Procurement Systems

D1392028

Procurement has long been, and remains, one of the most complex business processes. It can be approached in many different ways. Characterised by its novel cross-disciplinary project management approach, this new book covers more than the conventional themes of project solicitation and proposal evaluation. It builds on *Procurement Systems: A Guide to Best Practice in Construction* edited by Peter McDermott and Steve Rowlinson, and extends its content to make it relevant to the wider project management community.

Coverage includes general historical context issues, moving on to a practical discussion of different types of project and their procurement needs. This book focuses upon value generation through helping readers to design project procurement implementation paths that deliver sustainable value. It does this by showing readers how to facilitate project stakeholders to identify and articulate both explicit tangible and implicit intangible project outcomes. The book also discusses and provides cutting-edge research and thought leadership on issues such as:

- stakeholder management;
- ethics and corporate governance issues;
- business strategy implications on procurement;
- e-business;
- innovation and organisational learning;
- cultural dimensions and trust development;
- human resource development.

These issues are focused upon how procurement delivers value – a perspective of procurement that has been of increasing interest.

Derek H. T. Walker is Professor of Project Management at RMIT University, Australia.

Steve Rowlinson is Professor in the Department of Real Estate and Construction at the University of Hong Kong.

Also available from Taylor & Francis

Construction Safety Management Systems
S. Rowlinson   ISBN: 978–0–415–30063–6 (hb)

Occupational Health and Safety in
Construction Project Management
H. Lingard and S. Rowlinson   ISBN: 978–0–419–26210–7 (hb)

Procurement Systems: A Guide to Best
Practice in Construction
P. McDermott and S. Rowlinson   ISBN: 978–0–419–24100–3 (hb)

An Introduction to Building
Procurement Systems
J. Masterman   ISBN: 978–0–415–24641–5 (hb)
                                    ISBN: 978–0–415–24642–2 (pb)

Procurement in the Construction Industry
W. Hughes et al.   ISBN: 978–0–415–39560–1 (hb)

Profitable Partnering in Construction
Procurement
S. Ogunlana   ISBN: 978–0–419–24760–9 (hb)

Information and ordering details

For price availability and ordering visit our website **www.tandfbuiltenvironment.com/**
Alternatively our books are available from all good bookshops.

Procurement Systems

A cross-industry project management perspective

Edited by
Derek H. T. Walker and
Steve Rowlinson

Taylor & Francis
Taylor & Francis Group
LONDON AND NEW YORK

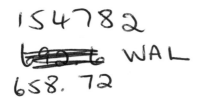
First published 2008
by Taylor & Francis
2 Park Square, Milton Park, Abingdon, Oxon OX14 4RN

Simultaneously published in the USA and Canada
by Taylor & Francis
270 Madison Ave, New York, NY 10016

*Taylor & Francis is an imprint of the Taylor & Francis Group,
an informa business*

Typeset in Sabon by
Newgen Imaging Systems (P) Ltd, Chennai, India
Printed and bound in Great Britain by
Antony Rowe Ltd, Chippenham, Wiltshire

British Library Cataloguing in Publication Data
A catalogue record for this book is available from
the British Library

Library of Congress Cataloging in Publication Data
Walker, Derek H. T.
    Procurement systems: a cross-industry project management
perspective / Derek Walker and Steve Rowlinson.
        p.cm.
    Includes bibliographical references and index.
    1. Industrial procurement. 2. Project management.
I. Rowlinson, Stephen M. II. Title.
    T56.8.W365 2007
    658.7'2–dc22   2007008896

ISBN10: 0–415–41605–1 (hbk)
ISBN10: 0–415–41606–X (pbk)
ISBN10: 0–203–93969–7 (ebk)

ISBN13: 978–0–415–41605–4 (hbk)
ISBN13: 978–0–415–41606–1 (pbk)
ISBN13: 978–0–203–93969–7 (ebk)

Contents

Contributors

Co-editors and chapter co-authors

Derek H. T. Walker (co-author of the Preface and Chapters 1 through 14) is Professor of Project Management at the School of Property, Construction and Project Management at RMIT University, Melbourne, Victoria Australia. He is Director of the Doctor of Project Management (DPM) programme http://dhtw.tce.rmit.edu.au/pmgt/. His PhD thesis is related to construction time performance management and this led him into further investigation of a range of project procurement, organisational learning and innovation aspects of PM. He has successfully supervised numerous PhD candidates in this area as well as having written several books, over 30 book chapters and in excess of 150 peer reviewed papers. Further details can be found on http://dhtw.tce.rmit.edu.au/. He is also editor of Emerald Insight's *International Journal of Managing Projects in Business*. His industry experience includes 16 years in the construction industry in the UK, Canada and Australia (including a two year period with a project planning software support and development organisation) with 20 years as an academic also providing consulting services to the construction and IT industry.

Steve Rowlinson (co-author of the Preface and Chapters 1, 2, 3, 4, 9 and 13) is a Professor in the Department of Real Estate and Construction at Hong Kong University and is involved actively in research and Doctoral supervision in the areas of procurement systems, construction management, occupational health and safety and ICT. He has been coordinator of the CIBW 092 working commission on Procurement Systems for over ten years now and has co-organised numerous conferences and symposia in this capacity. Steve has authored and co-authored more than 10 books and over 100 peer reviewed papers. He is an Adjunct Professor at Queensland University of Technology where he has a particular interest in international project management and construction innovation. Steve has acted as a consultant to, *inter alia*, Hong Kong Works Bureau, Hong Kong Housing Authority and Queensland Department of Main Roads and, as well as numerous consultancy reports, has produced over 100 expert reports in relation to construction site accidents and construction disputes over the past 20 years in Hong Kong. He is a member of the Institution of Engineers (HK), the Institution of Civil Engineers (UK) and a Fellow of the Royal Institution of Chartered Surveyors and is a keen golfer, skier and snowboarder. For more details see http://rec.hku.hk/steve/

Chapter co-authors

Mario Arlt (co-author of Chapters 1, 5 and 7) is a Senior Manager with SIEMENS Corporate Research, Inc. Princeton, NJ, USA and also a Doctor of Project Management (DPM) Candidate at RMIT University, Melbourne, Australia. Over the past 15 years he has actively managed and provided consulting services on small- to large-scale projects in Financial Services, Banking, Automotive, Pharmaceutical and High Tech industries. During the last four years he has established and managed a project management consulting group within the SIEMENS USA. He is a PMI Project Management Professional, Certified OPM3® ProductSuite Consultant and Assessor. His areas of expertise include project portfolio management, project recovery and PMO management.

Justin Stark (co-author of the Preface and Chapters 1 and 7) is a Senior Program Manager at TerraFirma Pty Ltd and was previously a Program Manager at Hewlett Packard Ltd, Melbourne, Australia. He is also a Doctor of Project Management (DPM) Candidate at RMIT University, Melbourne, Australia. He has extensive experience in managing a wide range of varied projects including infrastructure role outs and design (architecture), transition and transformation programmes, national and international data centre and office relocations as well as global SDLC programmes and pursuit management. He is PMP certified and has published a number of academic papers on subjects including ICT and Outsourcing Services as well as being an invited speaker on Asset Management throughout the Asia Pacific region. His areas of expertise are project recovery, offshore relocations, global programmes and PMO management; his thesis area is on Portfolio Optimisation within an Enterprise setting.

Lynda Bourne (co-author of Chapter 3) received her Doctor of Project Management degree at RMIT University where her research on defining and managing stakeholder relationships led to the development of a new project management tool, the *Stakeholder Circle*^{TM} (see URL http://www. stakeholder-management.com for more details). She has authored numerous papers, and is a recognised international speaker on the topic of stakeholder management, project communications and other related subjects. Professionally, she is the Managing Director of Stakeholder Management Pty Ltd and is responsible for the development and delivery of a range of advanced project and stakeholder management training courses; including courses for PMP, CAPM, PgMP and OPM3 ProductSuite accreditation. She was the first accredited OPM3 ProductSuite Assessor and Consultant in Australia, chaired the OPM3 ProductSuite Examination committee and was the inaugural winner of PMI's 'Project Manager of the Year' award (2003, Australia) and has contributed to a number of PMI standard development teams including the OPM3 and Program/ Portfolio Management Standards.

Michael Segon (co-author of Chapter 4) teaches at the Graduate School of Business at RMIT University, Melbourne, Victoria Australia. He received his PhD focusing on Creating and Implementing Ethics Systems in large organisations from the Queensland University of Technology in 2006. From 2003 to 2006 he was retained by KPMG Australia (Melbourne Office) to provide professional advice and consulting services in ethics and integrity.

James Norrie (co-author of Chapter 5) received his Doctor of Project Management degree at RMIT University and is currently the Director of and a Professor in the School of Information Technology Management (ITM) at Ryerson University in Toronto, Canada (www.ryerson.ca) where he teaches both graduate and undergraduate courses in project management and business strategy. He is the author of two books, several articles and speaks frequently around the world on topics related to project management with particular interest and expertise in enterprise project management (EPM), project portfolio management (PPM), strategic project selection and the use of the balanced scorecard as a project management tool.

Kersti Nogeste (co-author of Chapter 6) is Director and Principal Consultant of Project Expertise Pty Ltd (www.projectexpertise.com.au); received her Doctor of Project Management (DPM) degree from RMIT University, Melbourne, Australia; has managed successful projects and programmes of work in Australia and North America; is a doctoral supervisor, guest university lecturer and a board member of a regional hospital and health service. Using this combination of qualifications and experience, she applies a unique balance of rigour and relevance to helping her clients implement their organisational strategy via projects; ensuring alignment of organisational strategy, programmes of work and individual projects through to the level of project outcomes and outputs. Kersti's most recent roles have involved managing product development, IT&T infrastructure and company merger and acquisition (M&As) projects and programmes of work. In addition, Kersti is the author of a number of peer-reviewed journal articles and a regular speaker at national and international conferences.

Guillermo Aranda-Mena (co-author of Chapter 7) is currently a Lecturer in Property, Construction and Project Management at RMIT University, Melbourne, Australia. He holds a PhD in Construction Management and Engineering from the University of Reading and a Masters of Science in European Construction Engineering from Loughborough University of Technology, both in the United Kingdom. In 2003 Guillermo was appointed Post Doctoral Research Fellow at the University of Newcastle, Australia, working on a Cooperative Research Centre for Construction Innovation (CRC-CI) research project in Building Information Modelling (BIM) in collaboration with the Common Wealth Scientific and Industrial Research Organisation (CSIRO), Ove Arup and Woods Bagot Architects. He is currently RMIT principal investigator of four CRC-CI research projects including 'Business Drivers for BIM', 'Mobile Telcom in Construction', 'eBusiness Adoption in Construction' and 'Automated BIM Estimator'. He is currently supervising various Masters Theses and two PhDs. He is a Conjoint Academic to the Singapore Institute of Management, Singapore and the University of Newcastle, Australia. Publications and further details can be found on www.rmit. edu.au/staff/guillermo

Tayyab Maqsood (co-author of Chapter 8) is a Lecturer in Construction and Project Management in School of Property, Construction and Project Management at RMIT University, Melbourne, Victoria Australia. He is a Civil Engineer and has worked in Australia, Hong Kong, the UK, Thailand and Pakistan in various capacities as lecturer, project engineer and research associate over the last 11 years. His PhD investigated the role of knowledge management in facilitating innovation and learning in the construction industry. His PhD work is published in 22 refereed articles.

Fiona Y. K. Cheung (co-author of Chapter 9) recently completed her Master of Applied Science at Queensland University of Technology on the topic of Determinants of Effectiveness in Relational Contracting. She has also been working as research assistant with the CRC in Construction Innovation in Brisbane on a number of research projects associated with relational contracting and has co-authored a number of conference and journal papers. She is currently a PhD candidate at QUT focusing on relational contracting and sustainability in the construction industry.

Beverley-Lloyd-Walker (co-author of Chapter 10) undertakes teaching and research in the School of Management, Faculty of Business and Law, Victoria University, Melbourne, Victoria, Australia. Her PhD focused on IT-supported change in the Australian banking industry and its impact on bank performance. Her more recent research and teaching has maintained a strong interest and emphasis on strategic human resource management issues and she has co-authored with John Griffiths and Gary

Dessler *Human Resource Management*, published by Pearson Education Australia, the third edition of which is due for release in mid-2007. She has also contributed several book chapters on construction procurement and the role of IT in human resource management.

Helen Lingard (co-author of Chapter 10) completed a PhD in the field of occupational health and safety in the construction industry and worked as Area Safety Advisor for Costain Building and Civil Engineering (Hong Kong). She has lectured in occupational health and safety and human resource management at RMIT and Melbourne universities and provided consultancy services to corporate clients in the mining, construction and telecommunications sectors. Dr Lingard has researched and published extensively in the areas of occupational health and safety, work-life balance and human resource management. She has co-authored two books on *Human Resource Management in Construction Projects* and *Occupational Health and Safety in Construction Project Management*. She is writing a third book on the subject of *Managing Work-Life Balance in the Construction Industry*, to be released in 2007. She is Associate Professor (Construction Management) in the School of Property, Construction and Project Management, RMIT University.

Chris Cartwright (co-author of Chapter 11) has recently retired from Ericsson Australia after 35 stimulating years, the final 10 years taking responsibility for project management methods and competence. Driving the process improvement activities for the Ericsson Project Office the team was rewarded in October 2005 when the organisation was recognised 'best in class' globally within the group. He is currently completing his Masters Degree in Project management at RMIT. An active member of the Melbourne chapter of PMI he has spent time as a director and is currently managing a global team of 450 PMI members writing the second edition of the Project Manager Competency Development Framework, an ANSI standard, due for release in Q3 of 2007.

Peter Rex Davis (co-author of Chapter 12) is an Associate Professor in Construction and Project Management. He commenced teaching and research at Curtin in 1994. Prior to entering academia Peter amassed many years' experience in commercial construction, with both Government and corporate clients. Projects undertaken varied, and encompassed schools, hospitals, local government buildings and major developments of significant complexity and scope. He has professional experience in construction management in both the UK and Australia. He has a PhD that investigates the impact of relationship-based procurement. The doctoral programme was carried out at Royal Melbourne Institute of Technology (RMIT), Melbourne. Apart from his research into

Construction Procurement he actively researches teaching and was awarded a Curtin University Excellence and Innovation in Teaching Awards (EIT) in 2003.

Alejandro C. Arroyo (co-author of Chapter 14) has an MBA in Marine Resource Management, an MSc in Shipping and Ports, and is also a Doctor of Project Management (DPM) Candidate at RMIT University, Melbourne, Australia. He is focusing his thesis on an integrative project at transnational scale that is taking place today in South America, by involving knowledge management and communities of practice concepts within an ever-changing political and business environment. He is CEO and Consultancy Director of Buenos Aires-based Southmark Logistics SA (www.southlog.com) – a company focusing on project logistics across the Americas in the segments of mining, oil and gas, hydropower, infrastructure, transportation, marine environment and project sustainable development. He works at present on a number of large projects in the natural resources and infrastructure areas in the logistic and environment function, across the vast and complex geography of Latin America, the Caribbean and Alaska.

Preface

Derek H. T. Walker, Steve Rowlinson and Justin Stark

Chapter introduction

Project management (PM) theory and practice have undergone a transformation – from studying and understanding projects with tangible outcomes that are of finite-duration [such as delivery of constructed infrastructure or information technology (IT) products] to radically reflecting upon whether some types of project really exist at all in any concrete or conceptual sense (Hodgson and Cicmil, 2006). PM theory now extends to a broad range of project types from the highly concrete defined-time 'traditional' projects to projects that are completely ephemeral and intangible (e.g. events and change management initiatives). This book is mainly focused upon the nature of PM from a practical perspective. Definition of projects and project types are discussed in Chapter 2, and in Chapter 5 we discuss strategy as it is applied to projects and how projects integrate into the general management of an organisation. However, we refer to intangible projects throughout this book and recognise the existence and validity of less tangible projects.

Traditionally, PM theory and practice have poorly addressed the wider issues of procurement as being a key activity to achieve value for money. The USA-based Project Management Institute (PMI) has close to 250,000 members and for many decades has invested much effort in developing competency standards; however, it only recently expanded its focus on procurement as an important PM process. Indeed the latest version of the PMI body of knowledge (PMI, 2004: Chapter 12) devotes less than 30 pages to the process with much of the chapter's content being focused upon transactional issues such as letting contracts, contracts management and administrative aspects of contract closure. In research [using a survey of members from the UK Association for Project Management (APA) on its body of knowledge] procurement was seen as a commercial area and not a strategic one (Morris *et al.*, 2000). The APA is linked to the umbrella organisation, the International Project Management Association (IPMA), with over 40 PM associations as members (see www.ipma.ch/asp/) that help align and coordinate international PM standards. The evidence from the various associations' project management body of knowledge (PMBOK) reflects the tendency for project procurement thinking to be dominated by traditional PM experience gained over the past half century from construction, heavy engineering, aerospace, ship building and other industries, sectors. Recent work criticises PM professional bodies as not fully considering the complexity, indeed the reality, of projects as conceived by many PM practitioners (Cicmil, 1999; 2006).

Most PM practitioners' conception of PM is predominantly based upon an assumption of project delivery being an outsourced activity undertaken by the client (or client's agent) with a delivery chain comprising a design team, a main contractor who sub-contracted and supervised much of the detailed operational work and input resources (people, equipment and materials). This rigid and highly segregated form of procurement has attracted much criticism, centred on a failure of the prevailing system to capture value throughout the supply chain. There has been a series of UK government reports, for example, investigating the construction industry, and findings from these indicate that the traditional construction procurement system is a root problem, driving fragmentation and an adversarial project culture that fails to deliver best value (Murray and Langford, 2003). Several reports (Latham, 1994; DETR, 1998) that are particularly critical of the culture prevailing in the construction industry in the latter decades of the twentieth century illustrate a situation that is currently poor in numerous other PM sectors. The Standish group (1994) paints a similar picture of IT's and PM's failure to provide value despite suggesting a more recent 50% improvement in the situation. In their 2003 report they state that

> Project success rates have increased to just over a third or 34% of all projects. This is a 100% plus improvement over the 16% rate in 1994. Project failures have declined to 15% of all projects, which is more than half the 31% in 1994. Challenged projects account for the remaining 51%.
>
> (Standish, 2003)

In terms of procurement, Rooks and Snijder (2001) undertook a survey of 1,252 IT purchasing transactions of Dutch small- and medium-sized enterprises (SMEs) and found that only 28% of transactions for software, hardware and systems between IT providers and customers do not have some problems associated with them. Their findings show that 45% of transactions had inadequate documentation supplied, and 25% of transactions had 'incompatibility' or 'budget' issues. Although this was associated with the delivery of projects, we can (provided we have the linkage) relate the process of procurement (understanding deliverables, managing relationships, defining service levels and expectations, penalty clauses, liquidated damages, etc.) to the success rates. Thus, while project success or failure is caused by a variety of factors, clearly, selecting a procurement process that improves the likelihood of project success and value creation is an important goal that crosses many PM sectors. Strategising the procurement choice to fit the value propositions of stakeholders who can contribute to project success can be a vitally important element leading to improved PM processes.

Projects start with a procurement process that defines outcomes expected and the project scope, commissions the design to achieve identified outcomes, and assembles the means to deliver those project outcomes. Too often we see adoption of a default or one-size-fits-all procurement process. We attempt in this book to trigger a re-evaluation of the 'norms' because we believe that project procurement is ripe for reflection and adjustment to reflect the plethora of project types being undertaken and contexts in which the procurement process 'norms' are either inadequate or are simply counter productive. This short preface chapter introduces the book's purpose, aim, some founding definitions and the book's structure.

The value proposition for this book – its purpose and aims

The core issues identified by critics of the traditional procurement system (Kelly *et al.*, 2002; Langford *et al.*, 2003; Dalrymple *et al.*, 2006) recognise the need for value-for-money and improved relationship quality between team members in the project delivery system (Walker and Hampson, 2003). An emerging core interest in project procurement is supported and enhanced by the formation of the International Council for Building Research and Innovation in Construction (CIB) working commission W092 – Procurement Systems. W092 developed a focus on contracts management and forms of tendering for construction projects during the early 1990s. Over the past decade, however, it has embraced investigating the nature of how procurement choices affect organisational culture and working relationships (McDermott, 1999). This contextual discussion will be broadened later in this chapter. We will draw upon the construction industry as it is a reasonably mature PM environment or field of interest; however, we also draw upon many other PM industry sectors.

This book is primarily focused upon how project leaders can make and influence procurement decisions so as to realise a project that truly delivers value to the project stakeholders for project and organisational success. Particular attention is paid to the nature of 'value' in this process; in doing so, we introduce a range of intangible project outcomes, including ethical concepts, to help project managers consider and address relative, identified ethics issues when realising a project.

You will have mastered concepts discussed in this book when you can articulate how to improve the project procurement processes, to capture value and be capable of doing so by understanding the following:

- the fundamental meaning of delivering 'best value' for projects and programmes;

- the range of procurement options open to project managers and the interface between matching the strategy for value creation with an appropriate procurement delivery choice;
- cultural dimensions that show how project leaders can develop supply chain and competitor alliances and joint venture arrangements to synergise each partner's capacity to deliver project outcome value;
- how stakeholders can influence the value that a project can generate and how they can influence its realisation through a project by designing a procurement process that recognises their input and influence;
- how innovation and organisational learning can be incorporated into procurement processes to generate value;
- how value from e-business can be harnessed to effectively and efficiently procure projects;
- how PM leaders can balance cooperation and competition in procuring projects or being part of a group offering PM services;
- the impact that attracting the best talent to work on projects can be achieved through procurement choices that encourage and reward performance;
- how to use value measurement tools that provide better leading indicators of value generation and
- how to develop an appreciation for, and an awareness of, the role of ethics and being a good corporate citizen in delivering PM value.

The focus of the book is illustrated in Figure P.1. The key theme is procuring project value. Organisations tend not to want to merely source products, equipment, people or systems; rather they have a complex (often implicit)

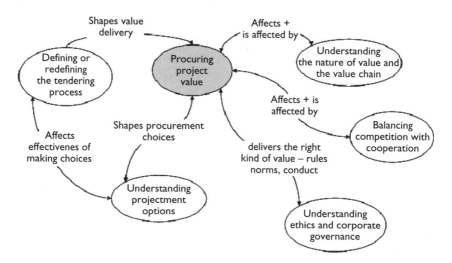

Figure P.1 Focus of the book.

need for a bundle of resources that help them generate and deliver something that will be valued by those receiving the project's product/service. Value can be explicitly and implicitly expressed. Understanding procurement options allows us to manage procurement choices and subsequent outcomes.

We can effectively define and redefine the procurement process to match the way that we and critical stakeholders perceive value so that we can focus on efficiency to reduce costs/time (doing the thing right) and effectiveness by focusing on value-based differentiation (doing the right thing). Our understanding of the value chain and the nature of value, affects our perception of value. We also find the need to balance cooperation and competition across project phases. Understanding ethics and corporate governance issues to ensure probity in project delivery is not only essential but it also affects our perception of project value.

The context of project procurement systems

This section provides an example of how the concept of procurement has been developing in a PM environment. When reviewing the PM literature such as the PMBOK (PMI, 2004) or the Morris and Pinto book (2004), for example, we see procurement viewed predominantly with a supply-chain management focus. Many IT or business process PM project managers mainly perceive procurement as a make-or-buy decision from a choice of in-house sourcing or from an outsourcing perspective. Some of the richer and more fine-grained research work into project procurement can be found in the construction management literature. It is for this reason that we will now summarise highlights from that wide body of work.

Jack Masterman produced one of the first texts in building procurement systems. (Masterman, 1992; 2002) and Rowlinson in Rowlinson and McDermott (1999) developed the work further and provided a definition of procurement systems based on nine years of the work of the CIB[1] Working Commission 092 (W092). Walker and Hampson (2003), among others cited in this text, brought relationship-based procurement systems more fully to the attention of the construction and PM fraternity. The evolution of this project procurement definition is repeated below and discussed in the subsequent sections of this chapter. W092 members and other academics in this field view a construction procurement system choice as a strategic decision that subsequently affects the whole construction process and that the procurement concept is a very broad issue. However, they recognised the need for appropriate procurement PM and administration processes to be followed regardless of which procurement strategy may be chosen. They describe the aims and objectives of CIB Working Commission W092 Procurement Systems (established in 1989) as follows:

- to research the social, economic and legal aspects of contractual arrangements that are deployed in the procurement of construction projects;

- to establish the practical aims and objectives of contractual arrangements within the context of procurement;
- to report on and to evaluate areas of commonality and difference;
- to formulate recommendations for the selection and effective implementation of project procurement systems and
- to recommend standard conventions.

This working group aimed to serve the needs of the international construction research community in times of change, as the 1980s had brought significant changes in the legal, economic and social structures of states in both Developing and Developed Countries. Privatisation, in its many guises, was on the political agenda not only in Western Europe and North America, but also in Eastern Europe (through the transitions from socialist to capitalist systems), and in Africa and Asia (through Structural Adjustment Programs). W092 has achieved a broadening of the procurement agenda, which is accepting a diversity of viewpoints, the recognition of the impact of culture and, overall, a contingency viewpoint.

The construction procurement concept had been poorly defined prior to CIB W092's formation and this is also reflected in other PM-related business activities. It is worth exploring the development of the procurement concept by this construction group as it is likely to be applicable to other PM situations. The CIB commission initially adopted a general definition of the procurement term as 'the act of obtaining by care or effort, acquiring or bringing about', from the standpoint that such a conceptualisation of procurement would raise awareness of the issues involved both in challenging generally accepted practices and in establishing new strategies. Mohsini and Davidson (1989: 86), attempted a more sophisticated definition – 'the acquisition of new buildings, or space within buildings, either by directly buying, renting or leasing from the open market, or by designing and building the facility to meet a specific need'. This led to a debate amongst commission members in Montreal in 1997 over the proposed definition: 'Procurement is a strategy to satisfy client's development and/or operational needs with respect to the provision of constructed facilities for a discrete life-cycle' (Lenard and Mohsini, 1998: 79). Thus, the commission's view was that the procurement strategy must cover all of the processes in which the client has an interest, indeed the whole lifespan of the building.

However, the usefulness of such definitions has been questioned. McDermott (1999), noted that existing definitions (at that time) while being useful for making comparisons of projects or project performance across national boundaries, were limited to developed market economies – a conclusion supported by Sharif and Morledge (1994) who have drawn attention to the inadequacy of the common classification criteria describing procurement systems (such as those discussed in Chapter 2 of this book) in enabling useful global comparisons. This criticism is exacerbated by procurement choice descriptions being used differently by different PM sectors: for example, in IT or product development – 'design and build' in construction terms is a valid and meaningful choice, yet in many other PM settings a component is often specified in broad performance terms by a main contractor; the supplier designs, manufactures and installs that element as being part of a supply chain. This is simply part of the make-or-buy decision discussed in Chapter 1. Thus, the whole procurement typology debate has been broadened in the construction industry context into one which covered a whole-of-life cycle to also include economic, political and cultural issues. Evidence of this appeared in Latham (1994: 5), who stated that 'some international comparisons reflect differences of culture or of domestic legislative structures which cannot easily be transplanted to the UK.'

McDermott further points out that key assumptions contained within the definitions, client choice and the availability of a range of procurement options, are irrelevant to Third World countries (McDermott et al., 1994). Following intensive debate, a working definition of procurement was developed by CIB W092 at its meeting in 1991, defining it as 'the framework within which construction is brought about, acquired or obtained' (unpublished document). McDermott argues that this definition serves a useful purpose as it is both broad, encouraging a strategic interpretation, and neutral, being applicable to developed and market economies.

W092 has a number of focus areas that are current and internationally relevant. Amongst these are development and privatisation; these are issues which, although having developed in the Western economies, have a high degree of relevance to Asia, where the rapidly developing economies of China and India have looked towards the private sector to meet some of their burgeoning infrastructure needs. Most countries in the world have moved towards privatisation and public–private partnerships (PPPs).

A commonly occurring project procurement theme is that of the role of culture in the shaping of procurement systems. In particular, the roles of trust and institutions are increasingly important. These issues will be discussed later in this book: these issues have arisen in both the West and the Asian economies for separate reasons. Of course, the nature of culture determines the nature of trust, just as history as well as culture determine the nature of the institutions within which procurement systems operate. This is a fruitful and an exceedingly interesting area of research.

Along with culture and institutions comes conflict. The move away from adversarial approaches to procurement and contracts, and a move towards collaboration and relationship management, can be seen to have permeated the systems in Europe, North America and Australia. However, such changes are also taking place in Asia, particularly in China, and the Japanese system has also embodied relationship management in its *Kaizen* approach, systematic incremental improvement (Imai, 1986), for many decades now. Thus, the work of W092 on an international basis has great relevance for research and dissemination, and along with procurement must come considerations of sustainability. Sustainability, a quadruple bottom-line view in terms of social, economic, environmental and safety and health issues, is an essential prerequisite for the continuance of development. Indeed, sustainability is embedded in procurement systems in the UK through the concepts of community benefit (Donnaly, 1999). In this section we have tried to introduce the reader to the broad field of project procurement that unfolds in this book.

Structure of the book

The book is presented in two main sections. The first section presents fundamentals of procurement. Each chapter in that section contains a short vignette, together with questions raised and links to useful resources to address those questions.

Chapter 1 provides a discourse on the nature of value, PM and procurement, and defines terms and the historical journey that has led from a lowest-cost tender to the sophisticated view expressed in this book, and the literature that has supported the development of this book. It discusses the 'make-or-buy' decision and the rationale for outsourcing and also the various outsourcing types.

Chapter 2 addresses the issue of the context of different types of project and how that impacts upon a choice of procurement methods to use. A discussion of relevant contract administration issues is also presented as they relate to generating and maintaining value within the context of procurement choices.

Chapter 3 views procurement from the stakeholder perspective. Stakeholders can add considerable value to projects as well as negatively threaten projects, draining management energy and causing disruption.

Chapter 4 includes important discussion of ethical issues relating to procurement. This chapter also discusses project and corporate governance and the impact of brand image upon those associated with a project.

Chapter 5 is important because it discusses the strategic decisions that anchor a project in a particular trajectory, the initial approval to proceed and the strategies that may govern the project throughout its life cycle.

Chapter 6 logically discusses performance measures and ways of value being perceived, and how to assess the effectiveness of project teams in a holistic manner. This chapter indicates how recent research has shown how hidden intangible value can be identified and linked to more tangible project outputs so that the true value of a project can be better articulated and monitored.

Chapter 7 investigates how e-business has affected project procurement choices and procedures. This chapter also discusses important facets of how information communication technology (ICT) is supporting project procurement and delivery processes and how it should be factored into a project procurement and delivery strategy.

Chapter 8 addresses the important and often neglected aspect of innovation, organisation learning and knowledge management (KM) that can add value to project procurement processes.

Chapter 9 brings forward critically important issues relating to cultural dimensions of procuring and delivering projects. These include the cultural environment of a project and how a procurement design can set it on a more positive trajectory. Project teams these days comprise cross-cultural groups, whether cross-national or cross-disciplinary. These issues are relevant to considering and designing an effective procurement and project delivery strategy.

Chapter 10 addresses a key issue relating to project delivery. Projects do not deliver themselves; people deliver them. Therefore, effective procurement choices should build in value that tempts the project organisations to attract the best available talent to deliver value. This is as relevant to both in-sourced and out-sourced projects.

In the second section, four chapters are presented which provide research results and case studies of organisations, some of whom have had their identities disguised. These are taken from the Construction Industry, the IT Industry and the Logistics Industry. They develop ideas on PM procurement practice through case studies involving alliancing, business transformation, process improvement and the establishment of a centre of excellence project. Each case study provides a description of the project context to help readers understand the issues, challenges and quandaries facing these projects in their procurement choices. Responses and/or solutions to these quandaries are then provided. Each aspect of the 'what' issues are given and explained in Chapters 2–5. Additionally, the 'how' and 'why' issues are addressed in Chapters 6–10.

We thank the generosity of those who assisted us in producing this book, the contributing authors, those who supported us while writing it, those who undertook the valuable editorial checking and administrative duties and those who provided information. Finally, we thank those authors whose work we cite. All value generated in any book or project such as this is developed from the baseline of previous literature developed by our academic and practitioner colleagues.

Chapter summary

This chapter introduced the book. Figure P.1 illustrates the focus of the book, being firmly on delivering value through a procurement system that builds in and designs in value as a fundamental proactive process. We introduced procurement in the context of PM and also traced the recent history of the evolving procurement field of study. We stress that we have to strategically design processes that drive a value agenda to effectively achieve value delivery in PM. We expect that this book will break new ground, triggering people to re-think procurement and how it can be used as a change-management agent. This can be achieved by introducing innovation, ensuring an ethical process, with due justice being a core value, so that value is generated for as many players as possible in the supply chain. We indicate how strategy and e-business trends can influence project procurement, and we also highlight new performance measurement tools that can address the needs of a range of stakeholders on a range of project types. Importantly, we address the issue of how to attract and retain the best talent to implement projects, be that from in-house or out-sourced teams. We believe that this book provides a fresh, engaging and strategic approach to the field of procurement that has for too long been viewed as being only transactional 'purchasing' and 'contracts administration' functions. We provide four stimulating case study examples of leading-edge procurement application where tools, transformation and relational engagement are key drivers of this fresh approach.

Note

1 CIB is the acronym of the abbreviated French (former) name: 'Conseil International du Bâtiment' (in English this is: International Council for Building). In the course of 1998, the abbreviation was kept but the full name changed to: International Council For Research And Innovation In Building And Construction. The CIB was established in 1953 URL is http://www.cibworld.nl/website/

References

Cicmil, S. (1999). 'Implementing Organizational Change Projects: Impediments and Gaps'. *Strategic Change.* 8(2): 119–129.

Cicmil, S. (2006). 'Understanding Project Management Practice through Interpretative and Critical Research Perspectives'. *Project Management Journal.* 37(2): 27–37.

Dalrymple, J., Boxer, L. and Staples, W. (2006). *Cost of Tendering: Adding Cost Without Value?* Clients Driving Innovation: Moving Ideas into Practice, Surfers Paradise, Queensland, 12–14 March, Hampson K., CRC CI.

DETR (1998). Rethinking Construction, Report. London, Department of the Environment, Transport and the Regions.

Donnaly, M. (1999). 'Making the Difference: Quality Strategy in the Public Sector'. *Managing Service Quality.* 9(1): 47–52.

Hodgson, D. and Cicmil, S. (2006). *Making Projects Critical.* Basingstoke, UK: Palgrave MacMillan.

Imai, M. (1986). *Kaizen: The Key to Japan's Competitive Success.* New York: McGraw-Hill.

Kelly, J., Morledge, R. and Wilkinson, S. (2002). *Best Value in Construction.* Oxford: Blackwell Publishing.

Langford, D. A., Martinez, V. and Bititici, U. (2003). 'Best Value in Construction – Towards an Interpretation of Value from the Client and Construction Perspectives'. *Construction Procurement.* 9(1): 56–67.

Latham, M. (1994). Constructing the Team, Final Report of the Government/Industry Review of Procurement and Contractual Arrangements in the UK Construction Industry. London: HMSO.

Lenard, D. and Mohsini, R. (1998). *Recommendations from the Organisational Workshop*. CIB W-92 Procurement – The Way Forward. The University of Montreal, 18–22 May, Davidson C. H. and T. A. Meguid, CIB, 1: 79–81.

Masterman, J. W. E. (1992). *An Introduction to Building Procurement Systems*. 2nd edn. London: E & FN Spon.

Masterman, J. W. E. (2002). *An Introduction to Building Procurement Systems*. 2nd edn. London: E & FN Spon.

McDermott, P. (1999). Strategic Issues in Construction Procurement. In *Procurement Systems A Guide to Best Practice in Construction*. Rowlinson S. and P. McDermott (Eds). London: E & FN Spon: 3–26.

McDermott, P., Melaine, Y. and Sheath, D. (1994). *Construction Procurement Systems – What Choice for the Third World?* East Meets West: Proceedings of CIB W92 Procurement Systems Symposium, Hong Kong, 4–7 December, Rowlinson S., University of Hong Kong, 1: 203–211.

Mohsini, R. and Davidson, C. H. (1989). *Building Procurement – Key to Improved Performance*. Contractual Procedures for Building: Proceedings of the International Workshop, Liverpool, UK, 6–7 April, D. Cheetham D. C., T. Lewis, and D. M. Jaggar, CIB, 1: 83–86.

Morris, P. W. G. and Pinto, J. K., Eds. (2004). *The Wiley Guide to Managing Projects*. Series The Wiley Guide to Managing Projects. New York: Wiley.

Morris, P. W. G., Patel, M. B. and Wearne, S. H. (2000). 'Research into Revising the APM Project Management Body of Knowledge'. *International Journal of Project Management*. 18(3): 155–164.

Murray, M. and Langford, D. A. (2003). *Construction Reports 1944–98*. Oxford: Blackwell Science Ltd.

PMI (2004). *A Guide to the Project Management Body of Knowledge*. 3rd edn. Sylva, NC, USA: Project Management Institute.

Rooks, G. and Snijders, C. (2001). 'The Purchase of Information Technology Products by Dutch SMEs: Problem Resolution'. *Journal of Supply Chain Management*. 37(4): 34–43.

Rowlinson, S. (1999). Selection Criteria. In *Procurement Systems A Guide to Best Practice in Construction*. Rowlinson S. and P. McDermott, Eds. London: E & FN Spon: 276–299.

Sharif, A. and Morledge, R. (1994). *A Functional Approach to Modelling Procurement Systems Internationally and the Identification of Necessary Support Frameworks*. East Meets West: Proceedings of CIB W92 Procurement Systems Symposium, Hong Kong, 4–7 December, Rowlinson S., University of Hong Kong, 1: 295–305.

Standish (1994). The Chaos Report (1994). Company research report. Dennis, MA: 14.

Standish (2003). Latest Standish Group CHAOS Report Shows Project Success Rates Have Improved by 50%, at http://www.standishgroup.com/press/article.php?id=2, 25 March.

Walker, D. H. T. and Hampson, K. D. (2003). *Procurement Strategies: A Relationship Based Approach*. Oxford: Blackwell Publishing.

Acknowledgements

A book such as this does not just happen. Our co-authors, for example, directly make a valuable contribution through their expertise and generously giving their time and effort to draft, re-draft and polish the chapters – we deeply thank them for that devotion and commitment. Behind the scenes are the reviewers, the copy editors and those in the publication chain that helped us by producing this tangible version of the book. Without their vital inputs of expertise and energy the quality of this book would be compromised. Authors need to discuss their work so we acknowledge and heartily thank our colleagues and friends who have provided moral support, tangible advice and valuable feedback throughout the writing phase. Our universities helped us by providing hardware and software and access to the library as well as moral support.

Finally, and with great gratitude, we all acknowledge and thank our social support structure, our family and friends, who provided the encouragement and support in so many ways to help make time by carrying out our share of the family work to allow us the time to think and write. Specifically, I (Derek) would like to deeply thank my partner Beverley Lloyd-Walker and my family including Georgina, David, Ruby our wonderful grandchild and Tony and Lee.

This book has been produced with the cooperation of CIB Working Commission 092 and is a CIB endorsed publication.

Derek H. T. Walker
Steve Rowlinson
September 2007

Chapter 1

Introduction and procurement fundamentals

*Derek H. T. Walker, Justin Stark,
Mario Arlt and Steve Rowlinson*

Chapter introduction

This chapter discusses procurement fundamentals from a PM context that sets the scene for the rest of the book. We argue that delivering value is the main purpose of PM and that value is generated by PM teams for customers and stakeholders including organisations involved in the PM process. We assert that a sustainable procurement process provides a way that value is generated for all parties concerned because envisioned value generation is a strong motivator of the required level of commitment and enthusiasm necessary to generate win–win rather than win–lose outcomes.

At the heart of value generation is the issue of who participates in this process. This brings to the surface the primary make-or-buy decision that involves determining which parties are best prepared, suited and equipped to undertake specific parts of a project. That leads to determining how to best procure the required resources whether they be internally or externally sourced. While a portion of the project will be inevitably undertaken by those in the lead or PM role position (even if that contribution is solely to coordinate activities), much of the project work will be outsourced. This chapter will focus upon the procurement fundamentals of the outsourcing rationale. It is often assumed as a 'given' without looking at this decision from a value generation and maximisation point of view.

This chapter starts by discussing the concept of 'value' and moves on to question and discuss what an organisation actually is and how it can deliver value. The next section focuses upon the nature of outsourcing and its derivatives. That leads into the issue of transfer pricing and how value generation is treated in a global context because the trend towards global outsourcing has led to often complicated taxation and value recognition issues. Privatisation is also discussed in this chapter because many of the major projects being undertaken by the private sector have been shifted through government policy in all market economies from the public sector to the private sector or through a partnership of public and private enterprises.

The nature of value

Value is in the eye of the beholder. Because the appreciation of value is governed by perception, we must start with the stakeholders who have a need that requires satisfying or addressing. Chapter 3 will explore the stakeholder dimension of defining and contributing to value. Robert Johnston (2004) argues that customer delight lies at the top end of a continuum. This ranges from dissatisfaction where fitness for purpose is manifestly absent, through compliance with minimum expected delivery performance, to that higher level where exceptional value is delivered. Delight is, therefore, the result of (excellent) service that exceeds expectations. However, this may come at an excessive cost to either the provider or recipient. His study was based on gathering over 400 statements of excellent and poor service gathered from around 150 respondents. Participants' phrases about excellent service provided by the respondents fell into four relationship-based categories:

1 delivering the promise;
2 providing a personal touch;
3 going the extra mile and
4 dealing well with problems and queries.

Characteristics of poor service were essentially the converse of excellence (Johnston, 2004: 132). The interesting part of this study, which related to customer service in general rather than procurement process in particular, was that while category one is transactional and satisfaction of the stated standard appears to suffice, the other three categories are about intangible and often poorly explicated standards. We could recognise value, therefore, as being not simply fitness for purpose at an agreed price in a timely manner but also as providing intangible deliverables for organisations that may include excellence in quality of relationships, leadership, learning, culture and values, reputation and trust. It is not easy to identify and prepare PM processes that have well-operationalised and developed measures to monitor and ensure effective delivery. However, Nogeste (2004) and Nogeste and Walker (2005) identify how this can be achieved. This aspect will be discussed later in more detail in Chapter 6. The point made here is that value is an amalgam of the 'iron triangle' of performance measures of time, cost and quality together with expectations and anticipated delivery of 'softer' often unstated needs. Traditional project delivery procurement systems may be adequate in defining tangible and defined outputs but they fall short in facilitating delivery of expected intangible and unstated outcomes. This brings in issues of expectations management as a valid aspect of project procurement processes. The extent to which these intangible outcomes are delivered is often associated with the level of customer delight and value.

Returning to the more tangible and defined elements of value, another way of viewing value for money is receiving a project or product at a reasonable price that reflects input costs and is delivered within a reasonable timeframe at the specified quality level. This concept of value relates to ethical and governance issues. Taking an extreme example, slave labour or using stolen materials could arguably deliver lower costs; however, this would be unfair and unethical. Issues relating to ethics and governance will be discussed in more detail in Chapter 4. Similarly, a process that ensures open and transparent availability of cost and scheduling information, for example, can provide a set of project governance property arrangements to satisfy fairness criteria. Thus, provisions for demonstrating value can be structured into procurement processes. Further, as will be discussed in Chapter 8, procurement options can be designed to facilitate innovation and/or organisational learning. Similarly, a focus on talent, both from the point of view of attracting talented project team members and developing capacity in teams and stakeholders, can also be designed into procurement choices. This is discussed in more detail in Chapter 10.

Another way of looking at value is to ensure that delivered projects meet strategic needs of stakeholders. An inexpensively delivered factory (strategically) located in the wrong place will not deliver competitive advantage and hence will have sub-optimal or negative value. For many change management or product development projects, making the correct strategic choice is part of the project procurement process – especially as many of these projects use organisation-internal markets to resource these projects. This aspect will be further discussed in Chapter 5.

New technology, particularly an e-business and information and communication technology (ICT) infrastructure, can offer new ways of configuring procurement options and can impact new processes and ways of both procuring and delivering projects. These innovative approaches can enhance value to be delivered and this aspect is presented in more detail in Chapter 7.

Much of the focus of this book, particularly Chapter 2, lies with exploring the range of procurement options open to deliver value in projects by matching these against the context of the project's need to be satisfied. The cultural dimension of both generating trust and commitment from all parties is explored in Chapter 9, and this is particularly important for relationship-based procurement options (Walker, 2003).

The nature of projects and project management

The PMBOK provides a useful definition of a project and project management. A project is defined 'as a temporary endeavour undertaken to create a unique product, service or result' (PMI, 2004: 5). Project management is said to be the application of knowledge, skills, tools and techniques to project activities to meet project requirements. Four main processes are listed: identifying requirements; establishing clear and achievable objectives; balancing competing demands for quality, scope, time and cost; and adapting the specifications, plans and approach to the different concerns and expectations of various stakeholders (PMI, 2004: 8).

Turner and Müller (2003) remind us that a project uses a temporary organisation to deliver its outcome and that there are transaction costs inherent in a project that do not appear in established ongoing production organisations. They also stress that much of the project manager's role is motivating team members, stakeholders and shaping agendas. They describe features of a project as follows (Turner and Müller, 2003: 2):

> The project aim is to deliver beneficial change. Features include uniqueness, novelty and transience. Pressures typically encountered are in managing to cope with uncertainty, integration and transience. Processes need to be flexible, goal oriented and staged.

Value in terms of projects is apparently more complex an issue than standard factory production of 'widgets' because the management and leadership intensity involves mobilising transient groups of people brought together for a limited time to work on specific projects; people within these groups often exit part way through the realisation of these projects for one reason or another. That is why this book includes aspects of stakeholder identification and management (Chapter 3), organisational learning and knowledge management (KM) (Chapter 8), talent management (Chapter 10) and cultural dimensions (Chapter 9) that make establishing an optimal framework for delivering value so important. As the collection of government reports reviewed and commented upon by Murray and Langford (2003) clearly indicate, the procurement systems traditionally used in the UK construction industry fail to address the complex reality of how projects can best deliver the more holistic concept of value noted earlier in this chapter.

The phases of a project need to be considered so that optimum value is delivered because a project often takes a long time (many years) to deliver. A typical project can be seen to comprise of four principal often overlapping phases. These are illustrated in Figure 1.1.

At the initiation phase the project needs are identified, clarified and prioritised along with expected benefits, the strategic impetus for the project and developing and challenging the business case for the project. This is where stakeholder input can provide significant project potential value. The appropriate involvement of stakeholders and the management of this process are discussed in Chapter 3 in more depth. The need for identifying stakeholder potential impact can be of crucial importance and add value by identifying often hidden opportunities for planning input

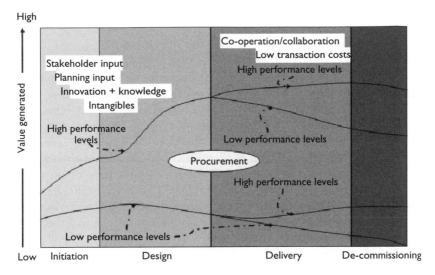

High

Value generated

Stakeholder input
Planning input
Innovation + knowledge
Intangibles

Co-operation/collaboration
Low transaction costs
High performance levels

High performance
levels

Low performance levels

Procurement

High performance levels

Low performance levels

Low   Initiation   Design   Delivery   De-commissioning

Figure 1.1 Project phases and performance.

and knowledge and can unearth hidden risks and threats (Bourne and
Walker, 2005a,b). Also, intangible value and benefits can be identified,
analysed and linked to the tangible project deliverables (Nogeste and
Walker, 2005). The design phase involves matching needs with proposals to
deliver the required value to meet the identified needs. Stakeholder
involvement, particularly in respect of prioritising needs and using value
engineering techniques to optimise value (e.g. see Green, 1999), can deliver
unexpected value. The design phase is usually followed by a process where
the successfully selected project team is mobilised and resources procured
to deliver the project. This is where sub-optimal procurement choices can
have a profound impact upon project delivery value. The final phase noted
above is de-commissioning and often a different set of additional stakeholders may be involved at this stage.

Figure 1.1 illustrates two trajectories of project value generation throughout the project phases. At the initiation and design stages stakeholder input
(including not only supply chain stakeholders but project end-users and
others affected by the project's existence) can provide critical definition and
perceptions of value that may shape the evolving design details. High-level
performance at this leading edge of the project time span is indicated by
identification of misunderstandings of value perceptions at as early a stage
as possible. These can include fully understanding the true needs driving the
feasibility of the project, contextual knowledge and matters relating to
planning and logistics. Without this kind of input from stakeholders, it is

common for project designs to start sub-optimally with high expense and rework being required to remedy this if discovered later. Worse still, this input may be realised too late to effectively incorporate ideas that could optimally realise project value. This is why the chosen procurement process can have such a significant impact upon the stage before committing the project to detailed design and tender. As illustrated above, the gap between poor and good performance at this stage can be significant.

The second part of the trajectories illustrated in Figure 1.1 relates to the delivery phase spilling over to de-commissioning. The point made here is that even if a poorly initiated and designed project has a top-level project team carrying out the project, it is unlikely to yield as much value as a mediocre or poor project team delivering a project where there was high performance initiation and design established. For example, Barthélemy (2001) identified a proportion of the hidden costs associated with outsourcing contracts, and their management and maintenance, which include:

- searching for and identification of a suitable vendor;
- transitioning services to the external service provider;
- managing the day-to-day contract and its service deliverables and
- the effort required once the outsourcing contract agreement has been completed or terminated.

Figure 1.1 clearly demonstrates that the value contribution lies not only in the way that stakeholders and project team initiators behave and contribute, but also in the timing of their contribution. It shows how the procurement phase, which often spans the design and delivery phases, is pivotal in generating project value.

Procurement choices should be geared towards maximising value. This involves not only considering which type of contract to use but how to engage stakeholders, how to link business strategy to select the project initially, how to define value both in its more tangible and easily explicated form, and the hidden tacit aspirations of those who are meant to benefit from the project. Procurement choices should be about balancing demands and responsibilities; protecting reputations of those involved in the project against using unethical practices; encouraging innovation, best practice and knowledge transfer where it can reap value; developing project governance and reporting criteria that highlight, protect and generate value; and they should aim to attract the level of talent that can deliver stakeholder delight. Clearly, this vision of procurement positions itself far from the traditional lowest-cost strategy. It requires a more intelligent treatment of how best to facilitate the project to generate sustainable value. If the word 'value' appears overused in this chapter it is because it has been largely absent from much of the project management and procurement literature. This book seeks to redress this situation.

The make-or-buy decision

At the heart of procurement is a fundamental make-or-buy decision. Chris Fill and Elke Visser (2000) argue that essentially, projects are triggered by a decision to realise a change within the organisation (a make decision) or to outsource this (a buy decision). Most projects have both make and buy elements to them. The interesting thing about the emerging and growing trend of organisations managing by projects is that we can look at outsourcing as being purely external (a buy decision) or we can choose to look at a project justification proposal to internally produce goods or deliver a service (a make decision). Also, each of these has an important impact upon management issues associated with the project lifecycle. Either way, these decisions essentially follow a similar process.

Calculation of costs relating to the delivery of transaction-based services has been the subject of extensive research. Transaction-cost economics practices (TCE) (Williamson, 1975; Chiles and McMackin, 1996; Robbins and Finlay, 1997) enable an organisation to quantify the costs associated with transaction-based services. Transaction-based services are defined as those services that can be quantified in a manner that allows them to be separated as individual items within a service delivery framework. This makes identification and cost analysis considerably easier than non-quantifiable services. Transaction-based costs are defined as the delivery of services that can be measured against a specified service level agreement. These activities are a significant contributor when measuring the total costs associated with outsourcing deals. TCE assumes that the costs are selective: those differing management choices will see a differing value for the costs of transactions and that the governance structure and organisations' interactions can affect costs. This may not be as relevant in outsourced deals, as the structure of the outsourced provider and the methodology used for management may not be within the direct control of the customer. There may be situations in which the customer has an expectation of specific interactions with the appropriate levels of the project management team for the type of contract or service level agreements that are in place. Although the governance structure of the delivery organisation will affect the relationship, it is the governance structure of the customer that can ultimately influence the ongoing service relationship. If organisations have the same governance structure, then costs will be reduced and cost management could be minimised, although both organisations may vary their governance structures to help reduce transaction costs.

The resource-based view (RBV) of the organisation assumes that organisations are a collection of people, processes, technologies, competencies (skills and knowledge), money and physical assets that together provide the means for firms to conduct business. Barney (1991) argues that for firms to gain and maintain sustainable competitive advantage these resources must be (compared to competitors) rare, valuable and inimitable. They must be deployed strategically in order to 'stay ahead' of competitors. Others, such as Conner and Prahalad (1996) argue that knowledge is a crucial resource and indeed, Prahalad and Hamel (1990) were early theorists who argued that a firm's core competencies lay at the heart of its value as a company. A further elaboration on this view includes the way that firms respond to turbulence and dynamic change in the market and these competencies (dynamic capabilities) are central to a firm's sustainability (Teece, Pisano and Shuen, 1997; Eisenhardt and Martin, 2000). Firms can (internally) view value as being their capacity to maintain or increase the most valuable of their stock of resources. Ideally this manifests itself in increasing revenues, physical or intellectual property, organisation competence or skills and knowledge that its people deploy in their daily job performance. Their attractiveness to clients and the market is partially determined by their choice of strategy to use their resources to generate value for their customers or clients.

Many managers and business leaders translate this strategic choice when delivering projects as a combination of undertaking parts of the project at least cost or arranging for others who can do so at lower overall cost. An often hidden cost that needs to be also considered is the transaction cost of arranging to sub-contract out part of a project. This can include the management energy (and opportunity cost of making more productive use of their time) in developing a tender brief, tendering, evaluating and awarding a contract and then administrating and ensuring that required quality is delivered. This 'transaction cost' as described by Williamson (1991) can be significant and the scale overlooked by those eager to outsource and subcontract parts of their project. Further, parts of a project will require resources of skill, competencies and capital/intellectual assets that the firm may not possess and so the obvious solution is to outsource these parts. Finally, there are contextual issues that inform a buy-make decision. An organisation may have all the required resources to not only undertake a project by themselves but may also wish to share risk, share resource demands or develop collaborative relationships with other firms. Fill and Visser (2000) identify reducing hassle as an outsourcing driver. Hassle may represent an element of risk management or binding up resources that could be more productively used elsewhere. They present a process of the make-or-buy decision as being determined by the inputs of:

1   the firm's strategy, and cost and business structures;
2   the transaction costs associated with a 'buy' rather than 'make' decision and
3   contextual factors that can include risks and future opportunities.

When a 'make' decision is made based on the above to undertake a part of a project in-house then not only should budgets and plans be presented to plan how to undertake the work and what resources are required at what times, but a business case should also be made to justify that 'make' decision. The business case should address risk elements, opportunities and provide contextual information, for example, there may be strategic, political or historical reasons why apparently non-cost-effective decisions may be made.

The make or buy decision, therefore, generally results in at least some part of a project being outsourced as no firm has all the requisite resources and skills to undertake an entire project under its own steam. Thus, the next relevant aspect to discuss is the rationale and forms of outsourcing. In this chapter we discuss the concept of outsourcing before entering into details of the various procurement options; that will be dealt with in Chapter 2.

The nature of outsourcing and its derivatives

Gay and Essinger (2000) see outsourcing as a process of transferring service delivery management responsibility to a third party for providing services that are governed by service-level agreements. This recognises the fact that the customer must take responsibility for the delivery of services to its organisation. Jacobides and Croson (2001) argue that effective service-level-agreement monitoring encourages the vendor to ensure that the delivery of service meets the contractual requirements. This implies that the management responsibility for the monitoring of services governed by the service-level agreement remains with the customer, especially when considering the successful delivery of services. The customer, in such situations, will need to rely on information provided both by reporting communications from the service provider and feedback from internal customers to be able to judge vendor service-level delivery, and to prompt any subsequent adherence action to ensure contractual service-level agreement performance. This information will then have to be compared to the contract between the service provider and the customer. Management reporting needs to be included in all outsourcing contracts and should reflect the service-level agreements and expectations that have been agreed upon. Measurement or reporting systems should be able to reflect the value associated with services, rather than 'management noise'.

Outsourcing can be thought of in one sense as sub-contracting activities when the firm has a contract for a project but is unable to internally absorb all the required work. In another sense it can be considered as undertaking part of a firm-internal activity or process. This may include IT support, or an element of production (perhaps manufacturing components or service delivery element such as logistics support). Outsourcing has become increasingly prevalent in facilities management (FM); for a more detailed treatment of that aspect (see Bröchner, 2006). A further subset of outsourcing discussed in this section includes out-tasking and privatisation.

Outsourcing and out-tasking

Project outsourcing naturally requires project management capabilities; whilst outsourcing of operational activities often involves a portfolio of cascading projects to deliver the outsourcing model. Some of these projects are directly related to the outsourcing act, such as shifting an operation or process from being internal to external delivery. Others are indirect, such as the development of an effective oversight or governance capacity for managing an outsourced venture, or may include change management activities that are nested inside or surround the actual outsourcing project.

Recent surveys in CIO Magazine (Bendor-Samual, 2002) and Deloitte Consulting (Landis, Mishra and Porrello, 2005) analysed the reason why large organisations outsource activities. In both studies about two thirds of the respondents named cost savings as the primary driver for their move towards outsourcing. However, as Deloitte's recent survey points out, 38% of the respondents which engaged in outsourcing ventures to save cost, experienced unexpected additional costs. Many outsourcing ventures did not deliver on other promises; the expected increased focus on core business was not achieved, critical processes were mislabelled as non-strategic and had to be in-sourced again, vendor risks had been underestimated and better access to quality resources were often overestimated.

While this by no means establishes that outsourcing is an unsuccessful model, it suggests that a solid business case is essential to the success of outsourcing activities. Such a business case should state the expected quantitative and qualitative benefits, both from a tactical and strategic perspective, as well as identify and estimate all related risks, in addition to establishing suitable metrics to measure the value-add of the outsourcing venture on an ongoing basis. A comprehensive business case should include all direct costs for the services being provided as well as the indirect costs associated with managing the supplier–customer relationships and those incurred from risk management activities.

Outsourcing should not be *the* strategic objective, but *an option* to achieve a strategic objective defined by management. Currently many variations of outsourcing can be observed. The following classification of outsourcing variations, based on six dimensions are presented in Table 1.1 (Stark, Arlt and Walker, 2006).

The *Activity Dimension* describes what organisational activity is to be outsourced and distinguishes between short- and long-term endeavours, such as projects and programs on one hand, and open-ended activities like the permanent execution of processes on the other hand. The *Geographic Dimension* relates to the locations included in an outsourcing scenario, and captures the geographic distance of the outsourcing partner. Furthermore, the *Legal Entity Dimension* defines the legal relationship of the partners of an outsourcing agreement, which can reside either within

Table 1.1 Outsourcing dimensions

Activity dimension • Projects • Programmes • Portfolios containing projects or programmes • Operations or business processes *Legal entity dimension* • Within one legal entity • In a different legal entity, but within a corporation • Outside the corporation *Engagement-temporal dimension* • Hands-off divesture • Experimental – spin-off • Experimental – growing partnerships to divest • Experimental – growing partnerships to potentially absorb the outsourced entity	*Geographic dimension* • Local/Regional/National (on-shore) • International/Global activities (near-shore or offshore) *Distribution dimension* • All activities are performed within the organisation (collocated execution); • Activities are distributed across multiple locations and partially executed in-house and outside (out-tasking) • All activities are performed outside the organisation (full outsourcing) *Mobilisation–demobilisation dimension* • Simple (no re-entry expected) • Moderate (re-entry possible) • Intense (re-entry probable)

(shared services model) or outside a corporation. The *Distribution Dimension* captures the degree that an activity is outsourced, which for a defined activity can range between co-location (0%) and full outsourcing (100%). The *Engagement-Temporal Dimension* classifies the partnering approach from the kick-off of an outsourcing relationship and will mainly impact procurement considerations for the outsourcing contract. Some of the successful outsourcing partnerships have moved through some experimental stages to a more hands-off model. Lastly, the *Mobilisation–Demobilisation Dimension* captures the impact of outsourcing an activity and may imply significant efforts, depending on the implied re-entry option: a full divesture with no plans for taking back the outsourced activity in the foreseeable future (simple), demobilisation of resources with possible re-entry (moderate) or probable re-entry (intense), which requires legacy capacity to do so.

As can be seen in Table 1.1, the six dimensions allow numerous strategic or tactical permutations of outsourcing scenarios. As an example for the mix of these dimensions, an organisation could decide to outsource a project component, such as the generic code development for a defined component of a software product, to a third party in an offshore location with little need for developing any long-term relationships or embedded learning product enhancement nor any need for complicated re-deployment or redundancy of existing staff.

Although outsourcing of services has been implemented by a number of government agencies for example, there are other procurement strategies that can be employed to meet the needs of large organisations. Out-tasking, selective delivery of expert services (often complete business processes) by an external party, can be used without committing excessive overheads and management. Recruitment agencies, payroll service providers and cleaning services are all prime examples of this style of procurement strategy. In these situations, a services contract is drawn up between the customer and the supplier for the delivery of specified services. It is then up to the service provider to ensure that the service is delivered to the contractual agreements. In these out-task scenarios, non-technical or specified skill set services are provided, and economies of scale for the delivery organisation should provide cheaper services to the customer (Domberg and Hall, 1995).

Out-tasking is a short-term operational decision that can be used to provide a service needed to continue operations or provide specialist skills that would not be required for long-term operations, or can be completely divorced from day-to-day deliverables. The risk of this procurement strategy is less than with outsourcing as there are lower switching costs, and often the skills provided are available freely within the market place. Out-tasking requires that a level of trust be established between the two organisations, but the level of relationship management and ongoing support is considerably less than with the outsourcing of services. Procurement is achieved by defining the service level required for specific tasks and sourcing them within the market place. For additional services, the incumbent may provide a more cost-effective service but the customer is able to freely negotiate within the market place.

At first glance, such a procurement strategy has the look and feel of an outsourced service, and the classification between outsourcing and out-tasking is a grey line. However, out-tasking is a specified deliverable so intellectual property would not change hands and the customer is more able to switch vendors in an easy and timely manner. With outsourcing, the cost of the service needs to consider the long-term implications of switching costs because when these contracts are terminated, it often occurs in an acrimonious manner.

Such an out-tasking scenario to a third party *may* yield certain benefits, although there is much literature to refute this as an activity that saves money and delivers (Treacy and Wiersema, 1993) unless there is an ongoing relationship with the supplier. It would be particularly suitable for a one-time, short-term outsourcing need for a clearly defined, and potentially rather commodified service, which requires specialist skills or could be provided by a third party at a substantially lower cost, in a shorter period of time, and/or of higher quality. The risk of this out-tasking strategy is lower than a full outsourcing scenario, due to lower switching costs associated with mobilisation and demobilisation of key resources and the likely wide availability of qualified resources in the market place to provide such (commodified) services. Out-tasking, as described in the above example, requires that a certain level of trust is established between the two organisations, but the level of relationship management and ongoing support is considerably less than with the outsourcing of an operation or process. The customer may procure additional services from the same provider or from within the market place due to the commodified nature of the underlying delivery models. Risks relating to the retention of intellectual property and the build-up of a potential competitor are limited, as the conscious decision is made to only out-task a part of the development effort and to focus on commodified activities. Outsourcing the development of further software components to other partners may increase the leverage while maintaining ownership of the overall design and architecture and securing the intellectual property of the overall product.

We suggest that outsourcing decisions should be a truly strategy-driven process, which could be executed as follows in Figure 1.2.

The stages can be summarised as follows:

1   *Define strategic objectives*: The clear definition of the objective and target variables, which will be used to measure the success of the taken action, is most critical for the success of an outsourcing venture. The relevance of this first step is often underestimated and will be further elaborated in this section.
2   *Define a suitable outsourcing model*: Based on strategic objectives, the six dimensions can be used for defining a suitable outsourcing model. Metrics defined in the prior step will help estimate the business benefits and set target values. Chapter 2 will explain this in greater depth.
3   *Validate the outsourcing model*: This validation step may include a number of stakeholders, which will be involved in the outsourcing projects within the organisation. It will further solidify the benefits and risks and could take into consideration benchmarking data for similar ventures within or outside the organisation.
4   *Define requirements and partner selection criteria*: As a result of the prior steps, parameters for a proposal request can be identified as well as vendor selection criteria.
5   *Select vendor*: Based on Stage 4, a vendor selection can be performed and negotiations will be initiated.

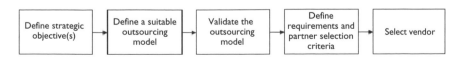

Figure 1.2 Outsourcing strategy.

Thus, the decision whether to use internal or external resources for executing an activity or service is strategic in nature and not merely a price-driven make-or-buy decision as long as services are typically not standardised to the extent where they are purchased and delivered like a commodity.

To successfully outsource activities, executive management needs to define the underlying drivers for such an action. To the extent that management can qualify and quantify the strategic objectives associated with the outsourcing activity, it should be possible to establish a solid business case and execute it. Such strategic objectives could be articulated as follows:

- being focused upon core business activities (metric: volume and support cost for activities) that are neither considered core competencies nor directly responsible for the organisation's competitive advantage;
- leveraging external subject matter expertise, for example, for non-core product components or one-time development or integration efforts (metric: in-house development staff hired or retained for non-core development);
- reducing labour-intensive processes and increase of cutting edge, customer-centric activities in locations which exhibit high labour cost and
- understanding the exit criteria for outsourced activities which will lead to a relocation of outsourced activities back to the parent organisation or reassignment.

However, when considering the outsourcing of services a series of additional risks should be taken into account. For example, successful outsourcing requires that the organisation ensures that not only the deliverables meet the specifications, but also the relationships and contractual requirements are to be measured and maintained (Cooke-Davies, 2002), which involve additional efforts and costs on both sides. The outsourcing contract needs to define a reporting and control mechanism with agreed upon metrics and audit capabilities such that delivery risk can be managed by the customer. Outsourcing activities often lead to a shift of primary skills sets from the customer organisation to the supplier, which may reduce an organisation's ability to manage the delivery from requirements definition to execution. For example, the outsourcing of a manufacturing activity will impact on the resident skill set for production engineering; the outsourcing of engineering capabilities may diminish the organisation's ability to develop new products. The outsourcing of a complete process is likely to impact on the detailed understanding of the process and hence the ability to manage, monitor and audit process compliance with the intended purpose, and so on. This risk of losing relevant skill sets requires mitigation and will lead to cost, which will impact on the business case and most likely, the procurement scenario.

Outsourcing often involves moving a defined set of skills from the existing organisation to an external party – for an example, see Willcocks and Choi (1995). Although this can be managed within the transition process, consideration must also be given to the management of the transition from a staffing perspective. Often government staff have been in roles for considerable lengths of time and the migration to a commercial operating environment is not successful. Staff benefits, expectations and organisational culture would need to be considered as part of the integration activities, while regulations, such as union agreements and government contracts and employment law can impact on the process (Lobez, 1999).

The procurement process for outsourced services needs to take into account the complete life cycle of the engagement. A number of large organisations often experience euphoric expectations at the inception of outsourcing agreements without fully understanding and managing the risks within this new approach to business. Organisations have to consider outsourcing agreements – as with any venture – as a temporary endeavour that, whether long-term or short-term, follows a life cycle, including contractual agreements and managerial contingency plans for the event of contract termination. As a part of such contingency plans, an organisation which has completely outsourced an activity will need to understand how it would proceed to provide that function either internally or through an alternative party if the contract was terminated.

Any termination clauses should ensure the least possible disruption of the activity, the protection of the intellectual capital required to continue the activity and the re-transfer (or in case of software and data assets the deletion) of proprietary assets, which were provided to the outsourcer under the agreement. A number of companies experienced difficulties in enforcing intellectual property rights in markets like China. Thus a potential risk management strategy that can be employed (besides the decisive pursuit of legal options) may include the aggressive hiring of key players of a former outsourcing partner, the protection of intellectual property (IP) or theft through selective out-tasking of only some or commodified components. Although it is not possible to include a re-allocation of key staff in case of a termination of outsourcing agreements, there is a need to define the intellectual property that would be provided in the event of any dissolution of the service delivered. In this context, intellectual property should include any hardware, software, customised processes and procedures that have been specifically funded by the customer and integrated into the operations of the services. It should not include any intellectual property that is used for other customers or would be commercially sensitive to the supplier. This definition, especially when considering the transfer of tangible assets, needs to be contractually defined with great care prior to the delivery of services; if not, the customer can be at the mercy of the supplying body, especially when considering mission-critical services or functions that support the core business of the customer.

Within this framework, the cost of maintaining a skeleton staff that is able to maintain the customer's IP and ensure that the services being delivered are appropriate and fit for their intended purpose is a procurement transaction cost expense that is often overlooked. Adding to the level of complexity of this procurement life cycle is the need to manage both the delivery of services and the relationships with the outsourced service provider.

For an operation, an outsourcing contract defines the services being delivered, functional and non-functional requirements covered as part of the service-level agreement, related cost and other terms and conditions. However, delivery of core services is not the only service required by the customer. The procurement life cycle within the outsourced contract needs to include facilities for delivery of additional projects and functions that may be required by the customer, as requirements change over time. Therefore, the customer organisation needs to understand its direct relationship with the service provider and how to engage when additional work is required. The internal customer management team needs to manage internal relationships so that service delivery is consistent and that shadow services are not provided by other, non-authorised external parties. This imposes an additional burden on the customer account team, but it also gives them an understanding of all the functions delivered within their organisation, and a process that allows the customer to manage costs.

Government bodies, however, may not be at the size required to successfully obtain economies of scale by outsourcing service functions. Within the Australian market place for example, the federal government attempted to overcome this limitation by clustering smaller government bodies to provide a larger 'carrot' for the outsourced providers. Although this was conceptually a good idea, it did not take into account the differing requirements of the previously independent government departments. As one of the drivers was cost reduction, this in turn led to lower service levels delivered to a number of the agencies, and drastically decreased customer satisfaction. In addition, the ongoing management of the relationship and delivery of services required that both the outsourced service provider and the clustered agency worked as partners to meet all parties' needs and expectations. From a procurement life cycle perspective, this is a complex task due to the differing requirements for each party.

Transfer pricing issues impacting on outsourcing decisions

The issues of taxation implications for outsourcing, and transfer payments for sections of a project undertaken offshore, have some complex implications that affect outsourcing decisions, particularly where these may involve service delivery, like decisions being made about where a global organisation's head office may be located to benefit from corporate taxation rates or deciding where to place a production facility. This is a wide area that cannot be dealt with in depth in this book; however, some discussion and consideration of project management procurement implication is needed.

Multinationals have the ability to deliver and produce services and products that span national boundaries and jurisdictions. When a multinational transfers goods and services within its subsidiaries located in different countries, where are its profits taxed? Government authorities, such as the US Internal Revenue Service or Australian Tax office, regularly challenge prices for transferring products and services across boundaries and under the banner of the same parent company. The OECD Transfer Pricing Guidelines, which were published in 1995, with subsequent updates every year up to 1999 were released[1] to assist corporations and governments in defining transfer pricing calculations (OECD, 2001).

Defined guidelines maintain the arm's length principle of treating related enterprises within a multinational group, and affirm traditional transaction methods as the preferred way of implementing the principle. Corporate tax administrations, taxpayers and enterprise organisations, all have a share in avoiding conflicting tax rules which may impact on the development of world trade.

Multinationals have the ability to utilise their size and influence to ensure that the most cost-effective procurement strategies are available; this may be based on differing procurement strategies, or utilisation of offshore services and manufacturing to maximise profitability for the parent company. For transnational organisations, this may include the provision of services, procurement of materials or staffing needs from the most cost-effective locations. In Australia, for example, state governments often compete with each other when dealing with transnational corporations by providing incentives to these bodies. Such incentives may include taxation breaks (especially with regard to state levies such as payroll tax) or incentives to maintain facilities, such as support/investments within factories settings (Mitsubishi Australia received substantial funding support to enable the manufacturing facility to remain open in South Australia) as reported by Kohler (2004). Obviously, government bodies believe that there is a positive financial outcome in providing incentives to enterprises to set up services within their states, and corporations and organisations would be remiss not to take advantage of such offers. Organisations utilise their purchasing power to generate the offers made by government organisations to maximise their return on investments and meet transnational stakeholder expectations.

Consider the implication of cost-of-living calculations. Organisations moved from the provision of repeatable services from areas with high cost of living requirements (e.g. London, Sydney) to lower-cost locations within the same country (Kent, Newcastle NSW). This re-allocation of delivery resources to lower-cost locations started the trend towards outsourcing but was done on a location basis, rather than an external, task-orientated focus. Hence, transnational organisations are able to utilise their market influence and purchasing power to drive down the costs of procuring services whilst obtaining grants or incentives from governments to locate their services within a specific location or environment.

The next stage of procurement management is the multinational corporation. Such companies are able to use their market influence to purchase the same goods and services from the most cost-effective location, regardless of where it is located in the world. Governments can counter such action by restricting imports via regulation (restricting certain imports, tariffs, import duties) but such practices are decreasing as the global economy moves towards a free trade environment.[2] In addition, countries with larger gross domestic products are able to influence global trade by the manner in which their currencies are managed. The US Dollar provides a staple yardstick for inter-country transactions, and some countries are able to support their exports by artificially undervalueing their currencies when compared to the US Dollar, which increases the quantity of exports (Ong, 2004).

Procurement, however, requires that there is a transaction in place. Goods or a service must be provided from the supplier to the consumer. Not unlike a customer purchasing a car, the multinational corporation is able to shop around and explore the market place. A consumer is able to obtain quotes and minimise the price that they would pay for the goods or service. Corporations take the same action in order to maximise their profitability and return on investments for their stakeholders. This includes, but is not limited to, purchasing in on-shore, near-shore and off-shore locations. Each of these locations has its own advantages and disadvantages.

On-shore locations are those in the same region or country as the parent organisation. This may be a conscience decision or a mandatory requirement. For example, the US Government would not allow for the delivery of sensitive (or any) services to support its deliverables from locations not defined as USA locations. Near-shore locations are becoming more popular as they have the advantage of being more cost effective (cheaper) but maintain a proximity to the home country. Examples are generally based on the location of the home country, but may also include Mexico or Puerto Rico when servicing the United States, or Malaysia when supporting Singapore. Near-shore locations may have the advantages of lesser problems with transportation costs, language requirements and time zones, but are not the cheapest alternative for an organisation to pursue. Off-shore locations (off-shoring) have become a buzz word with the services industry. This entails the delivery of services from locations remote from the home country. Languages issues, time zones and distance may be negatives to such strategies, but these are offset by the reduction in cost associated with such activities.

Each of these strategies can be addressed in two distinct ways. An organisation may purchase services from a specialist provider and then incorporate their deliverables into their value chain. This is a practice that Small and Medium Enterprises as well as Multinationals can utilise. Such a practice would not generally undergo any legal requirements or impinge on any legislative requirements. However, multinational corporations are in a position to either own off-shore delivery centres or purchase them if it suits their financial goals.

In such instances, there is a need to maintain a level of distance between the corporations working together. Jeffrey Owens from the OECD Centre for Tax Policy and Administration states the following in his on-line article[3] (Owens, 2006):

> Globalisation brings costs and benefits, even for the tax professional. The move towards a borderless world has opened up new opportunities for taxpayers to minimise their overall tax liabilities. Much of this tax planning is legitimate.

However, he adds the following cautionary note: 'Another all-too-common practice is for some large corporations to manipulate transfer prices between subsidiaries to artificially shift income into low tax jurisdictions and expenses into high tax jurisdictions' (Owens, 2006).

Transfer pricing plans and processes have changed considerably over the past 30 years, and the economic realities for dealing with differing jurisdictions with differing legalisation require careful planning and evaluation of the 'rules' within areas, where the multinational may operate. For example, in the 1970s, the management of transfer pricing was based on behavioural theorists. This required negotiated pricing plans, and was done between companies without external interference.

As governments understood that management of multinational corporations required a more hands-on approach to transfer pricing management, a different process for setting prices was utilised. Pricing management was defined on the basis of either perfect market conditions or imperfect market conditions (Ho and Lau, 2005). For perfect market conditions the pricing is based on market price. In simple terms this is defined as either the 'as is' or 'adjusted' market prices. The 'as is' price relates to that in which the commodity is available within the market place. For example, if the product needed is available at $100 within the market place; that is the internal transfer price which is used. The 'adjusted' market price needs to take into account economies of scale, profit margins and other internal costs that should not have to be double-counted when dealing with services available from within an organisation. Economically, it would make little or no sense to have overhead costs that are generally attributed to headquarters counted in two locations, hence the 'adjusted market price' may be less than that available within the market place and needs to take into account all costs associated with the delivery and production of services.

There are a number of associated costs that are identified as relating to the practice and process of outsourcing or out-tasking and therefore related to the management of transfer pricing. As provision of services from off-shore, near-shore or on-shore locations must be managed and maintained, in regards to whether the services are delivered internally via a subsidiary company or from an external source, the structure and expectations requirement in management of relationships is similar to outsourcing contract management. These factors are not generally included in the standard delivery costs invoiced for service rendered as they are known as hidden costs. As stated earlier, Barthélemy (2001) identified a proportion of the hidden costs associated with outsourcing contracts and their management and maintenance, which is worth repeating here:

1   searching for and identifying a suitable vendor;
2   moving services to the external service provider;
3   managing the day-to-day contract and its service delivery and
4   the effort required once the outsourcing contract agreement has been completed or terminated.

Any provider, be it internal or external, can help reduce costs that are identifiable and quantifiable. This can be by the manner in which services are offered and delivered, or by working closely with the parent organisation to determine needs, specific requirements and service-level agreement definitions that are suitable for the delivery of goods or services required by the organisation. It is not cost effective to deliver services beyond the needs or requirements of the customer, as these will generally incur additional costs and may not be beneficial to the customer's value chain, and will reduce profitability for the total organisation when considering the delivery as a component of the total value chain. Efficient delivery of service should be the aim of both parties to an external supports goods or services contract.

The list of identified hidden costs does not include all of the costs involved in managing external delivery. The costs involved in the management of the service contract should include both the costs incurred in external management (e.g. vendor–customer relationship management) but also the internal management costs. Within a large-scale enterprise, the finance department's role should include liaison with the other internal departments or business units, with a goal of ensuring that the profitability of the organisation is maximised; this leads directly to the management and definition of transfer pricing. As enterprises can consist of a substantial number of departments, this internal function could involve considerable time, effort and expense. Although not as pronounced in the smaller-sized, medium enterprises (<$100M contracts), internal management costs still need to be considered.

As IT management can be a highly complex process, it is important to have an internal liaison team that is able to determine what IT needs can be met externally, and those services that should remain within the organisation. The liaison team also needs to be able to look at IT objectively and determine if vendor proposals are effective, and if the organisation has the ability to utilise economies of scale to minimise additional IT costs. This becomes especially apparent when considering sharing resources within and outside the sponsoring organisation and in highly integrated value chains.

The ability to define where and when overhead costs are leveraged against any organisation tends to require careful corporate planning, especially when dealing with differing corporate tax rates across the world. Figure 1.3 depicts global corporate tax rates for 30 OECD countries (Warburton and Hendy, 2006: xxiv) and show both weighted by GDP at purchasing power parity in 2003 and unweighted averages for 2005. Warburton and Hendy (2006: 32) state that

> Unweighted averages are the simple average of a range of observations. Only unweighted averages are reported in the OECD's Revenue Statistics – the source of most of the data used in this report. This

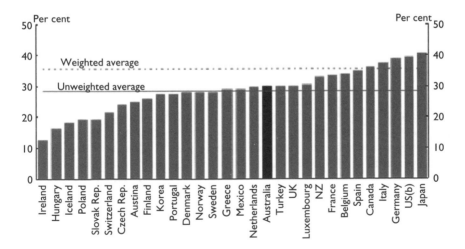

Figure 1.3 Statutory corporate tax rates OECD-30 2006.

Source: Warburtun and Hendy, 2006, See http://comparativetaxation.treasury.gov.au/content/report/downloads/CTRfull.pdf

means that each country's tax burden is given an equal weight in the average of all OECD countries' tax burden, regardless of whether that country is Australia, which in 2003 contributed 1.9 per cent of the total GDP of all OECD countries, or the United States, which contributed 36.3 per cent of total GDP. In comparison, weighted averages scale the tax burden of each country by that country's contribution to total GDP. Unweighted averages best reflect the diversity of experience in the OECD. In contrast, weighted averages may be useful if the observer wants to understand the proportion of economic activity relevant to a particular issue – for example, the rate of growth of OECD GDP is a weighted average.

Multinational corporations that have the ability to measure profitability in differing parts of the world would obviously prefer to register profits in the most economically suitable location. An example of such would be a corporation having its head offices in Ireland rather then the United Kingdom so as to minimise taxation costs.

To counteract this process, governments define their transfer pricing rules as requiring a number of tests in order to be approved. In Australia, for example, transfer pricing requires that each transaction be considered as an arms-length process. That is to say that there are two independent parties involved and the negotiated price is one agreed to between the parties. This agreement is then verified by following a four-step process (Bowman, 2006):

1   A functional analysis of the business is undertaken.
2   Publicly available data is used to determine an arms-length transfer price.
3   A comparison is then made to the materials lodged by the organisation.
4   An Year-on-year review of the process is undertaken.

The aim of this process is to ensure that commercial, realistic prices are set and used for transfer pricing requirements. Although initially onerous, once set, an organisation need not undertake every step in this process, but merely update any changes as defined. The issue with such a process is that it is self-regulating and the information is required only when a company is asked for the data to validate a claim or report. The primary requirement for organisations is to document the process and ensure that the data is maintained and validated on a regular basis.

Organisations do have a number of methods available to them when determining how to calculate the costs of goods and services and the subsequent transfer price used between organisations. These are based, not only on the method of production but also on the risk and rewards for each company (Aaron, Ackermann, Schuhmacher and Triplett, 1998). Arms-length pricing is generally used for manufacturing or non-routine deliverables. An example would be car engines built in one country but installed in a second by the same umbrella company. Non-routine deliverables would be activities such as business processing for sales and marketing, or business process management supporting multi-divisions within an organisation but which are delivered from an off-shore location. This may also include helpdesks, application support and other intangible deliverables that generally require an interaction between two individuals to resolve rather than a face-to-face meeting.

For organisations that are tightly integrated within their value chain, the methods of determination for transfer pricing become more difficult. Arms-length pricing models are not suitable when the determination of who is responsible for specific components of the delivery is not clearly identifiable (Aaron *et al.*, 1998). In such instances a profit-split process can be employed; however, this requires an understanding (and subsequent documentation) of the risk each party is taking. This can be calculated as the value of each organisation's investment in the project or the actual dollars placed at risk. Such calculations are more interesting as the risk and reward should be calculated in constant currency. Constant currency is an exchange rate that eliminates the effects of exchange-rate fluctuations and that is used when calculating financial performance numbers. Companies with major foreign operations often use constant currencies when calculating their yearly performance measures.[4] Although not immediately apparent, this provides for a baseline cost measurement that is consistent year over year and is well documented throughout the accounting profession.

As an example, the following statement is added to IBM results as reported to the Exchange and Securities commission and included within their press releases available from www.ibm.com/investor/2q06/2q06earnings. phtml:

> In an effort to provide investors with additional information regarding the company's results as determined by generally accepted accounting principles (GAAP), the company has also disclosed in this press release the following non-GAAP information which management believes provides useful information to investors.

IBM results:

- without non-recurring items,
- without divested PC business and
- adjusting for currency (i.e. at constant currency).

Arms-length transactions are the preferred method of calculation for transfer pricing, but they are not the only method available. As we have seen, the processes' needed to calculate transfer pricing for tightly integrated value chains becomes more difficult as the risk and reward measurements and the level of investment should be taken into account. For more complex transactions, organisations also have the option of purchasing a supplier and using them as an external supplier and to manage the taxation and profitability issues needed to maximise profitability for the parent organisation.

Transfer pricing requirements and profitability within each country do affect the financial impact and tax rates associated when making economic decisions with respect to how organisations structure their businesses. Service-based functions, such as call centres are delivered remotely and would not be subject to transfer pricing requirements. Rather, the delivery of the service and the subsequent profitability can be based around the pricing requirements. Does an organisation price its services based on where the service is delivered or on where it is provided? Is the profitability based on where the service is delivered from or where it is used?

Production costs are not the only consideration when dealing with production environments, such as 'widget manufacture'. For example, differential tax rates such as corporate taxations, import duties, withholding tax and tax credits will affect the overall costs associated with manufacturing. However, taxation issues can influence procurement choices and approaches for projects using a global supply chain.

Privatisation

Privatisation of government-owned organisations is a different strategy to outsourcing. For many government agencies, the possibility of privatisation is not available. It would not be possible to successfully privatise the total operations of the Australian Tax Office or the Public Heath system, as these are government services needed to successfully run the country. These agencies are candidates for outsourcing of services to support delivery requirements. Privatisation is a strategy that can be used to transfer publicly owned organisations into private hands (who are alleged to be better placed to effectively manage business activities). This includes functions such as utility companies – power, light and water – which are not only able to be regulated by the governments but also provide an opportunity for private organisations to reap benefits from technology upgrades and economies of scale.

Privatisation provides the government with the opportunity to focus on the core deliverables required to support the population but also to ensure that the essential, but not life-threatening, services are still provided; although recent experience in Australia has shown that governments have privatised functions that should be managed as non-profit organisations – for example, transport services and emergency assistance (Police, Ambulance and Fire-response numbers).

When considering privatisation, the sponsoring body must not only consider the immediate impacts of the move, such as reduction in public deficit, but also the long-term implications (Walker and Walker, 2000). There has been much debate within Australia over the sale of Telstra (Australia's government-owned Telecom provider) to private hands, see Daniel (2002) or www.citizensagainstsellingtelstra.com/. The debate should consider the long-term economic effect of selling one of the country's most profitable companies but the majority of discussion is on the upgrading of services to rural and disadvantaged areas. Although such debate is fruitful for those affected, it does not consider the reasons why the government-owned agency did not provide services to these areas despite being a wholly owned government agency. Such actions must be justifiable and it has now become a political debate as to how to require a private company to deliver services to a non-profitable area. By funding such activities, the government is reducing the net present value of the sale of Telstra and showing the inadequacies within the strategy.

Privatisation of service provision could be considered an effective strategy for the management of currently managed government services provided that the long-term effects of such behaviours are considered and any risks mitigated. Although most of us live in a capitalist society, the wholesale privatisation of utility government services needs to be considered in terms of the overall economic effect. This must include the immediate benefits of debt reduction, end-of-life maintenance and replacement costs being borne by non-public funding, and improvement in services as well as the long-term effects of having a guaranteed revenue stream removed from government hands, and the longer-term effect of not having overall control over the staffing and delivery of these services. Private companies' goals are centred upon maximising profitability for stakeholders; as a government agency its goals are where to provide a basic level of service for all consumers. Although these goals are not mutually exclusive, the provision of services for non-profitable areas and groups would not be maintained by the private operators unless government regulation and/or funding is provided. In this situation, what is the benefit to the private operator if they are still bound by management decisions beyond their control and are not able to work to maximise the profitability for their stakeholders?

Outsourcing and privatisation are two strategies that have been employed by governments to attempt to maximise the levels of service provided to the public. Outsourcing requires a long-term commitment to the provision of services by an external party. If the contractual details are managed from a strategic standpoint and consideration given for the longer-term requirements, such as additional service provision and termination clauses, then outsourcing can be used successfully by a government agency to provide a range of services needed so that the agency can concentrate on its core business. Care must be taken when drawing up such contracts as the management costs form a part of the overall cost of ownership. Consideration must also be given to switching cost, relationship management within and outside the agreements, as well as the cost of providing additional service. For example, what are the costs associated with having a third party provider deliver services to the agency that will then need to be managed by the outsourcer? What is the cost of transition? Finally, in a delivery model where multiple outsource providers are working together to deliver services such as infrastructure, network and application support, who is responsible for the overall service delivery?

Governments may start the privatisation process by separate legal entities, such as crown corporations, that generate profit and operate on a fully commercial basis prior to being sold as an entity or floated on the market as is the case with Telstra in Australia, or happened with British Airways in the UK. A non-government version of privatisation can be seen as the encouragement and development of spin-off companies or use of 'skunk-works' entities. These are purposeful and strategic decisions to encourage development through outsourcing experimental high-risk ventures to parts of an organisation that can be later divested or incorporated within the host organisation in a way that matches its culture and governance structures. Privatisation and developing spin-off companies can be seen as a high-level strategic outsourcing economic response.

Chapter summary

In this chapter we focused on value generation to stakeholders and how this may be achieved by organisations in a way that cascades value through the supply chain. The discussion in this chapter acknowledges that much of the PM work will be outsourced to a project supply chain, and so the value generation propositions for these groups involved in a project needs to be understood so that a win–win outcome can be not only achieved, but also through understanding value triggers throughout the PM supply chain, the generation of value can be designed into the procurement process. Later chapters in this book elaborate on a range of aspects of how value can be designed into the PM procurement process.

Vignette

OmniTrip is the organisation that won the tender for logistics support and management of the privatised integrated public transport system for Southland, a major city of 4 million people that until 10 years ago, had owned and operated a State-owned bus, tram, and train (both above and underground) public transport system. OmniTrip's contract was to provide services and maintenance of legacy transport assets and gradually replace these with new assets as part of Southland's government renewal and improvement policy. OmniTrip is a global organisation with experience in delivering similar services in 15 other cities in North America, Asia, Northern Europe and South America. Southland's Energy Efficiency Legislation requires phasing out hydrocarbon dependency and moving from the mix of current hydrocarbon diesel-fuelled 100-bus fleet to a hybrid hydrocarbon-electric fuelled fleet, moving towards sourcing of green-energy-rated electricity for the train system and phasing in in-ground power for the tram system, removing of overhead power cables for its 200-kilometre tram network over the next 20-year period and replacing the 600 existing trams with state-of-the-art energy-efficient trams. It also will replace the 150 trains with more energy-efficient and high-security and high-comfort trains.

Its coordination activities include: managing the procurement and maintenance of transport assets; staffing of all operations including facilities management of stations (owned by Southland); and developing a new generation of stored value smart card ticketing systems. Clearly, these complex operations represent a series of projects and programmes of projects in portfolios. In the interests of Southland's mandated best-value legislation, OmniTrip has decided that open, or even selective, tenders present best value alternatives. Provision of in-ground power for the tram system and removal of overhead power cables provide an illustration of the procurement choices facing OmniTrip.

The engineering for undertaking this task involves integration of many complex engineering groups: civil engineering, track engineering, signalling, environmental and safety engineering to name but a few. While OmniTrip's joint venture (JV) partner ComfortZone has undertaken similar work in three similar-sized cities in the world, each city has experienced different geological and social challenges so that scoping this phase of the overall programme for the tram network has proved difficult. ComfortZone could use its in-house skills gathered from its experience in the three cities or it could choose to outsource parts of the task to up to 10 of its overseas supply chain partners. Legal advice suggests that this type of project could benefit from the formation of a separate entity or use out-tasking because of the potential intellectual property (IP) to be gained in learning an improved way of delivering this kind of service and facility on many other similar projects. It also needs to consider capitalising on potential revenue from cross-business entities that can realise income from carbon-trading (see Chapter 4 triple bottom-line aspects) as more of its supply chain has recently joint ventured with legal entities that can maximise revenue from these carbon trade activities. Further, the business world has significantly changed over recent years, and the uncertainty associated with carbon-tax liabilities and trading in carbon credit off-sets has radically changed the notion of 'lowest cost' because costs and revenues associated with sustainability legislation (currently in place and anticipated) has radically affected procurement policies and structuring outsourcing agreements. The tram track works that OmniTrip and ComfortZone are currently grappling with represent a small part of the procurement decision-making policy that this JV is experiencing.

Issues to ponder

1   Describe three key aspects of value that you believe that OmniTrip offers Southland that its JV with ComfortZone enhances.
2   What top three issues relating to the make-or-buy decision do you see relevant to this case?
3   Discuss three main justifications that you would expect could be argued for and against the privatisation of the services originally offered by Southland to its citizens.
4   How does the uncertainty indicated above relating to sustainability potentially affect the concept of an organisation delivering best value?
5   Using Table 1.1 as a template, select three of the six outsourcing dimensions of procurement issues raised in this vignette and note what additional information should be provided to make a reasonable procurement decision about outsourcing.

Useful web resources

The following web sites accessed during 2006 should be of value to readers. These are only a few samples of interesting procurement-related web sites that illustrate procurement policies or strategies and tools and related information.

*   Commonwealth Procurement Guidelines January 2005 for the Australian Commonwealth Government Department of Finance and Administration, www.finance.gov.au/ctc/commonwealth_procurement_guide.html

- The Australian Procurement and Construction Council which is the peak council of departments responsible for procurement and construction policy of the Australian Commonwealth, State and Territory Governments. www.apcc.gov.au/
- Directory link to e-procurement tools can be found at www.builderspace.com/divisions/procurement-software.html
- The CIB web site provides references to many related conference proceedings, there are a number of related commissions follow links URL is www.cibworld.nl/website/

Notes

1 See URL www.oecd.org/document/34/0,2340,en_2649_33753_1915490_1_1_1_1,00.html to link to the OECD bookshop.
2 See www.fta.gov.au/ for more details and current updates on trends.
3 See URL www.oecdobserver.org/news/fullstory.php/aid/1945/Tax_in_a_border-less_world.html
4 See the web site www.investopedia.com/terms/c/constantcurrencies.asp for this and other explanation of other financial terms.

References

Aaron, D. M., Ackermann, R. E., Schuhmacher, C. W. and Triplett, C. S. (1998). 'Proposed IRS Regulations on Global Dealing: Ignoring Capital and Taxing it too?' *Ax Management Memorandum*. **39**(22): S322–335.
Barney, J. (1991). 'Firm Resources and Sustained Competitive Advantage'. *Journal of Management*. **17**(1): 99.
Barthélemy, J. (2001). 'The Hidden Cost of IT Outsourcing'. *Sloan Management Review*. **42**(3): 60–69.
Bendor-Samual, P. (2002). 'Outsourcing, Directions and Decisions for 2003'. *CIO Magazine – Advertising Supplement*.
Bourne, L. and Walker, D. H. T. (2005a). *Stakeholder Chameleon – Ignore at Your Peril!* PMI Global Congress 2005 – Asia Pacific, Singapore, 21–23 February, Newtown Square, PA: Project Management Institute: 8 pp.
Bourne, L. and Walker, D. H. T. (2005b). 'Visualising and Mapping Stakeholder Influence'. *Management Decision*. **43**(5): 649–660.
Bowman, C. (2006). 'A Mystery No More'. *Charter*. **77**(9): 56–58.
Bröchner, J. (2006). Outsourcing. *Commercial Management of Projects Defining the Discipline*. Leiringer, D. L. R. Abingdon, Oxon: Blackwell Publishing: 193–206.
Chiles, T. H. and McMackin, J. F. (1996). 'Integrating Variable Risk Preferences, Trust, and Transaction Cost Economics'. *The Academy of Management Review*. **21**(1): 73–99.
Conner, K. R. and Prahalad, C. K. (1996). 'A Resource-based Theory of the Firm: Knowledge Versus Opportunism'. *Organization Science: A Journal of the Institute of Management Sciences*. **7**(5): 477–501.
Cooke-Davies, T. (2002). 'The "Real" Success Factors on Projects'. *International Journal of Project Management*. **20**(3): 185–190.

Daniel, Z. (2002). 'Are There Benefits to Privatising Medibank Private?' *The Business Report.* Australia, Australian Broadcasting Corporation, Radio National, Saturday 4 May 2002. Available at http://www.abc.net.au/rn/talks/ 8.30/busrpt/stories/s550924.htm (accessed on 28 July 2007).

Domberg, S. and Hall, C. (Eds) (1995). *The Contracting Casebook – Competitive Tendering in Action.* Series The Contracting Casebook – Competitive Tendering in Action. Canberra: Australian Government Publishing Service.

Eisenhardt, K. M. and Martin, J. A. (2000). 'Dynamic Capabilities: What are They?' *Strategic Management Journal.* **21**(10/11): 1105–1121.

Fill, C. and Visser, E. (2000). 'The Outsourcing Dilemma: a Composite Approach to the Make or Buy'. *Management Decision.* **38**(1): 43–50.

Gay, C. L. and Essinger, J. (2000). *Inside Outsourcing: The Insider's Guide to Managing Strategic Sourcing.* London: Nicholas Brealey Publishing.

Green, S. D. (1999). 'A Participative Research Strategy for Propagating Soft Methodologies in Value Management Practice'. *Construction Management and Economics.* **17**(3): 329–340.

Ho, H. K. D. and Lau, T. Y. P. (2005). 'An Exploratory Study of Transfer Pricing Practices by Multinationals with International Affiliates'. *International Tax Journal.* **31**(4): 37–54.

Jacobides, M. G. and Croson, D. C. (2001). 'Information Policy: Shaping the Value of Agency Relationships'. *Academy of Management Review.* **26**(2): 202–224.

Johnston, R. (2004). 'Towards a Better Understanding of Service Excellence'. *Managing Service Quality.* **14**(2/3): 129–133.

Kohler, A. (2004). 'Mitsubishi Here to Stay: Phillips'. Australia, Australian Broadcasting Corporation (ABC), 24 May, http://www.abc.net.au/insidebusiness/ content/2004/s1113947.htm (accessed on 7 January 2007).

Landis, K. M., Mishra, S. and Porrello, K. (2005). Calling a Changing in the Outsourcing Market, The Realities for the World's Largest Organizations, Outsourcing Study. New York: Deloitte Consulting.

Lobez, S. (1999). 'The Brave New World of Contracting Out, Privatisation and Outsourcing'. The Law Report – Radio National. ABC Transcripts, Radio National, ABC Radio, 28 September, www.abc.net.au/rn/tranlist.htm#Law (accessed on 7 January 2007).

Murray, M. and Langford, D. A. (2003). *Construction Reports 1944–98.* Oxford: Blackwell Science Ltd.

Nogeste, K. (2004). 'Increase the Likelihood of Project Success by Using a Proven Method to Identify and Define Intangible Project Outcomes'. *International Journal of Knowledge, Culture and Change Management.* **4**: 915–926.

Nogeste, K. and Walker, D. H. T. (2005). 'Project Outcomes and Outputs – Making the Intangible Tangible'. *Measuring Business Excellence.* **9**(4): 55–68.

OECD (2001). *Transfer Pricing Guidelines for Multinational Enterprises and Tax Administrations.* Geneva: Organisation for Economic Co-operation and Development.

Ong, L. (2004). 'Is China's Currency Peg the Culprit behind US Economic Woes?', http://www.aseanfocus.com/asiananalysis/article.cfm?articleID = 711, 7 January.

Owens, J. (2006). 'Tax in a Borderless World. Achieving Tax Compliance is a Challenge Facing Governments the World over. Action can be Taken.' *OECD Observer.* October.

PMI (2004). *A Guide to the Project Management Body of Knowledge*. 3rd edn. Sylva, NC: USA Project Management Institute.

Prahalad, C. K. and Hamel, G. (1990). 'The Core Competence of the Corporation'. *Harvard Business Review*. 68(3): 79–91.

Robbins, H. and Finlay, M. (1997). *Why Teams Don't Work – What Went Wrong and How to Make it Right*. London: Orion Publishing Group Ltd.

Stark, J., Arlt, M. and Walker, D. H. T. (2006). *Outsourcing Decisions & Models – Some Practical Considerations for Large Organizations*. International Conference on Global Software Engineering ICGSE 2006, Costão do Santinho, Florianópolis, Brazil, 16–19 October, IEEE: CD-Rom Sessions, Research Papers I, GSE Management – 6 pp.

Teece, D., Pisano, G. and Shuen, A. (1997). 'Dynamic Capabilities and Strategic Management'. *Strategic Management Journal*. 18(7): 509–533.

Treacy, M. and Wiersema, F. (1993). 'Customer Intimacy and Other Value Disciplines'. *Harvard Business Review*. 71(1): 84–93.

Turner, J. R. and Müller, R. (2003). 'On the Nature of the Project as a Temporary Organization'. *International Journal of Project Management*. 21(3): 1–8.

Walker, B. and Walker, B. C. (2000). *Privatisation Sell off or Sell out? The Australian Experience*. Sydney: Australian Broadcasting Corporation.

Walker, D. H. T. (2003). 'Implications of Human Capital Issues'. In *Procurement Strategies: A Relationship Based Approach*. Walker D. H. T. and K. D. Hampson (Eds). Oxford: Blackwell Publishing: 258–295.

Warburton, R. F. E. and Hendy, P. (2006). *International Comparison of Australia's Taxes, Benchmarking Report*. Canberra: Australia Government Treasury, ISBN 0 642 74339 8: 449.

Willcocks, L. and Choi, C. J. (1995). 'Co-operative Partnership and "Total" IT Outsourcing: From Contractual Obligation to Strategic Alliance?' *European Management Journal*. 13(1): 67–78.

Williamson, O. E. (1975). *Markets and Hierarchies, Analysis and Antitrust Implications: A Study in the Economics of Internal Organization*. New York: Free Press.

Williamson, O. E. (1991). 'Strategizing, Economizing, and Economic Organization'. *Strategic Management Journal*. 12(Special Issue): 75–94.

Chapter 2

Project types and their procurement needs

Derek H. T. Walker and Steve Rowlinson

Chapter introduction

Early project management texts tended to assume that projects were substantially tangible and flowed along a well-trodden path. The early texts such as that edited by David Cleland and William King (1988) have very structured and traditional content that stresses techniques, examples from solid tangible projects such as those found in construction, advanced manufacturing, aerospace, with some minor attention directed towards IT projects from a programming and development perspective. The promise of business-development-type projects and intangible projects such as change management and process re-engineering has been acknowledged, but there was little written about these. That situation has been more recently remedied and there is a rapidly growing interest in these less-tangible-projects, and new texts have been written to reflect this; for example, the one edited by Peter Morris and Jeffrey Pinto (2004). That book, which will probably remain a 'must' for most project managers in the future has, for example, sections devoted to strategic portfolio programme management and supply chain management. That particular text clearly demonstrates the flowering of PM attention to being not only a part of general management theory but also of leading the way. Most management theory, particularly strategy development, have stressed the need for better implementation (Mintzberg *et al.*, 1998) but mainstream management texts generally fail to adequately grasp the PM nettle. We suspect and suggest that is because mainstream management literature has viewed implementation and 'operations' as a lesser form of intellectual pursuit than strategy and planning and so has generally overlooked the PM literature. We do not wish to appear as complaining about this perceived shortcoming; we observe this as a theory gap worth filling. We also assert that this gap can only be effectively filled by absorbing and reframing mainstream management theory to a project, programme and portfolio context. By doing so, we attempt to expand the interest in PM theory and practice. One inevitable outcome of this changing focus is that we must recognise the range of types of project and the management approaches that necessarily flow from that realisation. This chapter adds to that emerging perspective as well as providing a hindsight perspective on the forms of procurement options that have evolved over the decades to cope with a diverse range of project types and their realisation needs.

This chapter is thus presented in three broad sections and a vignette at the end of the chapter presents a scenario, and prompts relevant questions flowing from that scenario. The first section deals with a discussion on project types and their needs. The next section logically follows by outlining the various procurement options and how they may best fit the project type and need requirements. The third section focuses on writing and evaluating project proposals, and also contains a brief description of where contract administration fits into this picture. The chapter ends with a vignette and suggested activities and questions that readers can tackle to better frame their learning from this chapter.

Project types, forms and phases – influencing procurement choice

The PMBOK stresses the need for deciding what a project should deliver and how to plan to deliver that objective (PMI, 2004). There remains, however, a tension between taking a top-down and bottom-up approach to defining the scope of a project. The bottom-up approach essentially relies on a large number of well-understood and well-identified components that can be grouped into assemblies and these configured into subsystems, and thence into systems. A project becomes the summation of these systems that delivers a need. That need can have quite different systems that when brought together satisfy that need. Moreover, systems can include tangible parts such as a building or a weapon such as a ship/plane or missile delivery platform, as well as intangible systems such as training in the correct use of the weapon system, or commissioning a building including training in maintenance and facilities management, for example. Even these 'hard' or tangible projects have some 'soft' deliverables. Crawford and Pollack (2004) provide a useful discussion on this topic and link the concept of 'hard' projects or project objectives with objectivist scientific paradigms that believe that the truth lies 'out there' as a fixed reality. This approach led to a predominant systems view, while 'soft' projects relate more to subjective concerns, and is interpretive in nature, seeing sub-text and inexplicit needs as being highly important. This is a view that will be discussed further in this book, particularly in Chapter 3 relating to stakeholder management and in Chapter 6 relating to project performance measures.

The dichotomy is not complete as there has been an emerging use of a modified systems thinking approach (soft systems methodology or SSM) in project management, first developed by Checkland (1999) and used more recently to investigate project learning from experience and project learning histories (Maqsood *et al.*, 2006). Further, Crawford and Pollack (2004) identify a framework of seven dimensions that form the basis for a framework of understanding both hard and soft issues in analysing both hard and soft sides of the project. These are (Crawford and Pollack, 2004: 646) as follows:

1 the clarity of goals/objectives – how clearly these have been expressed;
2 the tangibility of goals/objectives – how well these can be defined, visualised or otherwise expressed;
3 success measures – the kinds of measures used to describe project success;
4 project permeability – the extent to which the project is subject to external risk and uncertainty influences;
5 number of solution options – the approach to exploring and refining goals;
6 degree of participation and practitioner role – the roles that team members adopt when managing a project and
7 stakeholder expectations – the view held of what is a valid expectation for the project that influential stakeholders believe to be reasonable.

They argue that this framework can be adopted by project managers when developing a strategy for delivering a project, and mapping issues against the seven dimensions so that they can avoid having a poorly balanced implementation and project definition strategy. If the expectation for the project is substantially 'hard' then there will be a clearly different PM approach to be appropriately adopted than for a 'soft' project. In their paper they illustrate their framework with three different case studies from public sector organisations, and conclude that the framework was useful in providing a structure to better analyse what went on in those projects to generate lessons learned. This kind of framework, based on experience and project histories gathered using the framework to provide a consistent way of making sense of project delivery performance, can be helpful at the project procurement design stage. For example, by improving understanding of hard and soft issues relevant to a particular project, a better fit for a delivery model, coordination of team members and communication media can be established.

During the beginning of the final decade of the twentieth century there was a vigorous debate about the value of and optimal form of developing a work breakdown structure (WBS) for a project in order that effective plans, monitoring, coordination and control could take place. Bachy and Hameri (1997), discuss the planning approach undertaken in a highly complex 10-year 'ordeal' as they call it, to develop a large expensive piece of scientific equipment – a Large Hydron Collider (LHC) – which allows scientists to look into the structure of matter. Their paper clearly favours the positivist paradigm. They advocated first developing a detailed and highly explicit product breakdown structure (PBS) or parts-list that defines the minutiae in an intensely bottom-up fashion. They then detailed an assembly breakdown structure (ABS) that describes how the parts should be assembled, and incorporated logistics and material availability and other relevant production and manufacturing data and information. They used the ABS detail information to develop a top-down approach tree-like representation of the project as a series of connected sub-systems or work breakdown structures (WBS) organised by function and component. The head of the WBS was the 'tangible project outcome' and a series of high-level components cascaded down into a breakdown of the work to be done. Detailed WBS work packages define suitable 'chunks' that they could provide a clear and simple description of what was to be done (objectives). An organisation breakdown structure (OBS) followed the WBS to define responsibilities and accountabilities and this, linked with the WBS, informed the way that resources would be consumed to deliver the project. The WBS and OBS formed the basis of the plan and in many ways is similar (but less detailed) to that used in the construction industry and other engineering applications often referred to as a global method statement and detailed method statement (Walker, Lingard and Shen, 1998). This approach is useful for integrating a range of plans, such as a risk management plan, an occupational health and safely plan, and an environmental plan (Shen and Walker, 2001) that can provide the basis for the project team to be sufficiently prepared for identified risks and have sufficiently modelled scenarios to be able to be in a position to cope with unexpected circumstances that distort the validity of planning assumptions (Walker and Shen, 2002). Lammers (2002) points out how complex the WBS on many projects can become and argues that the PBS also has a function of allowing tags or attributes to be linked to the WBS through the PBS so that a range of attributes can be stored in a data base that fully describes and defines the work packages. Turner (2000) recommends that work packages on a project of 1 year could have about 20 or so packages of work each delivering a project milestone with about 5–10 activities of work (each with resources and accountabilities clearly stated) within each work package that would take about 2 months to complete.

It is interesting to contrast the above with how PM and planning for control is handled in other industries where different paradigms prevail. The arts provide such a contrast as the view that prevails amongst creative PM teams is one of emergence rather than complete and rigid specification. For example, in sculpting an outcome such as the famous statue of David, the view allegedly expressed by Michelangelo was that he knew that the statue was inside the block of stone, he just had to liberate it (Briner, Hastings and Geddes, 1996). The live entertainment industry also provides another useful example. Hartman, Ashrafi and Jergeas (1998) provide an account of 14 entertainment project events in Alberta ranging from sports to music projects. It is clear from their account that while each event demands that they are completed on time, the mindset of the PM teams is quite different to the more traditional projects discussed in the preceding part of this discussion. They identify important major differences in PM practices that suggest different paradigm and value systems from construction of IT projects; for example: (Hartman *et al.*, 1998: 270). Artistic creativity is highly valued so that ephemeral and intangible tacitly 'understood' performance expectations are real outputs as well as outcomes that are expected – failure to deliver on expectations results in shame and loss of face. The way that work is defined and undertaken is fluid and undefined, as may be seen by the common saying about such projects, 'It will be OK on the night,' meaning that the culture forces people involved in project delivery at a range of levels in a 'hierarchy' to 'muck in' when required to ensure delivery on time. The authors (Hartman *et al.*, 1998) and others (DeFillippi and Arthur, 1998) stress that we have much to learn from such projects. It is interesting, from a procurement point of view, that many of these projects use volunteer labour or those who expect to sacrifice their present immediate potential return to gain 'experience' for expected future benefit not only from knowledge gained but also social capital such as trust and respect from those involved in the projects.

This leads us to pondering if there is a useful typology of projects that helps us to (1) identify different types of projects that would need a different delivery approach and (2) provide some guidance on how to design a procurement choice that best fits the need for the type of project. From the above discussion it seems obvious that a micro-management approach, a procurement choice with its inherent contract requirement and behaviour expectation inherent in the scientific project example (Bachy and Hameri, 1997), would be unenthusiastically embraced and delivered for the projects described by Hartman *et al.* (1998). Two sets of authors provide us with some guidance: Turner and Cochrane (1993) developed a goals and methods matrix that is useful in understanding the needs of different project types that can lead us to develop appropriate procurement strategies to meet the needs of these projects. Another way of looking at project types was presented by Shenhar and Dvir (2004). Both of these views shed light on our understanding of the impact that project type can have on procurement choice that encompasses contractual requirements which in turn may impose obligations that affect PM behaviour and culture.

Turner (1999) in his text book identifies a number of project types and PM styles that relate to these. In Turner's paper with Cochrane (1993) they identify four types of project based upon a 4-quadrant model of methods being well defined and goals being well defined (yes-no). Their typologies are as follows:

1   *Engineering projects* or 'earth' projects have both goals and methods solidly defined. Early published project case studies typically cite

construction, shipbuilding, aerospace and manufacturing projects as examples because they attract a 'scientific' view of operations management influence. The example presented by Bachy and Hameri (1997) typify this kind of project.

2   *Project development projects* or 'water' projects have poorly developed methods but well-developed goals. These are characterised as being somewhat fluid but structured in the way that a river, stream, lake or ocean naturally creates a boundary. The Bachy and Hameri (1997) case could have been looked at this way if the positivist paradigm of 'certainty of solution' (the research equipment design) was questioned as being merely 'the option'. However, it can be useful to view this kind of project this way if the procurement aim is to specify a fixed outcome (or set of outcomes) while leaving delivery methods flexible – rather than demanding any particular process of micro-management.

3   *Applications software development* or 'fire' projects have a well-defined methodology but poorly defined goals. The procurement emphasis may be directed towards requiring a particular methodological approach that is known (or reasonably assumed) to be successful while holding the end goal more fluid. The 'fire' could be seen as air that is focused somewhat like a flame generated by concentrating the rays of the sun through a lens.

4   *Research and organisational change projects* or 'air' projects have poorly defined methods and poorly defined specific goals. These are characterised as being illusive and generally invisible though these projects can be redefined – by including intangible goal elements through a process of linking intangible to tangible outcomes – see Chapter 6 and recent research in this area (Nogeste, 2004; 2006; Nogeste and Walker, 2005). The key to making these projects less difficult to deal with, is to either separate the outcomes into several phased projects or to fully link the tangible and intangible outcomes.

The key to success in using this typology is to recognise the realities of the type of project and to intelligently design a project governance structure, and through the procurement strategy require that this governance structure acknowledges the project type and ensures that an appropriate procurement choice delivers the appropriate project controls and goals/objectives/vision to match the type. Where necessary, projects may be segregated into several linked projects, such as one to deliver the output and another to deliver the outcome if a radically different paradigm means that one PM group is ill equipped to deliver both outcomes and that radically different styles and types of PM team are required.

Aaron Shenhar has devoted a considerable amount of energy in thinking about types of project and appropriate management styles. His co-authored book chapter (Shenhar and Dvir, 2004) builds upon a number of previous papers (Shenhar, Dvir and Shulman, 1995; Shenhar and Dvir, 1996; Shenhar, 1998; 2001; Shenhar, Dvir, Levy and Maltz, 2001) in which he extends the ideas identified above by (Turner and Cochrane, 1993). One of the extending features that Shenhar and Dvir identify is how pace as part of pace, complexity and uncertainty intervenes between the project product, task and environment and appropriate project management style. They develop what they call the Uncertainty, Complexity, Pace (UCP) Model (Shenhar and Dvir, 2004: 1267) where the Uncertainty dimension can be seen as a combination of novelty and technology. Their Novelty, Complexity, Technology and Pace (NCTP) framework was developed to guide project management style. The PM style can of course be influenced by procurement choices that favour an appropriate PM style. The framework has four dimensions.

Novelty is measured as derivative, platform or breakthrough. These describe the extent to which methods are well-defined. Derivatives are extensions of existing products or methods. Platforms are new generations and refinements in existing families of products or methods. Breakthroughs are paradigm shifts going beyond innovation to invention or significant reframing that develops a totally new way of looking at a problem.

Complexity is measured as assembly, system and array. Complexity refers to moving from an assembly to a system and array. An assembly involves creating a collective of elements into a component. A system, by contrast, involves a complex collective of entities into a new form, a re-configuration or re-framing of parts into a new whole with different characteristics to the pre-existing system. An array-type project radically shifts the paradigm. An array defence project may turn a set of physical network relationships into virtual ones where a radical new technology is introduced as the change agent.

Technology is classified as: low tech that relies on well-established technologies; medium tech that uses an existing technology base and incrementally extends it; high tech involves new technologies that may have been experimented with and tested in other contexts and so is 'new' in the applied context, but at least there is a reasonable body of knowledge of its impact and influences in other contexts; super high-tech projects are based upon new paradigms when the project was initiated.

Pace is perhaps the new concept in project types. Regular refers to an evolution as it happens with little sense of forced urgency. Fast/competitive projects are motivated by a sense of urgency so they do not follow a natural rhythm but are accelerated by force. Blitz/critical, as the tag implies, is driven by an acute sense of urgency and turbulence. The implication that these typologies present revolves, in procurement terms, around how to best encourage performance, accepting and trading risks, and developing a rewards and penalty structure that matches a project type to what can be otherwise developed by other PM teams. The image of flogging a dead horse may be crude but it is relevant in this context. Project sponsors need to recognise the project context, and match a risk and reward strategy as well as recognising the value generated by intellectual input of teams in different types of projects.

Selecting an appropriate management style involves assessing the nature of the project-external environment, including available resources, stakeholder expectations and so on, deciding upon the nature and characteristics of the end-product (deliverable) and the details of the task to be accomplished. These are represented by the 'shape' of the UCP model emerging from these influences. The response to the identified UCP risk profile should inform the project management style. The style involves how interactions between the project management team and its supporting supply chain will take place, the structure of the PM team and the way that they are integrated and coordinated, the tools that can be used to manage the project, the PM processes adopted and the type of people and the competencies, skills and motivations that they will bring to the project.

A badly structured risk acceptance provision in a procurement strategy can easily deter the 'best' potential project teams; clearly risk and rewards are linked. Even if a project sponsor relies heavily on the perceived motivation of the project team responding to an intriguing and engaging challenge, this may not prove effective in the presence of competing challenges. The degree of formality and micro-managing, the opportunities for divergent thinking and the trust that this requires, and the reliance on a stock of well-practised skills versus dynamic and emergent-experimental skills should frame the procurement choice, dependent upon the need of the project and the capacity of the PM team. This is an area of procurement and PM theory that is continually developing, where case studies contribute heavily to our understanding of a preferred way forward. When we look at just these two examples of ways of looking at project types (and there would be many others) we can see that project choice and appropriate PM approach is a complex and messy area where simple surveys of 'best practice' could be highly misleading.

We need to remember that projects are procured both internally (within an organisation) as well as externally (to an outsourced entity). The main drivers that influence procurement choice that we have seen in this chapter are related to the project characteristics (type of project), the project sponsors' expectations and their perceptions of what constitutes value, and the extent to which the rules and methods to deliver the project may be known or knowable. Project deliverables will generally have some mix of 'hard' and 'soft' deliverable elements and characteristics. These will link together with the environmental context that surrounds the project and shapes the project characteristics, which will also be affected by the complexity, uncertainty and delivery pace expected. An environmental factor affecting how a project may be characterised is the sponsor's expectations and value proposition because this will affect the expected role that governance, degree of formality and procurement and delivery processes expected to be used. The state of knowledge about governance rules and project delivery methods also influence the procurement choice. Further, in a paper by Shenhar and Wideman (2002), they construct a two-by-two matrix based upon a Y-axis of type of project work split into 'craft' or skills/training-based work and 'intellect' or education-creativity cells. Their X-axis cells relate to 'tangible' and 'intangible' types of product from the project. They cite traditional projects (construction, aerospace, shipbuilding, automotive), for example, as 'craft-tangible' and 'intellect-tangible' as a new mousetrap or development of a hybrid petroleum–hydrogen powered car. Examples of 'craft-intangible' projects could include updating procedures and processes, while 'intellect-intangible' projects could include producing a movie. This view is somewhat similar to the earlier hard and soft project idea.

Figure 2.1 illustrates the precursors to making a procurement choice based upon the literature, in particular ideas presented by (Turner and Cochrane, 1993; Shenhar and Dvir, 1996; 2004; Shenhar and Wideman, 2002). This not only moves beyond viewing project types simply as, for example, 'major expansion projects, development projects and operations and maintenance projects' but also includes the important project characteristic of the project's strategic purpose (Crawford et al., 2005) – because this factor may influence the extent to which it may be undertaken in-house as opposed to being outsourced.

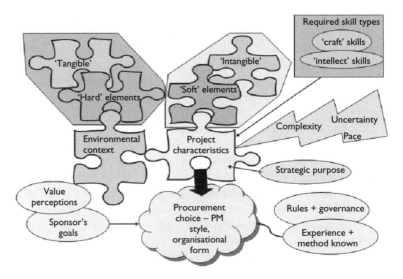

Figure 2.1 Project influences upon procurement choice.

In this book, we focus upon project classifications as being important to understand what a project may require for it to be delivered, and from that, the appropriate delivery procurement process that should be adopted. In Crawford *et al.* (2005: Appendix H) study report, they present a model of classification that links project classification purposes (for researchers, libraries, professional associations and organisations undertaking projects) with project attributes. Table 2.1 tabulates their 14 elements that they detail as project attribute categories.

While not all of these have been explained here, they do provide an indication of the diversity of project types and their demands in terms of finding an optimum solution to the procurement design and execution of their delivery. Clearly a 'one-size-fits-all' approach to procurement is unwise. Procurement systems should be designed to achieve the project objectives and deliver both tangible and intangible elements of perceived success – the value proposition perceptions indicated in Figure 2.1.

Having now set the scene for us to better understand what influences a project procurement choice and the preferred style and form of organisation that should deliver it; we now present a description of the types of procurement choices available. In this section, we will confine our discussion to forms of outsourced procurement choices. We discussed the make-or-buy decision in Chapter 1, so it is appropriate that we now discuss outsourcing procurement choices.

When making a procurement choice we need to be clear about the project phase being considered. The PMI's PMBOK defines project phases or stages as an initiation phase, several intermediate phases and a final phase (PMI, 2004: 21). Projects begin with a concept definition phase. A business or improvement need (concept) is identified, ideas are generated and justified. This phase involves developing the project idea based on satisfying some need, undertaking preliminary research into its likely scope, scale and viability (from a technical point of view), developing a business plan that tests its strategic and commercial viability, and sets the framework for further development. Successful passage through that gate permits a project definition stage to take place. In this phase the project vision, explaining its *raison d'être* and describing the preferred future that it will deliver, will be

Table 2.1 Project categorisation attribute

1 Application area/product of project	8 Project timing
2 Stage of life cycle	9 Uncertainty, ambiguity, familiarity
3 Stand-alone or grouped	10 Risk
4 Strategic importance	11 Complexity
5 Strategic driver	12 Customer/supplier relationship
6 Geography	13 Ownership/funding
7 Project scope	14 Contractual issues

articulated, and the design parameters will be established to create a clear boundary around the problem to be resolved so that it is achievable and understood. Project design may span several sub-phases such as a high-order conceptual design followed by (usually overlapping) a detailed design phase. There are often reality checks (stage gates) undertaken to test the validity of the project concept before the execution begins. Execution is preceded by a tendering or procurement phase where the team that will execute the project is selected, commissioned and mobilised. The execution phase, often on longer projects where the project deliverable is more than a highly tangible 'thing' that is simply designed and delivered, involves some element of design refinement and tuning coupled to the production process. The project is closed out when the deliverable is approved for acceptance, and there may be a further decommissioning phase considered where the 'spent' project need is dispensed with. Project phases and the processes whereby they are delivered are described in detail in methodologies such as PRINCE 2. For more detail refer to Bentley (1997).

Figure 2.2 illustrates the extremes of one way of looking at the choices available for project-outsourced procurement. One dimension indicates the degree of relationship between the outsourced entity and the project initiator. This relates to the closeness of the delivery team with the initiation group across the early phases of the project. With high-relationship situations, the delivery team will be intimately involved from the initiation stage in a joint problem-solving and advice-giving role. This may be channelled through a joint venture, alliance or other form of joint benefit-harvesting arrangement. At the other extreme, the relationship may be highly transactional with the project initiators simply buying in expertise to deliver the project with no joint benefit-sharing arrangements. Closely linked to the relational characteristics are those of culture, discussed in more depth in Chapter 9, most particularly, the degree to which trust,

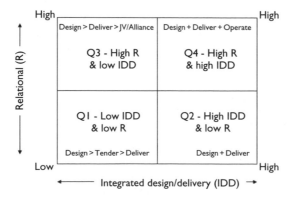

Figure 2.2 Broad procurement choices.

commitment and risk management strategies are interwoven. Each procurement choice carries with it different types of risk. The relational risk presents another consideration, particularly with alliances and joint ventures; this is where understanding, trust and culture implications become especially pertinent.

The design and delivery dimension refers to the degree of collaboration that occurs from the start of the initiation phase between the designers of the solutions and those who deliver it. In the construction industry, and many other manufacturing-type industries, the design group is quite separate from the delivering group, and the deliverer is selected on an open-tender method. The builder often tenders on a given design and builds strictly according to the specifications. Another transactional but more integrated form of this is where the project initiator develops performance specifications and the builder designs and constructs to that performance specification. In a more highly relational basis the entity that designs and builds also works with the project initiator to operate the facility and then hands that back to the initiator at some specified future time in an agreed manner (e.g. in perfect working order or completely dismantled and removed). In this latter case, the relationship is long term and ongoing due to governance provisions to ensure that when the close-out date for handover occurs, the expectations of the project conditions are ensured. A project sponsor with concerns about internal project delivery competence risk, for example, can deal with this by outsourcing – as discussed in Chapter 1, however, there needs to be a competence in the outsourcing process to do this. This risk can be managed by retaining the design processes, or both design and production can be outsourced, or even the entire operational management, as happens in the case of Build Own Operate and Transfer (BOOT) projects.

Procurement choices

Having broadly sketched out the options we will now discuss procurement choices in more detail.

Procurement is about the acquisition of project resources for the realisation of a constructed facility. This is illustrated conceptually in Figure 2.3, which has been adapted from that produced by Austen and Neale (1995) and discussed by Rowlinson in Rowlinson and McDermott (1999). The figure clearly illustrates the facility as the focal point around which a whole series of resources, players and institutions coalesce. Central to this model is the client's own resources that are supplemented by the project participants – in the example of the construction industry, the consultants and the contractors delivering the project along with the suppliers and subcontractors. However, these players are partially constrained by the 'institutional side'. The model clearly illustrates the need for the acquisition of resources in order to realise the project and turn it into a facility. This acquisition of

Figure 2.3 Procurement choices.

Source: Adapted from Austen and Neale (1995).

resources is part (and only a part) of the procurement system. This part of the system can be referred to as the contract strategy – that is, the process of combining these necessary resources together. The contract strategy is not the procurement system but only a part of it; the rationale behind this view is that the procurement system involves other features – such as culture, leadership, management, economics, environmental, ethical and political issues – which are all dealt with in this book.

   In reviewing this model it can be seen that there are institutions, stakeholders and players and all of these entities interact in order to both develop the delivered facility and also to constrain the decision-making process within bounds set by each element of the system. These bounds can be brought about by institutional or regulatory constraints or by the limitations inherent in the client's own human resources or the consultants' skill and experience levels, for example, for construction projects. The project is metamorphosed into the constructed facility. Thus, one can see the continuous process of development from an idea or a need to create a facility, and which essentially encompasses, and is constrained and at the same time facilitated by, the client organisation, its mission and its values. Hence, in procurement terms, when one looks at client resources one is in fact not just assessing the client's experience and knowledge but also its mission, values and organisational culture. Thus, the simple schematic model belies the complex managerial, cultural and organisational systems underpinning the whole of the process.

In addressing the contract strategy issue the majority of writers have, sometimes inadvertently, reverted to systems and contingency theory viewpoints and emphasised the importance of integration throughout the whole project process and within the project participants as keys to success. Hence, one is drawn back to Ireland's view (1983) on the almost insignificant differences between nominally different contract strategies. The issue that needs to be addressed in determining the crucial issue of contract strategy is the mechanism for ensuring integration, coordination and active collaboration between the project participants. This is achieved by not only devising an appropriate organisational form and set of conditions of contract but also by focusing on the soft infrastructure which underpins these notionally hard concepts of structure and contracts. Thus, the need to develop a team spirit, appropriate organisation culture and a 'best-for-project' attitude are all part and parcel of this human relations infrastructure, which ensures that the adopted strategy has a reasonable chance of success. Cheung *et al.* (2005) indicate how such an approach can be developed and implemented in a wide range of contract strategies with success.

Figure 2.4 illustrates examples drawn from the construction industry, and we use the quadrants referred to in Figure 2.2 – more detail about the definition of the types of contract can be found in Walker and Hampson (2003b). However, we briefly describe these for the benefit of readers unfamiliar with the construction industry terms.

The *traditional* approach (Q1) to procuring projects involves discrete design development, tender, contract award and construction delivery phases. Each phase is, in theory, separate and distinct. A client begins the process by approaching the principal design consultant – generally the architect for building projects or a design engineer for engineering projects – who then develops the design to as close to 100% complete as possible before inviting tenders. Often refinement and amendment of design details are undertaken during the project delivery phase to resolve many design issues left incomplete and unresolved. Tenders may be invited on an open-tendering basis that allows anyone to tender for the project, or on a closed- or pre-qualified-tendering basis – restricted to those who have met pre-tender qualification criteria such as demonstrated financial soundness and relevant project experience.

– Traditional approach
– Accelerated traditional
– Management contracting
– Construction management
– Project management
– Design and Build/Turnkey
– BOT 'family'/PPP

Q1 Delivery

Increasing integration
of design and
construction/delivery

Q4 Delivery

Figure 2.4 Contract strategy.

The main criticism of the traditional approach has been that it invites a confrontational approach over disputes arising out of contract variations, the unresolved design issues and an entire *claims industry* has developed out of this to advise contractors on how to claim for extra work, and client representatives on how to counter such claims (NBCC, 1989). The traditional approach imposes role-rigidity upon all parties. It remains mainly because most contractors and clients are familiar with it and so it often becomes a default approach. However, it removes the contractor from the design development phase and thus much management and constructability information and knowledge is lost, with serious cost and relationship risk consequences (Francis and Sidwell, 1996; McGeorge and Palmer, 1997). The contractor is closest to the workface with intimate knowledge of the production process and supply chain delivery players and so has much technical and relational knowledge about how best to meet design specifications in a cost- and time-effective manner (Francis and Sidwell, 1996). This can reduce or obviate contract claims that have been highlighted as resulting from the impact of poor design coordination and subsequent design changes to make design details workable (Choy and Sidwell, 1991; Yogeswaran and Kumaraswamy, 1997). Relationship aspects are impaired under the traditional procurement system because the contractor is answerable to the principal design consultant with no formal direct access to the client to suggest improved design for constructability.

Early contractor involvement in the design development process can be facilitated using non-traditional procurement methods that allow contractor expertise to be made readily available to the design team. This 'buildability' or 'constructability' advice is crucial to the development of design solutions that maintain value in terms of the quality of product as well as providing elegant solutions to production problems (Francis and Sidwell, 1996; McGeorge and Palmer, 1997). Under *Construction Management* (CM) (a term synonymous with 'management contracting') the contractor acts as consultant builder providing significant advice on the practicality of the design and expected construction methods (Sidwell and Ireland, 1989). The CM also provides services for construction planning, cost control, coordination and supervision of those who have direct contracts with the owner to carry out operational work, usually for a negotiated fee as an *agent* or *directly* take responsibility for a guaranteed maximum price – this is similar to a traditional approach but with the benefit of interacting closely with the client and design team. Under the agency CM model, trade or work package specialists physically undertaking the work would do so under separate contracts with the client, being coordinated, supervised and managed by the CM team.

Advantages of using the CM approach are as follows:

- reduced confrontation between the design teams and the team responsible for supervising construction;

- early involvement of construction management expertise;
- overlap of design and construction;
- increased competition for construction work on large projects due to work packaging and splitting the construction activities into more digestible 'chunks';
- more even development of documentation;
- fewer contract variations;
- no need to use *nominated* trade contractors (those specifically demanded to be used by the client) and
- public accountability.

The *project management* (PM) procurement delivery approach emerged during the 1960s and 1970s, arising out a move towards multidisciplinary design practices and in-house project teams undertaking government projects. The traditional client representative, generally an architect, led the design team with the major role of coordinating the design activities and overseeing the construction operations as a non-executive project manager (Walker, 1993). Variations in responsibility, and remuneration forms varying from a fee for non-executive PM services to a guaranteed fixed price with contractual relationships between PM and contractor teams (Barnett, 1998/9) have added to the complexity of this approach, with the PM's role ranging from 'adviser' to 'executor'.

The main difference between CM and PM approaches is that the PM team generally takes responsibility for *design coordination* as well as *supervision* of those responsible for delivering the work packages. If these responsibilities are not supported by a contractual arrangement, or where the PM team advises a client willing to take a more active management role, the PM will take an *agency* position using a power base of persuasion and expert knowledge to influence others. Major decisions are directly presented to the client, if an executive, or addressed to a project board of directors for formal approval if this group is charged to represent a client's interests. The PM team has a contractual arms-length relationship with both the design teams and the construction team in 'pure' PM arrangements. If the 'adviser' role is adopted, then the PM managerial influence will take the form of persuasion rather than authoritative direction.

A *design and construct* (D+C) approach allows a client to contract the D+C organisation to manage both the design and construction processes as a single point of contact. This takes the traditional approach further upstream by contracting out the design and construction (delivery) to a set requirement specification. There may have been preliminary design information developed to indicate a generalised design solution or the design brief may be left fairly for the D+C contractor to offer proposals. In *combined project management and construction management* (PM/CM) procurement options, an organisation undertakes to represent the client in leading the design team, and undertakes the management of the construction process including providing construction advice during design development. The PM/CM entity may act as adviser (in which case the managerial links are persuasive rather than directive) or may undertake the work under a contractual arrangement in which it carries financial risk. In a D+C arrangement, this team will hire both the design team members and the construction management team members, either within the design and construction company entity or as sub-contractors. The design team may be sourced from in-house staff or, as more frequently, sourced from the pool of design consultants undertaking a variety of procurement forms (Walker, 1993). In many cases, the design and construction contractor subcontracts or forms a joint venture with design firms. It can be appreciated that PM/CM and D+C blur quite substantially as procurement process options, so that a PM/CM entity that carries financial risk is really a D+C contractor.

The client will often commission a design consultant to develop a brief for a D+C approach that will have a range of possible conceptual design-stage information to move forward from. Thus the client has the opportunity to work with the design team to develop a brief to a stage in which it can test the market for proposals that will develop a project solution based upon the concept design. A selection procedure is then put in place to evaluate D+C proposals. However, designers from D+C groups submitting ideas and developed plans have the opportunity to provide innovative solutions and to take the concept and provide their own footprint on the result. This has several potential advantages, it

- provides for innovative solutions to be tested by the market;
- often results in a cost-effective solution;
- combines the expertise of both design and construction professionals and
- allows the client to have a single point of contact with whoever manages the project if it wishes, or in the case of PM/CM, can maintain contractual control and essentially use this group as its internal design and construction group.

The relationship experience for project teams moves towards treating design and construction supervision entities as being contributing partners in an enterprise to deliver a project solution that combines the skills of both design and construction groups. Dulami and Dalziel (1994) undertook a study of 37 managers of the construction team of whom 22 were involved on traditionally procured projects and 15 on D+C projects. They suggest that D+C improves project team integration through better cooperation and communication between the design and construction teams than is the case under traditional procurement approaches. As design and construction contractors are primarily builders, they have constructability input that can be of significant value, in much the same way that CM consultants can offer this advantage. This is just one example of many research studies across the globe into the effectiveness of moving from a traditional approach to a D+C approach.

Gunning and McDermott (1997: 221) provide insights into how D+C operates in Northern Ireland through interviewing 20 client organisations and 30 contractors. They found that

- competitive tendering is often based on minimal design work or site investigation, but with estimates inflated to cover consequent risks;
- life-cycle costing receives limited consideration by contractors in pricing, particularly with poor client briefs;
- the quality of design and construction is related to the adequacy of the client brief and the professionalism of the project team. This professionalism can be strongly challenged by commercial pressures and poor communication and
- clients are often ill-informed about the implications of the design and build approach, and about the importance of a comprehensive brief for the designer and contractor.

A variant of D+C is *Novation*, in design and construction, D+C responsibility is undertaken by one legal entity responsible for project execution. The client starts along a traditional route by commissioning design consultants to develop a design concept or advanced concept solutions (perhaps up to 25% of design finalisation) to enable D+C tenderers to bid for the project. These design solutions, though substantially complete at the point of novation, are then improved upon by absorbing constructability advice while maintaining design intent and integrity. The intellectual property of the design team, as well as the design team themselves, then takes on a new form (novating) into a D+C contractor's organisation to further develop the design solution initially developed by the client and design team. This provides the design solution to be novated and open to contestability by the construction team's expertise and buildability advice in terms of systems or more specific detailed design information. This can, however, pose serious problems for clients who have a more advanced design concept in mind than might be the case for the design and construction option, however. The means to meet the spirit and essential content of the design solution can be addressed by the entity that successfully takes over a D+C role to *novate* the design solution and take responsibility for the CM role in realising the project. This can represent a useful option, enabling clients to influence the project design to a greater extent than is possible in the standard D+C option.

The client's risk can be reduced because the contractor takes over design development after novation while accepting pre-novation design assumptions.

This arrangement allows the contractor to substantially fine-tune the design to take account of its competitive advantage including constructability issues. For example, in the A\$200 million-plus Melbourne Cricket Ground (MCG) Great Southern Stand project in Australia, the structural design was changed from cast-in-place to precast concrete in response to an overheated cast-in-place concrete market and advantages inherent in an off-site fabrication structural design solution. Additionally, novation allows fast tracking and a fixed price; however the client forgoes some flexibility of making design changes without incurring potential cost penalties. The contractor, by embracing the pre-novation design, cannot claim against design omissions or errors thus accepting contractual risk for the design. The design team, who frequently risk being simply 'passed' onto the contractor, may see their 'baby' amended to a level felt unacceptable by them. For some design professionals this may pose important issues of design integrity and ownership that may be difficult to accept.

Generally, novation is seen as a D+C variant. The client commissions a design team to undertake a partially complete design instead of each D+C company separately developing a design to a proposal stage. Upon successful negotiation of the construction contract, the design team and its design is passed to the successful bidder. The difference between novation and D+C is principally one of client control and influence. The client has more influence and control to shape the desired outcome in novation than with traditional D+C.

Turnkey procurement systems provide for the company supplying D+C services to also finance the project. The client makes staged or periodic payments towards the value of work completed for PM, CM and D+C work. In turnkey projects, however, the company is generally paid upon completion of the commissioning and testing. Literally, the client pays for the project and gets the key to gain access to the project. While this shares some similarities with 'BOT family' projects (described next) the contractor does not undertake to operate the constructed facility. The turnkey approach suits many clients who wish, for tax or other financial purposes, to make a payment only upon delivery of an acceptable product.

The BOT family of *total package* procurement options allow a client's project need to be met by an entity that contracts to design, build, operate, own for some period of time and transfer the facility back to the owner. In the *BOO, BOT, BOOT* the 'B' represents the word 'build', the first 'O' as 'operate' and the second 'O' as 'own' and the 'T' as 'transfer'. This option has been in operation for decades with many examples of these projects in most countries of the world. Readers can refer to texts (Walker and Smith, 1995; Smith, 1999) for more details on this approach. BOT requires that 'contracted parties must accept the conventional wisdom that risk should be assumed by the party within whose control the risk most lies. A major function of the BOT arrangement is, therefore, to recognise and provide a mechanism for the assignment and management of those risks' (Walker and Smith, 1995).

With the BOT 'family' of procurement options an alliance or joint venture group forms to provide a facility for a client for which the client makes a concession agreement to fund the facility until that facility's ownership is transferred to the client. This arrangement is more common for infrastructure projects than buildings because the concession allows for tolls or other payments to be made by end-users to cover the cost of both procuring the facility and its operation. Extensions of this idea have been cited in which buildings have been renovated and leased back on this basis and others where the facility is required to be removed and the site returned to an acceptable environmental standard. One advantage of this arrangement is that it extends the ideas of constructability further to embrace life cycle cost effectiveness. If the entity proposing the design solution is responsible for maintaining and operating the facility then they will have the incentive to reduce long-term costs and thus develop a highly cost-effective product over the product life cycle (Smith, 1999).

The BOT entity undertakes financing, design and construction as well as operation and so the client is taking no direct cost risk other than the possibility that the facility does not meet its needs or that the concession agreement is unsatisfactory (Smith, Merna and Grimsey, 1994). Legal and financing costs – establishment costs – can be considerable for the BOT entity, and should be compared with the client's legal and finance costs in undertaking the project in other forms or options. This option is unlikely to be viable for small-scope projects however. Governments are increasingly using this option for hospitals, prisons and other projects previously undertaken through other procurement scenarios outlined in this section (Grimsey and Graham, 1997). A new variant of this form is where a government will joint the partnership by injecting equity or in-kind resources (perhaps land or an existing facility that needs upgrading) in a public/private partnership or PPP (Akintoye, Beck and Hardcastle, 2003).

There has been recent criticism of these (and by inference other BOT family delivery systems – for further details readers can refer to (Fitzgerald and January, 2004; Hodge, 2004). Experience with BOT/BOO/BOOT schemes reveals some notable failures. Generally these have been based on failures of trust and/or communication. One of the most prominent failures has been the Bangkok Second Stage Expressway. The competitive tender phase took two years to find a successful bidder. Part of the complexity derived from the land-procurement process to allow the project to proceed. At the date when the work was supposed to start only 1% of the land required was available. That problem was overcome, but as the completion of the construction phase neared, other fundamental problems regarding the operation emerged. Toll rates were decreased instead of being increased to fund the project. This led to loans being frozen in response to financiers' concerns about the financial sustainability, and then arguments broke out over the right to operate the expressway. The project environment subsequently dissolved into acrimony amongst accusations of corruption and widespread political interference. Thus, both relationship and cost risks were severely affected by the Thai government changing its mind (Ogunlana, 1997; Smith, 1999). Clients and stakeholders involved in such projects can at times be severely affected by the integrated nature of finance, design and construction. Other PPP and BOOT projects with questionable outcomes are discussed in this book in Chapter 4 under procurement and the financial bottom line, and dealt with in depth by Flyvbjerg, Rothengatter and Bruzelius (2003: 14).

Why has this range of contract strategies historically developed within, for example, the construction industry? One of the reasons is that each participant wishes to obtain as much financial advantage and other value as possible from the project procurement process. Because the process takes place in a competitive market, there is a cycle of fluctuating pressure on prices at all times and commensurate downgrading in the priority of relationship and occupational health and safety (OHS) issues. If the client forces (through the market or design requirements) a lower-cost solution, then this downward pressure on costs often forces the contracting participant to look to alternative means to recoup profit. This often leads to a claims-conscious behaviour, and can also stimulate reductions in quality, functionality and OHS performance. Consequently, the client and its advisers are forced to exert greater surveillance over the contractor in order to minimise the effects of this behaviour. This results in a vicious cycle of negative behaviour as well as increasing project transactions costs. The result may be poor relationships on the project, poor OHS performance, poor quality and 'claimsmanship'.

One way of avoiding this problem, which is based around a traditional procurement contract strategy, is to adopt alternative contract strategies as indicated in Figure 2.4. Other forces are at work in bringing new contract strategies to the market place. The competitive nature of the market forces organisations to innovate if they wish to grow and secure market share. Hence, as the construction industry typically has been a conservative and traditional industry, there has been, in the past, ample scope for the introduction of new and innovative strategies. However, until the late 1960s such attempts at innovation were quite rare, especially in the UK construction industry. With the advent of more experienced and sophisticated clients, there has been an opportunity for leading contractors of the industry to explore new routes. Also, as buildings have become technically more complex, and clients managerially more sophisticated, there has been an increasing recognition that the conventional (traditional) approach to procurement is inadequate. Recently, modern management concepts, that is, business process re-engineering and partnering, have taken root in client organisations. The construction industry has experienced these concepts at second-hand, when working with a major client and first-hand in the rush to reorganise in the face of declining markets. Consequently, factors have combined to force the construction industry into the position where it has to change to survive and the trends indicated in Figure 2.4 have been a consequence.

Much of the literature in this area uses terminology such as the traditional approach, design and build, build-operate-transfer, management contracting, and so on. In order to clearly define the types of arrangements in which the trust, ethical, relationship management and OHS issues and strategies described in this book arise and are implemented, a generic taxonomy of organisational forms is required, as described in Rowlinson and McDermott (1999). The function of this taxonomy is to provide a clear and simple description of construction project organisation forms which, when taken with other contract strategy variables, uniquely define a strategy that is further clarified when put in the context of the overall procurement system.

On top of all of this structure are the concepts of relationship management and the development of collaborative and cooperative working relationships (Walker and Hampson, 2003a). Rowlinson, Cheung, Simons and Rafferty (2006: 77) describe these concepts as follows: Relationship management is 'a system that provides a collaborative environment and which provides a framework for them to adapt their behaviour to project objectives. It is about sharing resources and experiences, exposing the "hidden" risks.' Like all relationship approaches, trust between alliance partners is important because it creates an opportunity and willingness for further alignment, reduces the need for partners to continually monitor one another's behaviour, reduces the need for formal controls and reduces the tensions created by short-term inequities. Thus, the modern view of the procurement system is one predicated on issues of trust, collaboration and ethical behaviour, rather than the traditional view of structure and legal frameworks. This paradigm shift in how procurement is viewed has laid the foundation for an alternate, client and stakeholder-focused approach which is the essence of the ideas presented in this book.

A more recent development in the construction industry has emerged for a number of sophisticated (i.e. construction-procurement-experienced) clients where they have experimented with relational-based procurement approaches. Several definitions and ways of looking at partnering and alliancing are shown below; these have been developed to view partnering and alliancing as defined by Rowlinson, Cheung and Jefferies (2006: 6–7) from a research study into this aspect, working closely with two major Australian Government Construction clients.

Traditional contracts operate under AS2124 or similar, and are adversarial in nature; the contractors having been selected mainly on the basis of lowest price in a competitive bid and the superintendents see their role as gatekeepers, safeguarding the client's interest.

Partnering has been implemented by putting a partnering agreement on top of the traditional contract and encouraging contractor, consultant and client to proactively address project risks, identifying them before they affect the project, and take action, jointly agreeing to manage the risk. The problem with this approach is that superintendents have continued to see their role as gatekeepers rather than as team members, and contractors have kept one eye on the conditions of contract and claims, whilst going through the partnering process.

Alliancing has been implemented, in the main, through a management or cost plus contract where pain and gain are shared between the parties to the contract and a team approach is engendered in many instances by locating all team members in the same office. Thus, the traditional, adversarial roles have been removed by adopting a different form of contract and way of working out issues, particularly in the area of RFIs.[1]

Relational contracting embraces and underpins various approaches, such as partnering, alliancing, joint venturing, and other collaborative working arrangements and better risk sharing mechanisms. Relational contracts are usually long-term, develop and change over time, and involve substantial relations between the parties. The characteristics of relational contracts and construction contracts are summarized in the report.

For readers who are interested in partnering in construction there is a wide range of papers and academic literature on this topic (e.g. see Matthews *et al.*, 1996; Smyth, 1999; Uher, 1999; Akintoye *et al.*, 2000; Bresnen and Marshall, 2000; Walker and Hampson, 2003a). For those interested in the difference between partnering and alliances, a pertinent comment was made by Walker that project alliances ensure that partners in an alliance sink or swim together, that is, benefits or penalties are shared using an agreed arrangement based upon project performance rather than individual players' performance. This ensures that alliance participants have the incentive to focus their efforts upon helping each other to make the project a success rather than their own involvement being a success (Walker and Hampson, 2003a). This whole issue of alliancing and relationship-based procurement will be dealt with in Chapter 9 on culture and in the two case study chapters, Chapters 12 and 13.

Writing and evaluating procurement proposals

The aim of writing a good project proposal that has the best chance of generating value is one that clearly states the scope, scale, quality expected and relational aspirations that will govern the manner in which the project will be delivered. It is also important that the project be linked to the strategic needs of the organisation (Archer and Ghasemzadeh, 1999; Artto *et al.*,

2001; Morris and Jamieson, 2004). Disappointment generally flows from unrealised expectations. Yet if these expectations are not well expressed and formalised in some way acceptable to the parties engaged, then the gulf between what was delivered and what was expected can lead to litigation, bad feelings and adverse impacts upon reputations.

Working out what is expected to be done

Each project should have a clear vision. The dictionary definition of vision is 'the ability to think about or plan the future with imagination or wisdom' (Oxford, 2001: 2066). Kotter (1995) describes vision in terms of something that helps clarify the direction in which to proceed – this makes sense as the word has implications of the sense of sight. Christenson and Walker (2003) argue that a project vision is complex because projects use multiple temporary organisations, each with its own cultures and subcultures. They argue that a project vision should have the following characteristics:

1   It must be understood: It must capture the core purpose, preferred future state and essence of the project objectives, its *raison d'être*.
2   It must be motivational: It must make a convincing case for following the project vision concept that can be internalised by project stakeholders and that provides a compelling value proposition.
3   It must be credible: It must be consistent with stakeholder cultures or sub-cultures to appeal at the assumptions and values level so that the vision statement cultural artefact resonates with them.
4   It must be demanding and challenging: It should be proactive to facilitate teams to work smarter and more effectively, perhaps identifying stretch goals.

Lynn and Akgün (2001) in a paper on project vision, deploying various types of innovation, demonstrate the value for vision support and stability, and argue that a true vision should be a beacon on the hill that guides project teams delivering the vision. Where clear and stable vision is important and different from specification is that vision articulates the preferred future state rather than expected delivered object. Many projects involving rapidly changing technologies or which experience other kinds of unsettling turbulence (such as dynamic influence of stakeholders on complex inter-linked sets of projects) are forced to make changes in design that would suggest instability. In such circumstances a strong vision can prevail over detailed project design (Christenson and Walker, 2004). The envisioning process is an evolving one as it moves from a highly tacit knowledge object, into a more explicit form before being internalised as a cultural artefact.

The starting point for a project should be a compelling vision that arises out of the identified project need. Once that has been established by a project sponsor, and is tested by being approved in principle by an organisation's senior management, then more detailed needs and requirements analysis can take place. This would involve close interaction with influential and knowledgeable stakeholders (as is described in more detail in Chapter 3) and a combination of performance measures that reflect both tangible and intangible needs (see Chapter 6 for more details). The process is one of bubbling up as well as cascading down. Ideas for projects with a somewhat hazy vision will bubble upwards as needs are pragmatically identified. These are then tested and validated by higher levels of organisational authority for strategic fit and made to show how the project's vision can enhance value for the organisation. This gestation period varies depending upon the urgency of the need and the processes available to encourage and test these ideas. Many projects are currently viewed as parts of programs or portfolios of logically linked change initiatives (see, for example, Section II of [Morris and Pinto, 2004] for 12 chapters that discuss this aspect of managing projects in more detail than can be dealt with here).

The project initiation process will involve a project sponsor who most likely will shepherd the initial steps and, depending upon the project scope and scale, may have a dedicated person or team to drive the feasibility stage. At the early stage a feasibility study is undertaken that should outline the purpose (vision) of the project, its strategic, business and technical case that will be tested at the conceptual and functional level. For many large-scale projects, such as a petrochemical facility or advanced weapons systems, this process becomes a complex and intensive project in its own right. When the feasibility is established and approved more detailed plans for procurement can take shape. This triggers the mandate to proceed with the procurement process for appointments that include a project manager, designer advisers (even if this may be to provide a schematic design rather than a more developed design to be used for tendering on) and a host of other vital advisory roles that will develop the detailed project brief. It is at this stage that procurement-form choices are made. A project governance board (Project Executive Board) will be developed as more work towards undertaking the project proceeds. This board will include key stakeholder representatives and senior organisational management representatives whose roles, responsibilities and accountabilities should be well defined by this stage. This board will often expand as the project develops and may appoint advisers at various stages to assist them.

Strategic needs analysis generally provides the basis for a project brief to be developed (Barrett and Stanley, 1999; Smith and Jackson, 2000). It is beyond the scope of this book to delve in detail into the briefing process that will generate the terms of reference for the project, but only to highlight that this occurs well before project concept material is developed into a project specification that may form the basis of tender documents.

Suffice it to say that in line with the bubbling up of ideas as part of the brief development process that concerned and closely impacted stakeholders can best be marshalled to develop the needs analysis part of a project brief. Stuart Green and various colleagues (Green, 1996; 1999; Green and Simister, 1999) have developed approaches to group decision making, evolving a design brief including their involvement in value analysis and brainstorming design ideas. One approach is the use of metaphors which allows tacit information and knowledge to bubble up into a more recognisable and tangible form. Many great design ideas started out as a metaphor; the Sydney Opera House as sails on the harbour for example. Another approach is the use of Soft Systems Methodology (Checkland, 1999) which uses 'rich pictures' as cartoon-like images to help unearth emotions about what is taking place in messy complex situations (such as developing a design brief through brainstorming). For a more detailed description of some case studies of how the briefing process has been organised for some construction and projects, readers can refer to (Green and Lenard, 1999). The outcome from this process should be a Project Initiation Document. Bentley (1997: 27) PRINCE 2 states that 'a Project Initiation Document (PID) is an extension of the Project Brief to include details of the Project Management Team and risk analysis, plus a refinement of the Business Case and the Project Plan.'

This chapter is concerned with choices that can be made to procure resources. This section relates in general terms to the process of how to frame proposals and to evaluate them. To procure project resources the following need to be established:

1   The scope and scale of the project so that boundaries between it and other projects or parts of a program are clear – this helps to limit or eliminate scope creep.

2   The composition of the project components specified in a way so that quantity, quality and performance measurement is clear – this helps the deliverer to understand what is to be delivered, and how to scale and scope the deliverable so that it functionally and aesthetically meets stated objectives without either providing what may be viewed as sub-standard or super-standard (excessive).

3   What constitutes value for money – with the value proposition known, a more intelligent deliverer may be able to offer insights into how to deliver the expected need using less resources with derived benefits amicably shared.

4   What the 'rules of the game' are, procurements and practices that govern the way that the work is measured, valued, paid for, how changes to contractual conditions are handled, behavioural expectations (ethical standards etc.), constraints imposed by both external and

internal conditions (perhaps workflow considerations) and a range of issues relating to contracts administration; these issues may be the cause of cultural clashes between parties if not clarified at the outset.

5   Who takes responsibility for various risks – the guiding principle for effective risk management that delivers value is that those who can best manage a particular risk should take responsibility for it and be recompensed accordingly.

6   How disputes may be resolved, processes and escalation plans – these recognise the limits of authorisation at various levels in both teams and seek to design a speedy and effective dispute-resolution system and

7   The incentives and penalties that may apply.

Procurement choices will in essence provide a range of templates and options that define the Figure 2.2 dimensions of the nature and closeness of the relationship, the degree of separation or the integration of design and delivery as well as how perceived risks are managed.

As we aim in this chapter to present a guide to the sorts of things that should be present in a project brief, we do not intend to provide either contract templates, proposal writing shells or specific tips and hints on writing a request for a tender or a proposal response. There are a number of books on the market that concentrate on that aspect (Bartlett, 1997; Freed *et al.*, 2003; Nickson, 2003; Blackwell, 2004; Sant, 2004; Frey, 2005).

Working out how to do it

If the issues discussed in the section above have been adequately thought out and carefully considered, the project scope, scale and preferred 'rules of the game' can be established. We also discussed and presented in this chapter a number of outsourced project procurement approaches and some of these can be adapted for insourcing (i.e. competitively seeking within organisation groups to be chosen to realise the project). So, we have a situation where a choice now has to be made between the various available options. In fact, there are two types of choice; first, the choice of procurement form, and second, the choice of which tenderer should undertake the projects. Both choices can be reduced to achieving best value. Best value for the approach so that it meets the needs, with the minimum negative impact, of compromises about the inevitable conflicting demands of attributes that can not be always simultaneously achieved while managing complementarities to maximise synergies. Best value in selecting a PM team to deliver a project may also involve similar dynamics.

The literature reviews a number of ways of approaching how this choice can be made – based upon multiple-factor analysis (Kumaraswamy, 1996; Hatush and Skitmore, 1997; Kumaraswamy and Walker, 1999). Palaneeswaran *et al.* (2003), for example, identify five specific and one general potential value vectors that influence best value. These are stated and expanded upon as follows:

1   *Time* is similar to pace as described earlier by Shenhar and Dvir (2004: 1267); that is, time to deliver a project so that it is usable;
2   *Budget* naturally is important and relates to the funding required, that is, money, but also can relate to cash-flow (pace of expenditure) as well as resource categories (may be a consideration of recurrent and capital or 'bucket A' and 'bucket B' which is linked to political as well as practical constraints);
3   *Life cycle* considerations move beyond the current capital outlay issue to the timing of commitment and availability (e.g. assuming the intro-duction of technology improvement at some future time in the project);
4   *Environment*, both from an impact and cost perspective, though this can be expanded to aspects of the social as well as environment parts of the triple bottom line (3BL), as will be further discussed in Chapter 4;
5   *Image and esteem* – this consideration brings in important intangible aspects that will be discussed further in Chapter 6 and
6   *Other factors* – these are highlighted by Palaneeswaran *et al.* (2003) as including some of the social 3BL elements as well as quality issues and health and safety obligations, with an 'upgradeability' issue added that can be generalised to being 'more strategically-tactically optimised' so that it leaves future options and flexibility open.

These value vectors are useful to helping us visualise how complicated the decision process actually is. It becomes clear that what is required is not an A or B choice on some kind of net present worth as advocated by purely economic considerations but a carefully considered matrix of needs that have metrics built to allow comparisons to be made. Palaneeswaran *et al.* (2003) advocate a four-stage framework that is relevant to choosing a procurement method as well as choosing a successful proposal. In stage 1, 'vital' mandatory aspects are filtered based upon set established bench-marks; options can be filtered on a simple go/no-go basis using questions designed for simple yes/no answers. One useful example of a filtering system that narrows the field is the Performance Assessment System PASS that has been successfully used in Hong Kong construction (Kumaraswamy, 1996). In stage 2, the essential best-value parameters are assessed across the options to produce a set of parameters and their relative score contribution. Taking an 'all are equally weighted' easy decision is avoided as well as tuning these to options presented. The weighting matrix should be rigorously linked to the project context to provide a model that describes the 'ideal' situation. If, for example, time-to-market is a key driver and this takes up 50% of the score weighting and cost only represents half of the time consideration then it should attract only 25% of the weighting. It is advisable that the weighting regime is agreed by a representative group of stakeholders so that it matches and is in harmony with the project vision. Care must be taken to ensure that the weighted attributes are independent so that the cause–effect relationship intended between the choice of weighted attributes reflects reality rather than being a poor proxy that is only an indicator rather than a driver (Chang and Ive, 2002). In stage 3, with the evaluation criteria fixed, each option can then be scored for each attribute that has been weighted. This should take cognisance of the incremental value-adding of these attribute scores for each choice. Stage 4 involves undertaking sensitivity analysis. This should reveal how (the precise circumstances) one option could be better than others, and this can be tested and explored and form the basis of developing tie-breaking (or near ties) to ensure that the decision-making process is coherent and that, in the case of choosing between proposals, any basis for negotiations between close proposals is based on a defendable and logical basis. This last stage demands transparency and a credible and contestable approach being made to retain the confidence of all that it is a real rather than a sham process.

Finally, consideration needs to be made about the actual cost of tendering if a poorly or suboptimal tender call and submission process is adopted (as is widely the case currently) in terms of what does the value of the tendering process add to the organisation tendering, those tendering and society that ultimately pays the cost in a variety of direct and indirect ways as indicated by Dalrymple *et al.* (2006). Making these kinds of decisions can be expensive and resource hungry, and it is both unfair and unwise to impose a system in place where inappropriate costs and resources are expended in making the choice.

Working out how to administer it

Contract administration is an important 'detailed' part of the process. Because it is less glamorous than 'decision making' it tends to get poorly attended to in projects. Contract administration can generally be seen as being part of the PM monitoring and control, coordination and governance processes.

Monitoring and control revolves around the 'iron triangle' of cost, time and quality. Administration relies on standards being clearly stated, measures being practical and applicable, and processes being in place that allow the information and knowledge generated to be used effectively. These days, this means that there needs to be an interface between information communication technologies (ICT) that support gathering, transfer, analysis and reporting of information. All this can and should be used in a knowledge management (KM) context that leverages experience to lead to continuous improvement. It would make sense if procurement systems required ICT advantages to be taken advantage of so that organisational learning can take place (Walker and Lloyd-Walker, 1999). In some contexts an Enterprise Resource Program (ERP) approach links and forces a supply chain to share information through the ERP (Akkermans *et al.*, 2003). One way of advancing current procurement administration practices would be to extend value in terms of control and coordination so that the project brief and tender documentation require the recognition of information and knowledge as a valuable resource; this can be built in through the procurement system if it is indeed recognised as a valuable attribute.

Contract administration can also be seen to influence the way that negotiation takes place. While this is closely linked with cultural aspects relating to alliance-type projects in particular, discussed in Chapter 9, it is worth flagging here. This can result in the procurement process stipulating, for example, how dispute resolution can painlessly take place in a cooperative rather than an adversarial manner, as often happens in partnering and alliancing procurement forms.

Chapter 6, Project Performance Measures and Procurement, will deal with performance measures and the administration of projects to highlight and properly record achievements in satisfying often intangible or poorly expressed needs.

Chapter summary

In this chapter we discussed different ways of understanding how a project can be viewed so that a project procurement choice can be guided by a project typology. We then discussed a range of options that present themselves based on the degree of collaboration and integration between supply chain parties and their relationship. We highlighted some important themes including understanding the role of project vision and objectives as well as understanding the nature and characteristics of project delivery methods and how project context richly influences the setting that demands an intelligent project procurement choice. These themes are developed further in this book. Our point of departure from many available books and texts relating to project procurement and contract administration is our focus on the context of project type and its influence upon procurement choice – we argue that this gap needed to be addressed.

We also discussed how project procurement proposals could be written, evaluated and how contract agreements can be administered. Detailed discussion of those project procurement aspects from this chapter is limited and intended to be of secondary focus to us because project proposal writing and evaluation as well as project contract administration are topics that have been widely addressed in the construction management literature (Hughes *et al.*, 1998; Murdoch and Hughes, 2000; Hughes, 2006; Lowe and Leiringer, 2006) and general business literature (Bartlett, 1997; Freed *et al.*, 2003; Nickson, 2003; Blackwell, 2004; Lewis, 2005).

Vignette

OmniTrip is the organisation that won the tender for logistics support and management of the privatised integrated public-transport system for Southland, a major city of 4 million people that until 10 years ago, had a public sector owned and operated bus, tram and train (both above and underground) public-transport system described in the vignette in Chapter 1. OmniTrip's contract was to provide services and maintenance of legacy-transport assets, and gradually replace these with new assets as part of a Southland's renewal and improvement government policy. Its coordination activities include: managing the procurement and maintenance of transport assets; staffing of all operations including facilities management of stations (owned by Southland) and developing a new-generation, stored-value smartcard ticketing system.

OmniTrip clearly has a portfolio of projects ranging from highly tangible ones such as taking down tram power lines and replacing them with an in-ground third rail power access that is still in a stage of improvement from existing systems in place[2] of those designed by the French firm Alstrom.[3] This project falls within the tram hardware delivery programme in the tramways portfolio. Although it represents a construction element (physically building the new tram lines and de-commissioning overhead power lines) it also has a large systems-engineering component because the third rail idea requires a lot of services of engineering integration including a *WiFi* network to allow commuters to have full broadband internet access for an additional charge on their smartcard ticketing system. The stored-value smartcard development system is another project within the tram hardware and software program. Training, development, stakeholder management and other people–systems integration and engagement projects fall within the *organware* program.

Each project and programme within the tramways portfolio has different characteristics that demand different procurement decisions and solutions. Some aspects such as the civil-engineering works can be scoped to be tendered either to a range of local small-scale construction contractors or to one of the several global facilities management and maintenance contractors such as Transfield[4] or Thiess.[5] Other aspects of the service projects, that is, developing customised stored-value smartcards software that links to customer relationship management

(CRM) application that has vast data-mining potential (subject to data privacy legislation and ethical considerations) may lead to JV, alliance or other IP-sharing procurement approaches. OmniTrip aims to maximise its value generation and business development potential through its procurement choices and interaction with its upstream and downstream supply chain. OmniTrip also has been considering a BOOT-type scheme for acquiring the trams based upon fares gathered.

Issues to ponder

1   Considering only the tram-track refurbishment project, select three key aspects of value that you believe that OmniTrip could identify to make a procurement choice that maximises its value delivered by its downstream supply chain.
2   What procurement options do you think best suit the value propositions identified by you under point 1 above? What justification can you make for your choice?
3   Discuss three types of difficulty that you think OmniTrip may encounter in using a BOOT-type option for acquiring the tram component.
4   How might the identified difficulty be different if an alliance or JV arrangement were used instead of a BOOT-type procurement option?
5   From the above and Chapter 1 vignette information, what would you say is OmniTrip's vision for this portfolio of projects? What top-five ranked selection criteria may be used when assessing procurement submissions from its downstream supply chain partners?

Notes

1 Request for information (RFI).
2 See the Bordeaux tram system for example http://en.wikipedia.org/wiki/Ground_level_power_supply
3 See http://en.wikipedia.org/wiki/Alstom
4 See www.transfieldservices.com/index.htm
5 See www.thiess-services.com.au/home/index.htm

References

Akintoye, A., Beck, M. and Hardcastle, C. (2003). *Public-Private Partnerships: Managing Risks and Opportunities*. Oxford: Blackwell Science Ltd.
Akintoye, A., McIntosh, G. and Fitzgerald, E. (2000). 'A Survey of Supply Chain Collaboration and Management in the UK Construction Industry'. *European Journal of Purchasing & Supply Management*. 6(3–4): 159–168.

Akkermans, H. A., Bogerd, P., Yucesan, E. and van Wassenhove, L. N. (2003). 'The Impact of ERP on Supply Chain Management: Exploratory Findings from a European Delphi Study'. *European Journal of Operational Research*. 146(2): 284–301.

Archer, N. P. and Ghasemzadeh, F. (1999). 'An Integrated Framework for Project Portfolio Selection'. *International Journal of Project Management*. 17(4): 207–216.

Artto, K. A., Lehtonen, J.-M. and Saranen, J. (2001). 'Managing Projects Front-End: Incorporating a Strategic Early View to Project Management with Simulation'. *International Journal of Project Management*. 19(5): 255–264.

Austen, A. D. and Neale, R. H. (Eds) (1995). *Managing Construction Projects – A Guide to Processes and Procedures*. Series Managing Construction Projects – A Guide to Proceses and Procedures. Geneva: International Labour Office.

Bachy, G. and Hameri, A. P. (1997). 'What has to be Implemented at the Early Stages of a Large-Scale Project'. *International Journal of Project Management*. 15(4): 211–218.

Barnett, A. M. (1998/9). 'The Many Guises of a Project Manager'. *The Australian Institute of Building Papers*. 3(1): 119–134.

Barrett, P. and Stanley, C. (1999). *Better Construction Briefing*. Oxford, UK: Blackwell Science Ltd.

Bartlett, R. E. (1997). *Preparing International Proposals*. London: Telford.

Bentley, C. (1997). *PRINCE 2 A Practical Handbook*. Oxford, UK: Butterworth-Heinemann.

Blackwell, E. (2004). *How to Prepare a Business Plan*. London: Kogan Page.

Bresnen, M. and Marshall, N. (2000). 'Building Partnerships: Case Studies of Client-Contractor Collaboration in the UK in Construction Industry'. *Construction Management and Economics*. 18(7): 819–832.

Briner, W., Hastings, C. and Geddes, M. (1996). *Project Leadership*. Aldershot, UK: Gower.

Chang, C.-Y. and Ive, G. (2002). 'Rethinking the Multi-Attribute Utility Approach Based Procurement Route Selection Technique'. *Construction Management and Economics*. 20(3): 275–284.

Checkland, P. (1999). *Systems Thinking, Systems Practice*. Chichester, UK: John Wiley & Sons Ltd.

Cheung, F. Y. K., Rowlinson, S., Jefferies, M. and Lau, E. (2005). 'Relationship Contracting in Australia'. *Journal of Construction Procurement*. 11(2): 123–135. *Special Issue on 'Trust in Construction'*.

Choy, W. K. and Sidwell, A. C. (1991). 'Sources of Variations in Australian Construction Contracts'. *The Building Economist*. December: 24–29.

Christenson, D. and Walker, D. H. T. (2003). *Vision as a Critical Success Factor to Project Outcomes*. 17th World Congress on Project Management, Moscow, Russia, 3–6 June: On CD-ROM.

Christenson, D. and Walker, D. H. T. (2004). 'Understanding the Role of "Vision" in Project Success'. *Project Management Journal*. 35(3): 39–52.

Cleland, D. I. and King, W. R. (1988). *Project Management Handbook*. New York: Van Nostrand Reinhold.

Crawford, L. and Pollack, J. (2004). 'Hard and Soft Projects: A Framework for Analysis'. *International Journal of Project Management*. 22(8): 645–653.

Crawford, L., Hobbs, J. B. and Turner, J. R. (2005). *Project Categorization Systems.* Newtown, PA: Project Management Institute.

Dalrymple, J., Boxer, L. and Staples, W. (2006). *Cost of Tendering: Adding Cost Without Value?* Clients Driving Innovation: Moving Ideas into Practice, Surfers Paradise, Queensland, 12–14 March, Hampson K., CRC CI.

DeFillippi, R. J. and Arthur, M. B. (1998). 'Paradox in Project-Based Enterprise: The Case of Film Making'. *California Management Review.* **40**(2): 125–139.

Dulaimi, M. F. and Dalziel, R. C. (1994). *The Effects of the Procurement Method on the Level of Management Synergy in Construction Projects.* East meets West – CIBW92 Symposium, Hong Kong University, 4–7 December 1994, Rowlinson S. R., University of Hong Kong: 53–59.

Fitzgerald, P. (2004, January). Review of Partnerships Victoria Provided Infrastructure. Final Report to the Treasurer. Melbourne, Growth Solutions Group: 42.

Flyvbjerg, B., Rothengatter, W. and Bruzelius, N. (2003). *Megaprojects and Risk: An Anatomy of Ambition.* New York: Cambridge University Press.

Francis, V. E. and Sidwell, A. C. (1996). *The Development of Constructability Principles for the Australian Construction Industry.* Adelaide: Construction Industry Institute.

Freed, R. C., Freed, S. and Romano, J. D. (2003). *Writing Winning Business Proposals: Your Guide to Landing the Client, Making the Sale, Persuading the Boss.* New York: [Maidenhead], McGraw-Hill.

Frey, R. S. (2005). *Successful Proposal Strategies for Small Businesses: Using Knowledge Management to Win Government, Private Sector, and International Contracts.* Boston, MA: Artech House.

Green, S. D. (1996). 'A Metaphorical Analysis of Client Organizations and the Briefing Process.' *Construction Management and Economics.* **14**(2): 155–164.

Green, S. D. (1999). 'A Participative Research Strategy for Propagating Soft Methodologies in Value Management Practic'. *Construction Management and Economics.* **17**(3): 329–340.

Green, S. D. and Lenard, D. (1999). 'Organising the Project Procurement Process'. In *Procurement Systems. A Guide to Best Practice in Construction.* Rowlinson S. and P. McDermott (Eds). London: E & FN Spon: 57–82.

Green, S. D. and Simister, S. J. (1999). 'Modelling Client Business Processes as an Aid to Strategic Briefing'. *Construction Management and Economics.* **17**(1): 63–76.

Grimsey, D. and Graham, R. (1997). 'PFI in the NHS'. *Engineering Construction and Architectural Management.* **4**(3): 215–231.

Gunning, J. G. and McDermott, M. A. (1997). *Developments in Design & Build Contract Practice in Northern Ireland.* CIB W-92 Procurement Systems Symposium 1997, Procurement – A Key To Innovation, The University of Montreal, 18–22 May 1997, Davidson C. H. and T. A. Meguid, CIB, **1**: 213–222.

Hartman, F., Ashrafi, R. and Jergeas, G. (1998). 'Project Management in the Live Entertainment Industry: What Is Different?' *International Journal of Project Management.* **16**(5): 269–281.

Hatush, Z. and Skitmore, M. (1997). 'Criteria for Contractor Selection'. *Construction Management and Economics.* **15**(1): 19–38.

Hodge, G. A. (2004). 'The Risky Business of Public-Private Partnerships'. *Australian Journal of Public Administration*. **63**(4): 37–49.

Hughes, W. (2006). 'Contract Management'. In *Commercial Management of Projects Defining the Discipline*. David Lowe and Roine Leiringer (Eds). Abingdon, Oxon: Blackwell Publishing: 344–355.

Hughes, W., Hillebrandt, P. M. and Murdoch, J. R. (1998). *Financial Protection in the UK Building Industry: Bonds, Retentions, and Guarantees*. London, New York: E & FN Spon.

Ireland, V. (1983). *The Role of Managerial Actions in the Cost, Time and Quality Performance of High Rise Commercial Building Projects*. PhD thesis. Sydney: University of Sydney.

Kotter, J. P. (1995). 'Leading Change – Why Transformation Efforts Fail'. *Harvard Business Review*. **73**(2): 59–67.

Kumaraswamy, M. M. (1996). 'Contractor Evaluation and Selection: A Hong Kong Perspective'. *Building and Environment*. **31**(3): 273–282.

Kumaraswamy, M. M. and Walker, D. H. T. (1999). 'Multiple Performance Criteria for Evaluating Construction Contractors'. In *Procurement Systems. A Guide to Best Practice in Construction*. Rowlinson S. and P. McDermott (Eds). London: E & FN Spon: 228–251.

Lammers, M. (2002). 'Do You Manage a Project, or What? A Reply to "Do You Manage Work, Deliverables or Resources"'. *International Journal of Project Management*. **20**(4): 325–329.

Lewis, H. (2005). *Bids, Tenders, and Proposals: Winning Business Through Best Practice*. Sterling, VA: Kogan Page.

Lowe, D. and Leiringer, R. (Eds) (2006). *Commercial Management of Projects Defining the Discipline*. Series Commercial Management of Projects Defining the Discipline. Abingdon, Oxon: Blackwell Publishing.

Lynn, G. S. and Akgün, A. E. (2001). 'Project Visioning: Its Components and Impact on New Product Success'. *The Journal of Product Innovation Management*. **18**(6): 374–387.

McGeorge, W. D. and Palmer, A. (1997). *Construction Management New Directions*. London: Blackwell Science.

Maqsood, T., Finegan, A. and Walker, D. H. T. (2006). 'Applying Project Histories and Project Learning Through Knowledge Management in an Australian Construction Company'. *The Learning Organization*. **13**(1): 80–95.

Matthews, J., Tyler, A. and Thorpe, A. (1996). 'Pre-construction Project Partnering: Developing the Process'. *Engineering, Construction and Architectural Management, Blackwell Science*. **3**(1–2): 117–131.

Mintzberg, H., Ahlstrand, B. W. and Lampel, J. (1998). *Strategy Safari: The Complete Guide Through the Wilds of Strategic Management*. London: Financial Times/Prentice Hall.

Morris, P. W. G. and Jamieson, A. (2004). *Translating Corporate Strategy into Project Strategy*. Newtown Square, PA: PMI.

Morris, P. W. G. and Pinto, J. K. (Eds) (2004). *The Wiley Guide to Managing Projects*. Series The Wiley Guide to Managing Projects. New York: Wiley.

Murdoch, J. R. and Hughes, W. (2000). *Construction Contracts: Law and Management*. London, New York: E & FN Spon.

NBCC (1989). *Strategies for the Reduction of Claims and Disputes in the Construction Industry – No Dispute*. Canberra: National Building and Construction Council.

Nickson, D. (2003). *The Bid Manager's Handbook*. Aldershot: Gower.

Nogeste, K. (2004). 'Increase the Likelihood of Project Success by Using a Proven Method to Identify and Define Intangible Project Outcomes'. *International Journal of Knowledge, Culture and Change Management*. **4**: 915–926.

Nogeste, K. (2006). Development of a Method to Improve the Definition and Alignment of Intangible Project Outcomes with Tangible Project Outputs. Doctor of Project Management, DPM, Graduate School of Business. Melbourne: RMIT.

Nogeste, K. and Walker, D. H. T. (2005). 'Project Outcomes and Outputs – Making the Intangible Tangible'. *Measuring Business Excellence*. **9**(4): 55–68.

Ogunlana, S. O. (1997). *Build Operate Transfer Procurement Traps: Examples from Transportation Projects in Thailand*. CIB W-92 Procurement Systems Symposium 1997, Procurement – A Key To Innovation,, The University of Montreal, 18–22 May 1997, Davidson C. H. and T. A. Meguid, CIB, **1**: 585–593.

Oxford (2001). *The New Oxford Dictionary of English*. Oxford, UK: Oxford University Press.

Palaneeswaran, E., Kumaraswamy, M. and Ng, T. (2003). 'Targeting Optimum Value in Public Sector Projects Through "Best Value"-Focused Contractor Selection'. *Engineering Construction and Architectural Management*. **10**(6): 418–431.

PMI (2004). *A Guide to the Project Management Body of Knowledge*. Sylva, NC: Project Management Institute.

Rowlinson, S. and McDermott, P. (1999). *Procurement Systems: A Guide to Best Practice in Construction*. London: E & FN Spon.

Rowlinson, S., Cheung, F. Y. K. and Jefferies, M. (2006). *A Review of the Concepts and Definitions of the Various Forms of Relational Contracting, Brisbane*. CRC in Construction Innovation http://www.construction-innovation.info/images/pdfs/2001–004-A_Industry_Booklet.pdf,2002-022-2A-01 (accessed on 7 January 2007).

Rowlinson, S., Cheung, F. Y. K., Simons, R. and Rafferty, A. (2006). 'Alliancing in Australia – No Litigation Contracts; A Tautology?' *ASCE Journal of Professional Issues in Engineering Education and Practice: Special Issue on 'Legal Aspects of Relational Contracting'*. **132**(1): 77–81.

Sant, T. (2004). *Persuasive Business Proposals: Writing to Win More Customers, Clients, and Contracts*. New York: Amacom.

Shen, Y. J. and Walker, D. H. T. (2001). 'Integrating OHS, EMS and QM with Constructability Principles when Construction Planning – A Design and Construct Project Case Study'. *TQM, MCB University Press, UK*. **13**(4): 247–259.

Shenhar, A. J. (1998). 'From Theory to Practice: Toward a Typology of Project-Management Styles'. *Engineering Management, IEEE Transactions on*. **45**(1): 33–48.

Shenhar, A. J. (2001). 'Contingent Management in Temporary, Dynamic Organizations: The Comparative Analysis of Projects'. *The Journal of High Technology Management Research*. **12**(2): 239–271.

Shenhar, A. J. and Dvir, D. (1996). 'Toward a Typological Theory of Project Management'. *Research Policy*. 25(4): 607–632.

Shenhar, A. J. and Dvir, D. (2004). 'How Projects Differ, and What to Do About It'. In *The Wiley Guide to Managing Projects*. Morris P. W. G. and J. K. Pinto (Eds). New York: Wiley: 1265–1286.

Shenhar, A. J. and Wideman, R. M. (2002). *Toward a Fundamental Differentiation Between Project Types*. http://www.maxwideman.com/papers/differentiation/differentiation.pdf (accessed on 27 July 2006).

Shenhar, A. J., Dvir, D. and Shulman, Y. (1995). 'A Two-Dimensional Taxonomy of Poducts and Inovations'. *Journal of Engineering and Technology Management*. 12(3): 175–200.

Shenhar, A. J., Dvir, D., Levy, O. and Maltz, A. C. (2001). 'Project Success: A Multidimensional Strategic Concept'. *Long Range Planning*. 34(6): 699–725.

Sidwell, A. C. and Ireland, V. (1989). 'An International Comparison of Construction Management'. *The Australian Institute of Building Papers*. 2(1): 3–12.

Smith, A. J. (1999). *Privatized Infrastructure – The Role of Government*. London: Thomas Telford.

Smith, J. and Jackson, N. (2000). 'Strategic Needs Analysis: Its Role in Brief Development'. *Facilities*. 18(13/14): 502–512.

Smith, N. J., Merna, A. and Grimsey, D. (1994). *The Management of Risk in BOT Projects*. Internet 94, 12th World Congress on Project Management, Oslo, Norway, 9–11 July, Lereim J., International Project Management Association, 2: 173–179.

Smyth, H. J. (1999). 'Partnering: Practical Problems and Conceptual Limits to Relationship Marketing'. *International Journal of Construction Marketing*. 1(2): 1–14 online version http://www.brookes.ac.uk/other/conmark/IJCM/issue_02/010202.pdf.

Turner, J. R. (1999). *The Handbook of Project-based Management: Improving the Processes for Achieving Strategic Objectives*. London: McGraw-Hill.

Turner, J. R. (2000). 'Do You Manage Work, Deliverables or Resources?' *International Journal of Project Management*. 18(2): 83–84.

Turner, J. R. and Cochrane, R. A. (1993). 'The Goals and Methods Matrix: Coping with Projects With Ill-defined Goals and/or Methods of Achieving Them'. *International Journal of Project Management*. 11(2): 93–102.

Uher, T. (1999). 'Partnering Performance in Australia'. *Journal of Construction Procurement*. 5(2): 163–176.

Walker, A. (1993). *Project Management in Construction*. London: Blackwell Science.

Walker, C. and Smith, A. J. (1995). *Privatised Infrastructure – the BOT Approach*. London: Thomas Telford.

Walker, D. H. T. and Hampson, K. D. (2003a). 'Enterprise Networks, Partnering and Alliancing'. In *Procurement Strategies: A Relationship Based Approach*. Walker D. H. T. and K. D. Hampson (Eds). Oxford: Blackwell Publishing: 30–73.

Walker, D. H. T. and Hampson, K. D. (2003b). 'Procurement Choices'. In *Procurement Strategies: A Relationship Based Approach*. Walker D. H. T. and K. D. Hampson (Eds). Oxford: Blackwell Publishing: 13–29.

Walker, D. H. T. and Lloyd-Walker, B. M. (1999). 'Organisational Learning as a Vehicle for Improved Building Procurement'. In *Procurement Systems: A Guide to Best Practice in Construction.* Rowlinson S. and P. McDermott (Eds). London: E & FN Spon: 119–137.

Walker, D. H. T. and Shen, Y. J. (2002). 'Project Understanding, Planning, Flexibility of Management Action and Construction Time Performance – Two Australian Case Studies'. *Construction Management and Economics.* 20(1): 31–44.

Walker, D. H. T., Lingard, H. and Shen, Y. J. (1998). The Nature and Use of Global Method Statements in Planning, Research Report. Melbourne: The Department of Building and Construction Economics, RMIT University.

Yogeswaran, K. and Kumaraswamy, M. M. (1997). 'Perceived Sources and Causes of Construction Claims'. *Journal of Construction Procurement.* 3(3): 3–26.

Chapter 3

Stakeholders and the supply chain

Derek H. T. Walker, Lynda Bourne and Steve Rowlinson

Chapter introduction

Projects serve the needs of stakeholders by ensuring that their expectations and needs are realised. Project management does not occur in a vacuum but requires an infusion of enthusiasm and commitment powered by the full range of project stakeholder energy sources in an energy grid that can develop a positive or negative trajectory. The key to effectively harnessing this force is for project managers to know how to connect into this organisational grid and how to identify tipping-point key stakeholders and their value propositions. Project managers are unlikely to deliver project success without paying attention to the expectations and needs of key influence drivers and the diverse range of project stakeholders that may cumulatively exert a significant impact on the perception of project success. A project that does not meet the expectations of influential stakeholders is not likely to be regarded as successful, even if it remains within the original time, budget and scope.

Effective project managers require keen analytical and intuitive skills to identify high-impact and cumulative-impact stakeholders and work with them to understand their expectations to influence project success. This facilitates managing a process that maximises stakeholder-positive input and minimises any potential detrimental impact. The authors argue that project managers need to be able to engage more effectively with the hidden reservoirs of power that are exercised by project stakeholders in the interaction between individuals in their social networks.

In Chapter 4 we discuss business ethics as well as how considering the triple bottom line (3BL) can help us identify a broader set of stakeholders who may be otherwise left out of the project development process resulting in valuable perspectives, insights and support being ignored. Chapter 4 also discusses project governance and the importance of an active and supportive project sponsor – a key and influential stakeholder. Chapter 5, in discussing strategy, explores elements of the 'politics' of decision making surrounding the initiation and development of projects, and in particular how the choice of projects to be realised should be determined. Whenever terms such as 'politics' or 'strategy' are used there is an implicit recognition that a fundamental issue to be first resolved is identifying stakeholders and influence-shapers by assessing their potential impact upon the way that the project could proceed. This chapter links with Chapter 6 because effective performance measurement systems can only be developed by understanding the project purpose and what benefits are expected to be delivered to whom – that is, stakeholders. Chapter 8 discusses innovation and learning and how the upstream supply chain (clients) as well as the downstream supply chain (suppliers and sub-contractors) have wisdom and knowledge to offer, so identifying them and their potential impact is also of vital importance. Chapter 9 explores the cultural dimension of project procurement, and this is highly relevant to stakeholder management. Chapter 10 concerns attracting and retaining the most talented teams to help deliver project success; again this is relevant to managing project team stakeholders.

This chapter is presented in three broad sections with a vignette at the end of the chapter that prompts relevant questions. The first section deals with stakeholder theory because this forms the basis of our understanding of which stakeholders should be focussed upon, to fairly and effectively apportion attention and consideration of their issues, needs and potential contribution. The section that follows discusses types and sources of power, influence and concepts of trust and distrust, and from this describes how stakeholder management links into these concepts with an upstream focus. The ability to understand the often hidden power and influence of various stakeholders is a critical skill for successful project managers. Stakeholders can be a considerable asset, contributing knowledge, insights and support in shaping a project brief as well as supporting its execution. Project managers welcome any tools that can help them identify and visualise stakeholders' likely impact, and advances their ability to address the often-thorny problem of stakeholder relationship management. The third section then places the first two sections in context with management of the downstream supply chain with a value focus.

Stakeholder theory

Successful completion of project deliverables is critically dependent upon relationship management skills, amongst these the need to achieve project objectives that fully address stakeholder expectations throughout the project life cycle (Cleland, 1999: chapter 6). However, one major task that needs to be undertaken in developing a project's strategic aims is to identify stakeholders in order to develop a project brief that best addresses their often-conflicting range of needs and wishes.

Identifying stakeholders

Stakeholder theory offers a number of perspectives and expectations that stakeholders may hold. *Social science stakeholder theory* tends to focus around concepts of justice, equity and social rights having a major impact on the way that stakeholders exert moral suasion over project development or change initiatives (Gibson, 2000). Readers may wish to reflect upon the Chapter 4 ethics section as being relevant here. Thus one prevailing view is that a stakeholder is someone affected by a project and having a moral (and perhaps a non-negotiable) right to influence its outcome. This view is very broad and its consequences unmanageable because there are so many ways in which a project can impact on a very wide range of people – from affecting a business environment through to other more physical or social dimensions that relate to quality-of-life issues.

Instrumental stakeholder theory holds that stakeholders and managers interact and their relationship is contingent upon the nature, quality and characteristics of their interaction (Donaldson and Preston, 1995). In this view, the identification of stakeholders is more concerned with their instrumentality, agency capacity, or being vectors of influence. This implies a need for negotiation, and expected reactions ranging from standoff to mutual adjustment, depending on such intermediate variables such as trust and commitment, and motivational forces (being harmonised or in conflict).

Jones and Wicks' (1999) *convergent stakeholder theory* holds that stakeholder actions and reaction to change leads to project managers needing to develop mutual trusting and cooperative relationships with their stakeholders. Consequently, their actions should be based on ethical standards – see Chapter 4. By meeting both objectives, organisations can gain competitive advantage. This accords with 3BL principles where performance success is defined as meeting financial bottom-line performance measures as well as environmental and social responsibility performance measures (Elkington, 1997).

What becomes clear is that 'legitimate and valid' stakeholders need to be identified and their power and influence mapped so that their potential impact on projects can be better understood. Appropriate strategies can then be formulated and enacted to maximise a stakeholder's positive influence and minimise any negative influence. This becomes a key risk-management issue for project managers to avoid many of the project failures detailed in the literature, for example by Morris and Hough (1993).

Briner *et al.* (1996) identified four sets of stakeholders: client; project leader's organisation; outside services and invisible team members. Cleland (1995: 151) recognised the need to develop an organisational structure of stakeholders through understanding each stakeholder's interests, and negotiating both individually and collectively to define the best way to manage stakeholder needs and wants. He identifies several clusters of stakeholders from the supply chain. Stakeholders have also been described as *The ones who hold the beef* (Dinsmore, 1999) or those who have an interest. Effectively managing these stakeholders is essential at all phases of the project from 'initiation' to 'closeout' (Cleland, 1995).

It becomes necessary to consider what a stakeholder's stake actually is when trying to define what his/her needs or requirements are, or how he/she could impact the project. A *stake* could be an Interest, a Right or Ownership. An Interest is a circumstance in which 'a person or group will be affected by a decision, having an interest in that decision.' A Right is either a 'legal right when a person or group has a legal claim to be treated in a certain way, or to have a particular right protected' or a 'moral right'. Ownership occurs 'when a person or group has a legal title to an asset or a property' (Carroll and Buchholtz, 2000: 65). Most project stakeholders will have an Interest, many will have a Right – people with a disability or citizens with a right to privacy, and some will have Ownership – as in workers' right to earn their living from their knowledge or shareholders in an organisation.

The definition of *stakeholder* that will be used in this chapter is: Stakeholders are individuals or groups who have an interest or some aspect of rights or ownership in the project, and can contribute to, or be impacted by, either the work or the outcomes of the project.

Figure 3.1 illustrates stakeholders in four groups: upstream stakeholders, comprising the paying customer and end users of the product/service; downstream stakeholders who include suppliers and sub-contractors; external stakeholders are often ignored and comprise the general community and independent concerned individuals or groups who feel that they will be impacted by the project and its outcomes, invisible stakeholders who engage with the project team in delivering the ultimate project benefit but whose

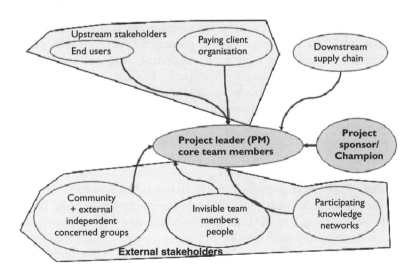

Figure 3.1 Stakeholder types.
Source: Adapted from Walker, 2003: 261.

cooperation and support is vital for project success, and also the knowledge network that interacts with the project delivery team in a variety of ways; and finally there is the highly visible project stakeholder group, comprising the project sponsor or champion as well as the project delivery team.

People naturally tend to form knowledge networks to share and re-frame knowledge that they routinely or occasionally use. History provides many such examples of learning communities, for example, the medieval guilds of Europe, and more recently clusters of people in knowledge-sharing networks centred around a particular skill forming 'tech clubs' or communities of practice (COPs) (Wenger *et al.*, 2002). A COP shares knowledge and skills and sustains its members through obligations to exchange knowledge, providing access and accessibility to shared insights and knowledge about work practices (Wenger *et al.*, 2002: 4). This hidden stakeholder group is often ignored, and yet COPs provide a significant source of influence and referential support that project managers can tap into.

Apart from the stakeholder groups identifiable by their more obvious connection with projects, there are clear and major groups that are invisible but whose cooperation and support is vital for project success. These groups would include family–support networks – this has family-friendliness (work–life balance) and workplace implications discussed in greater depth in Chapter 10 – but it also includes communities of practice and other social networks. People naturally tend to form knowledge networks to share and re-frame knowledge that they routinely or occasionally use.

Identifying stakeholders influence

Cleland (1999: 151) offers a process for managing stakeholders: identifying appropriate stakeholders; specifying the nature of the stakeholder's interest; measuring the stakeholder's interest; predicting what the stakeholder's future behaviour will be to satisfy him/her or his/her stake and evaluating the impact of the stakeholder's behaviour on the project team's latitude in managing the project. He also provides practical advice on how to do this. Most stakeholder groups and individuals, however, are external and hence many project management sub-processes are impossible to achieve for these stakeholders. Many organisationally internal individuals lie outside the boundaries of authority available to project managers. Cleland (1999: 175) offers, after the first step of identifying stakeholders has been achieved, a simple way to visualise stakeholders and their likely impact and influence. The approach is simply to list stakeholders along one axis of a table, list the significant stakeholder interest along another axis of the table and to then indicate the perceived magnitude of their interest. This simple idea is illustrated in Table 3.1.

This idea can be expanded using concepts derived from risk management. Risk assessment can be undertaken using a probability-impact analysis;

Table 3.1 Stakeholder interest intensity index (VIII)

Stakeholder interest	Stakeholders vested interest intensity index (VIII) value									
For colleaques and COP	I	2	3	4	5	6	7	8	9	10
Develop team's skill base	VH	H	N	N	L	VL	H	VH	L	N
Enhance workplace environment										
Family-friendly policy										
Demonstrated lessons learned										
Exemplar of better practice										
High-profile/strategic project										

Notes
Vested Interest (v) levels 5 = Very high, 4 = High, 3 = Neutral, 2 = Low, I = Very low.
Influence impact levels (i) 5 = Very high, 4 = High, 3 = Neutral, 2 = Low, I = Very low.
Vested interest-Impact Index (VIII) = $\sqrt{(v*i/25)}$ e.g. if Vested Interest (v) level = 4 (high) and
Influence impact levels (i) then VIII = $\sqrt{(4*4/25)}$ = $\sqrt{(16/25)}$ = 0.80 = high.

however, we need to distinguish between not only the size of impacts and their probability of occurring, but also the nature and timing of feasible responses to such risks (Ward, 1999). Ward and Chapman (2003) later argued that risk is perhaps a misleading term and that uncertainty would be a better term to use.

As a first step in assessing the potential impact of a stakeholder interest in terms of contributing to project success, the product of an interest-strength and its influence-impact potential may provide a useful form of visualising these two dimensions of stakeholder interest. From the stakeholder perspective, they have a vested interest in the project's success that varies in intensity from very low to very high. Also, the impact of that interest can be assessed in terms ranging from very high to very low. This provides a means by which a stakeholder interest intensity map can be developed. It can also be segmented as seen above and can be applied to a sub-set of stakeholders. In this illustration we are illustrating collegial and COP interest. This could be useful in designing strategies for maximising collegial support and commitment to project success and developing success criteria measures. The 'impact' part of the index relates to the power that these individuals may have to exert influence. Their influence is bounded by their source of power.

Figure 3.2 illustrates the process of influence shaping through social networks. With an organisation or entity is a number of opinion-shapers and they tend to belong to several social groups. For example, Group 1 may have affiliates through university classmates and alumni and Group 2 may represent members belonging to a professional association (or indeed any type of 'club'). Mentoring and seeking validation from reference groups can lead, for example, a sponsor to refer to a key network link who then seeks information, knowledge and advice from network colleagues. This helps to explain how opinion-shapers outside any organisation can exert a hidden (though not necessarily sinister) force that contributes to or results in firm

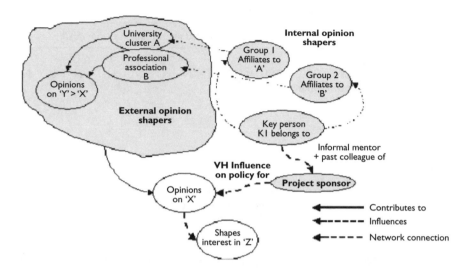

Figure 3.2 Influence mapping.

impressions and perceptions being formed about issues. In this way, we can see that tools that help us visualise influence and impact are pivotal in any stakeholder management approach.

Supply chain members as stakeholders

Figure 3.1 illustrates upstream and downstream stakeholders in a supply chain. Supply chain management will be discussed in more detail later in this chapter, but here we are merely establishing the theoretical justification of viewing both upstream and downstream supply chain partners as valid stakeholders. Supply chain members working in well-integrated projects share access to systems, knowledge and motivation to cooperate as seamlessly as possible to the extent that they appear to be one organisation rather than an integrated flow of separate organisations.

In Chapter 7 we discuss IT tools used by supply chains, such as Enterprise Resource Planning (ERP), and other e-business programs including web portals and customer relationship management (CRM) systems. These systems are intended to connect organisations seamlessly in terms of data and information flows to facilitate efficiencies. Bessant *et al.* (2003) argue that supply chains should be viewed as mechanisms for upgrading and transferring *appropriate practice*, and they explain this term to include experimentation to improve and fine-tune processes and ways that they cooperate to deliver projects. They also mean appropriate practice to include joint learning as experimental improvement initiatives, as well as effective joint problem-solving activities. In Chapter 8 we highlighted and discussed the implication of users, clients/customers as well as supply chain members being generators of practical knowledge and sources of innovative improvement ideas (Von Hippel, 1988; Leonard-Barton, 1992; 1995; Cavaleri, 2004; Cavaleri and Seivert, 2005).

Bessant *et al.* (2003) undertook a series of interviews in six detailed supply chain case studies in: the semiconductor industry equipment; production of tubular structures for the oil and gas industry; the computing equipment industry; two different chains in the chemicals industry; and the aerospace industry. Their findings have particular relevance to stakeholder management. First, they found that one leading partner in the supply chain needs to take on the role of coordinator. This is a traditional role of the project management team. However, in the construction industry, there has a been long-term criticism about the way that smaller sub-contractors and suppliers have been treated, being often left to pick up any time slack lost through poor scheduling or consideration of production and delivery logistics (Latham, 1994; DETR, 1998). While upstream supply chain partners are often well considered and their reaction and impact is well thought through, small downstream supply chain partners are frequently poorly consulted and engaged with for joint problem solving so that avoidable delays and costs are incurred because of problems with planning, logistics and production.

Bessant *et al.* (2003: 182) highlight four supply chain learning (SCL) themes from the six case studies that they consider require more detailed analysis and development:

1 the importance of implementing SCL on a platform of 'good practice' supplier management (and the need to review such programmes to add the SCL dimension);
2 the concept of supply chain coordination or 'governance' and the roles which can be played by different actors in the SCL network;
3 the role of facilitation and the skill sets and enabling toolkit which permit effective learning networks of the kind reported here to evolve and
4 the processes through which a shared learning agenda (and related 'curriculum', assessment frameworks, etc.) can be developed. Early evidence suggests this needs to take place at a sector or supply chain level – for example, via business associations.

The above themes are consistent with the general literature, for example – Spekman *et al.*, 2002; Sherer, 2005 and Maqsood *et al.*, 2007. To achieve this kind of improvement, these stakeholders need to be carefully considered as part of a broad stakeholder engagement approach to procurement.

A stakeholder management system example

This section will focus on explaining the process and antecedents to a stakeholder engagement, and a management tool developed by one of the authors for her doctoral thesis (Bourne, 2005). The basis of stakeholder identification, impact assessment and engagement strategy planning relates to how the stakeholder relationship with the PM team will be managed.

Underlying concepts

Three underlying concepts govern this relationship – trust, power and commitment. French and Granrose (1995) define relationships in the following way:

- 'Exploitation' – one person uses another to achieve his/her own selfish objectives without considering any benefit to the other;
- 'Reciprocity' – two persons are each using the other in a way that ensures each partner benefits. In this type of relationship there is a sense of stability and balance, absent from exploitation relationships. These relationships are based on rewards and 'give and take' and
- 'Mutuality' – this relationship is beyond exploitation and reciprocity. The two parties treat each other not as means but as themselves, by taking an interest in the other's goals and needs.

'Mutuality' is the superior of these three relationships. Whether they are organisations working to form partnerships or organisations dealing with employees, each party must have 'mutuality' as their goal. The concept of organisations working with their employees or with other organisations in less than superior exploitative relationships is one where it will be more likely to breach ethical bounds because the idea of mutual benefit is ignored or not understood. The minimum that any stakeholder engagement strategy must aim for is 'reciprocity', but by definition 'mutuality' will ensure the building and maintenance of robust and successful project relationships (French and Granrose, 1995).

Trust, commitment, power and stakeholder management

A special edition on trust in the 1998 'Special Edition Issue of the Academy of Management Review volume 23 number 3' provides some useful literature to understand the concept. McAllister (1995) reported on results from a quantitative study of 194 managers that provides empirical evidence to support the model developed as proposed by Mayer *et al.* (1995) and Rousseau *et al.* (1998).

Figure 3.3 illustrates the model developed by Mayer *et al.* (1995). They identify three factors that support trustworthiness: ability, benevolence and integrity. Ability means the capacity to do something, benevolence refers to intentions and integrity refers to coherence between what is promised and what is delivered. These factors are in turn modulated by the trustor's

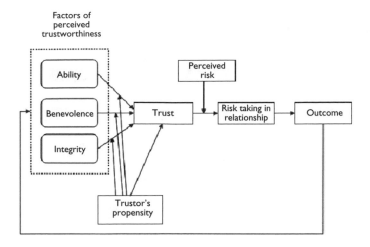

Figure 3.3 A model of trust building.

Source: Mayer *et al.*, 1995: 715.

(i.e. the person who provides property and creates a trust) propensity to place trust in the entity to be trusted. There must be a perceived risk that the trustor may be taken advantage of, so that a sense of vulnerability is instilled in the trustor. This risk test will have an outcome in the view of the trustor that either validates trusting or indicates that the trust was misplaced. This becomes a cycle in which the assessment continues until the trustor feels that the experienced perceived vulnerability is unbearable. This represents a kind of trust bank process described by Walker and Hampson (2003a: 199).

Inkpen and Beamish (1997) introduce the idea that stability of the relationship also has some impact, and this makes sense in a PM context with changes in trust strength occurring during different project phases when parties have shifting ability to perform various trust tasks as their influence changes with the importance of their involvement. Rousseau *et al.* (1998) note that there are different types of trust. Institutional trust remains fairly constant as it is perception of the institution's record of trustworthiness and, as discussed in Chapter 4, this is often wrapped up in perceptions based upon ethics and governance. They also describe calculative trust that merges into relational trust as tests of trust yield results to which the trustor and trustee adjust their perceptions.

Figure 3.4 illustrates a model modifying ideas put forward by Rousseau *et al.* (1998) combined with Mayer *et al.* (1995) and Inkpen and Beamish (1997) adapted to a project phase continuum. The interesting point from this is that research by Rousseau *et al.* (1998) indicates that integrity was

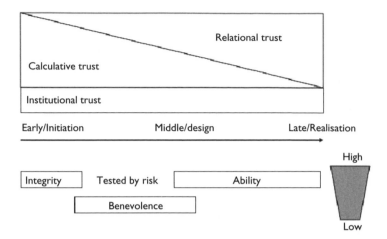

Figure 3.4 Trust in phases.

the prime factor initially determining trust but as testing continues benevolence takes over followed by ability being the dominant driver, all occurring as a project moves through its initial, often chaotic, initiation phase through more stable times as design leads to production and realisation of the project objectives.

The final modifier of trust that we wish to discuss, is the presence of both trust and distrust simultaneously. Lewicki *et al.* (1998: 445) developed a model that combines trust with distrust in a two-by-two matrix illustrated in Figure 3.5.

The range of trust and distrust helps explain motivations and perceptions that govern views about trustworthiness of partners in an alliance, a hands-off contract or in a well-integrated supply chain.

Figures 3.3–3.6 help explain some of the underlying forces that help shape trust. Figure 3.6 combines Meyer and Allen's (1991) commitment theory with Maslow's (1943) motivation theory. The lowest form of commitment is *compliance* where the minimum possible standard is volunteered, often related to legal or rule-based requirements. *Continuance* commitment shares similarities with compliance in that the motivation is to satisfy basic needs such as an individual's need to make a living, a firm's need to make a profit or provide a service that is its *raison d'être*, and so on. Once that need is substantially met the commitment will dissipate and there may also be easily available substitutes that can replace the object of that need so in this sense that 'can' of commitment is weak or transitory. *Normative* commitment is more reciprocal (ought to) and is motivated by social needs and obligations. This kind of commitment is wrapped up in

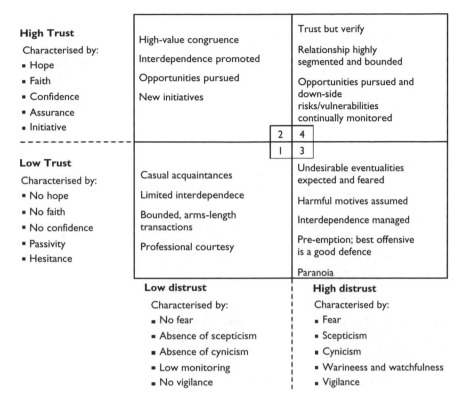

High Trust Characterised by: • Hope • Faith • Confidence • Assurance • Initiative	High-value congruence Interdependence promoted Opportunities pursued New initiatives	Trust but verify Relationship highly segmented and bounded Opportunities pursued and down-side risks/vulnerabilities continually monitored
	2	4
	1	3
Low Trust Characterised by: • No hope • No faith • No confidence • Passivity • Hesitance	Casual acquaintances Limited interdependece Bounded, arms-length transactions Professional courtesy	Undesirable eventualities expected and feared Harmful motives assumed Interdependence managed Pre-emption; best offensive is a good defence Paranoia
	Low distrust Characterised by: ■ No fear ■ Absence of scepticism ■ Absence of cynicism ■ Low monitoring ■ No vigilance	**High distrust** Characterised by: ■ Fear ■ Scepticism ■ Cynicism ■ Warineess and watchfulness ■ Vigilance

Figure 3.5 Trust and distrust.
Source: Lewicki *et al.*, 1998: 445.

loyalty and emotional facets of culture that are stronger, can be extremely strong, and are much harder to displace. The strongest form of commitment is *affective* because it is 'want-to' commitment based on a motivation of self-actualisation and/or ego needs. The desire, of course, can be illusionary and fade and in that sense we could argue that normative commitment is stronger. However, affective commitment can move people to contribute beyond expectations.

Commitment levels can also relate to the sources of power deployed during the relationship. Greene and Elfrers (1999: 178) outline seven forms of power that follow from the power literature reported in a number of leadership texts (Hersey *et al.*, 1996; Yukl, 2002).

1   Coercive – based on fear. Failure to comply results in punishment (*position power*);

2   Connection – based on 'connections' to networks or people with influential or important persons inside or outside organisations (*personal + political power*);

3   Reward – based on ability to provide rewards through incentives to comply. It is expected that suggestions be followed (*position power*);

4   Legitimate – based on organisational or hierarchical position (*position + political power*);

5   Referent – based on personality traits such as being likeable, admired, and thus able to influence (*personal power*);

6   Information – based on possession of or access to information perceived as valuable (*position, personal + political power*) and

7   Expert – based on expertise, skill and knowledge, which through respect influences others (*personal power*).

The nature of power and influence, the sources of this power, and the way in which it is used to contribute to or manipulate cooperative relationships underpin all procurement strategies and the relationships that develop from these. It is interesting that a number of books have appeared providing advice on the use of power to undermine the competitor and to win against a perceived enemy. The works of Machiavelli and Sun-Tzu are among the most prominent. A recent book on power and its use – which features ideas

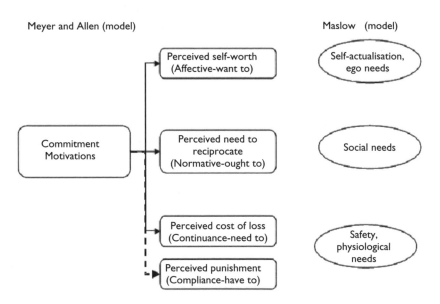

Figure 3.6 Compliance and commitment.

Source: Adapted from Meyer and Allen, 1991 and Maslow, 1943.

from the Machiavelli, Sun-Tzu and others – relates to winning power and holding power for personal gain, and not to achieve a goal that is shared by others (Greene and Elfrers, 1999). Positional power, however, is the least effective of the three outlined in building commitment to shared objectives, win–win outcomes and constructive dialogue, whether in resolving differences or building shared understanding. Project managers need to be aware of the types of power that people can wield to influence the opinions and actions of others. Power, trust/distrust and commitment are closely linked with project phases that impact on this project delivery dynamic as the criticality of supply chain partners shifts with their contribution level.

Stakeholder management – the Stakeholder Circle™ tool

While Table 3.1 provides a useful visual representation of stakeholder interest intensity it can be made more informative by employing a greater degree of graphical imagery such as an influence map or social network map based on an organisation's formal structure and showing who has strong or weak influence in the project environment. Project stakeholders may have deep (extensive) or shallow (limited) influence in terms of their network of others that may be proxies for their interest. For example, an individual with weak influence on the project driving power force may have very deep and strong influence on another individual or group that may in turn have a very strong influence on the project power source. Information about relationships may come through interviews, formal and informal documentation, or the 'grapevine'. Astute project managers keep their antennae active constantly, and know when and how to use such influence maps to achieve success through others who may be able to influence the outcomes.

Figure 3.7 illustrates how project managers view various stakeholders, those they deal with through looking upwards, downwards, sideways and inwards. This model was developed to describe the skills set needed by a project manager (Bourne and Walker, 2004). Dimension 1 relates to knowledge of how to look forwards and backwards to apply correct PM techniques – in the case of this chapter, this would include the appropriate strategy to draw upon when making decisions with stakeholders. Dimension 2 relates to the knowledge of relationships of how to look inwards, outwards and downwards, which is also relevant to the power school of strategy as it includes the ability to manage relationships with key influencing stakeholders. Dimension 3 skills relate to considering and ensuring that political influence and lobbying is addressed by looking sideways and upwards. This is what Bourne (Bourne, 2005), and Bourne and Walker (2004: 228) refer to as 'tapping into the power lines' of project stakeholder influence.

Following from techniques previously discussed that map stakeholders and their influence patterns, a visualisation of stakeholder power and impact can now be constructed.

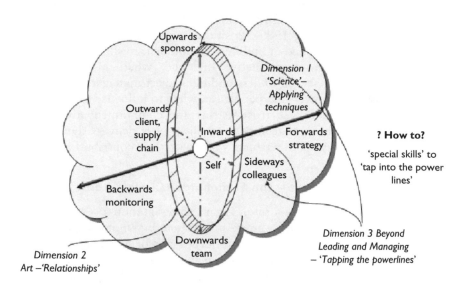

Figure 3.7 Tapping into the power lines.

The *Stakeholder Circle*^TM is both a methodology for stakeholder management and a software tool. The methodology will be described in detail later in this section. The software tool is a relational database that guides the team through steps for data input and prioritisation assessment and then calculates each stakeholder's relative importance. The *Stakeholder Circle*^TM tool develops a 'map' of the project's stakeholder community to facilitate decisions about the amount of effort the project team should allocate to managing the relationship with any given stakeholder (see Figure 3.8).

Figure 3.8 illustrates the concept (referred to as the *Stakeholder Circle*^TM) that one of the authors developed, based on the idea that a project can really only exist with the full consent of its stakeholders (Weaver and Bourne, 2002). The methodology and tool were developed as part of a doctoral thesis (Bourne, 2005) and recently outlined in the PMI journal (Bourne and Walker, 2006). The tool has since become commercialised (see URL www.stakeholder-management.com for more details). Key elements of the *Stakeholder Circle*^TM are: concentric circle lines that indicate distance of stakeholders from the project or project delivery entity; the size of the block, its relative area, indicates the scale and scope of influence; and the radial depth can indicate the degree of impact (Bourne, 2005; Bourne and Walker, 2005).

Patterns and colours of stakeholder entities indicate their influence on the project – for example, orange indicates an *upwards* direction – these stakeholders are senior managers within the performing organisation that are

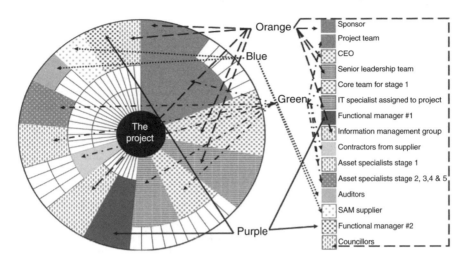

Figure 3.8 The *Stakeholder Circle*^TM tool.

necessary for ongoing organisational commitment to the project; green indicates a *downwards* direction – these stakeholders are members of the project team; purple indicates a *sidewards* direction – peers of the project manager, essential as collaborators or competitors and blue indicates *outwards* – these stakeholders represent those outside the project such as end users, government, 'the public', shareholders. The final colour coding is dark hues and patterns for stakeholders internal to the organisation and light hues and patterns for those external to the organisation.

This depiction of the stakeholder community represents the project's key stakeholders as assessed by the project team. In the *Stakeholder Circle*^TM for the Asset Management Project below, the most important stakeholder has been assessed as the Sponsor: this stakeholder appears at the 12 O'clock position; followed by the project team as the second most important and the CEO as third most important.

This tool can be very useful for project managers trying to understand and remain alert to the nature of stakeholder impact. The model has been tested through research conducted by one of the authors, and presented at Project Management Institute (PMI) chapter meetings and conferences on several continents – in each case the feedback indicated its resonance with practising project managers as a useful tool, so we illustrate it here.

The *Stakeholder Circle*^TM methodology consists of five parts: step 1 – identify; step 2 – prioritise; step 3 – visualise; step 4 – engage; step 5 – monitor.

Step 1 – identify stakeholders

First, the project stakeholders are identified and then categorised into groups indicating how they may influence the outcomes of the project: upwards for senior managers; downwards for members of the project team; sideways for peers of the project manager and outwards for other stakeholders outside the project – such as government, users, unions. The definition of what each individual or group *requires from the project* as well as a definition of the *significance to the project* of these individuals or groups must be agreed and documented at this stage. This concept is based on the idea of mutuality as discussed earlier in this chapter. This exercise is conducted by workshops with individuals who are familiar with the project deliverables and constraints, and with the organisational structure (and the organisational politics). It may be useful to use a metaphor to visualise stakeholder characteristics. Shelly (2007) uses the concept of an organisational zoo in which individuals are described in terms of a series of attributes that links them to particular animal behaviours – this can be both an amusing and highly enlightening exercise in considering not only stakeholder characteristics but their influences, power bases and habits and behaviours.

Step 2 – prioritise stakeholders

Next, prioritisation of these stakeholders is undertaken by considering three factors that can assess the relative importance of stakeholders.

- Power – is their power to influence significant or relatively limited?
- Proximity – are they closely associated or relatively remote from the project?
- Urgency – what is their stake? Are they prepared to go to any lengths to achieve their outcomes?

A simple definition of *power* used in the prioritisation workshops: it is based on the stakeholder's relative power to terminate the project. It is rated by the workshop participants on a scale of 1–4, where 4 is 'high capacity to formally instruct change (can have the project stopped)'; and 1 is 'relatively low levels of power (cannot generally cause much change)'.

Proximity as used in this methodology is self-explanatory. The team must rate the stakeholders on a scale of 1–4, where 4 is 'directly working in the project (team members working on the project most of the time)'; and 1 is 'relatively remote from the project (does not have direct involvement with the project processes)'.

Urgency can be viewed as having two attributes: time sensitivity and criticality. Based on these conditions, the methodology requires workshop participants to rate stakeholders on a scale of 1–5, where 5 is 'immediate action is warranted, irrespective of other work commitments' and 1 is 'there is little need for action outside of routine communications' (Mitchell *et al.*, 1997: 867). In projects where these ratings cannot be simply applied, the methodology supports a breakdown of the process into two subsidiary sets: 'vested stake' (how much 'stake' does the person have in the project's outcome?); and 'perceived importance' (likelihood to take action, *positive or negative*, to influence the outcome of the project). The ratings can then be combined in the software to give the overall *urgency* rating.

Step 3 – visualise stakeholders

The data from the previous steps is transformed into the *Stakeholder Circle*: one example has been described in Figure 3.8. The *Stakeholder Circle*™ will be different for each project and for each phase of the project – the relationships that visualisation shows will reflect the project's unique relationships.

Step 4 – engage stakeholders

The fourth part of the *Stakeholder Circle*™ tool methodology is centred on identifying engagement approaches tailored to the expectations and needs of these individuals or groups. The top-15 stakeholders, defined as being the most important and influential for the project, should receive special attention, but engagement strategies for all stakeholders must be developed. Their value proposition (what they require from the project) will often include intangible outcomes such as enhancement of personal or organisational reputation, and satisfaction of a measure in an individual's key performance indicator (KPI) set, that is, for delivery of project benefits.

The first set of this analysis involves identifying the level of interest of the stakeholder(s) at five levels: from committed (5), through ambivalent (3), to antagonistic (1). Next step is to analyse the receptiveness of each stakeholder to messages about the project: on a scale of 1–5, where 5 is – direct personal contacts encouraged, through 3 – ambivalent, to 1 – completely uninterested. If an important stakeholder is both actively opposed and will not receive messages about the project, he or she will need to have a different engagement approach from stakeholder(s) who are highly supportive and encourage personal delivery of messages. The 5 by 5 matrix (see Figure 3.9) thus developed will become the engagement baseline that is the starting point for measuring the effectiveness of the communication activities of the project.

Based on each stakeholder's engagement strategy, a communication plan will be developed, consisting of: specific messages or message forms (reports); how messages will be delivered; by whom; whether formal or informal, written or oral; at what frequency. The frequency and regularity

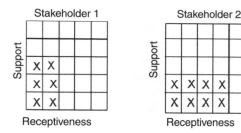

Figure 3.9 Stakeholder engagement profiles.

of delivery of these messages will vary with the level of support and receptiveness of the stakeholder as well as the stage of the project. The messenger need not just be the project manager; other members of the project team may be more appropriate to deliver the message.

Step 5 – monitor effectiveness of communication

Once the Communication Plan has been developed and team communication responsibilities allocated, the principal communication points must be included in the project schedule. Including communication in the project schedule means that team communication activities will be reported regularly at project team meetings. Regular Stakeholder Review meetings, similar to Risk Review meetings will maintain the currency of the project's stakeholder community, or provide information about changes in that community that will cause the project's stakeholders to be re-assessed, re-prioritised and re-developed as a new *Stakeholder Circle*^{TM} (community).

Re-assessment of the engagement matrix of project stakeholders is an essential part of the project-review processes, whether by regular team meetings, reviews or in response to other unplanned events around the project. In the case of a stakeholder that was first assessed as actively opposed and uninterested in receiving project messages, an engagement strategy and communication plan should be developed to change the engagement matrix to (say) neutral for both support and receptiveness. If on re-assessment, the engagement profile has not improved, this lack of change will provide the evidence that the current communication is not effective: a different approach must be taken. This is illustrated in Figure 3.10. On the other hand, achievement of the expected improvement as shown on the new matrix is evidence that the engagement strategy is effective and the communication is achieving its intended objectives.

Figure 3.9 shows the levels of support and receptiveness of two stakeholders. The engagement strategy must adapt to this profile and be

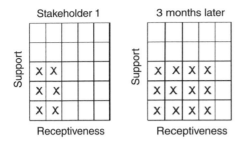

Figure 3.10 Monitoring communication effectiveness.

reflected in the communication plan. If at a later re-assessment the profiles have not changed as expected, a new communication plan must be developed and implemented.

Figure 3.10 shows the effects of comparing a stakeholder's engagement profile over time. The comparison of the new profile with the baseline shows that while the stakeholder is more receptive to messages about the project, the level of support for the project is unchanged. This result should trigger a re-assessment of the communication plan for this stakeholder.

For further details of two case studies on the use of the tool refer to (Bourne and Walker, 2006) and for substantial detail refer to Dr Bourne's doctoral thesis (2005).

Maintenance of the stakeholder community

The process of identifying, prioritising and engaging project stakeholders cannot be a once-only event. Stakeholders change as they move within the organisation or leave it, or as their relative importance to the project and their power and influence within the organisation changes. As the project moves through the project life-cycle or implementation stages, different stakeholders may have more or less impact on the project. The process may have to be repeated in whole or in part many times. An essential part of the methodology is the repetition of the process of the methodology and building of the *Stakeholder Circle*™ when any of these events occur.

Strategy relating to the 'who, what, when and how' of delivering the tailored messages defined for the important stakeholders must be converted into action. The communication plan should be part of the project schedule and thus reported on through team meetings and regular reports.

Value of the methodology

The benefit of this methodology and tool is derived from the analysis process itself as participants of the workshops discuss potential project stakeholders and their needs and potential contributions. These discussions and related negotiations about agreements on ratings of stakeholders enables all project team members to share their knowledge of the individuals and groups being assessed, as well as knowledge of the organisation and its politics. Additional benefits come from the ease with which key stakeholders' influence on the project can be judged once the diagram is complete. To be most effective, the assessment should be updated regularly as the project progresses through the phases of the life cycle or as the stakeholder community changes to reflect the dynamic nature of project relationships.

A methodology that provides a simple, relatively time-efficient process for the identification of key stakeholders is a useful adjunct to the project-planning processes. The methodology also supports a logical process to allow the project manager to decide which of the project's stakeholders to focus effort on, since it would be impossible to attend to the needs and expectations of all stakeholders. The use of a process that steps the project team through analysis of the expectations of the project stakeholders, and the best means to ensure their support of the project provides another benefit to the project manager. Managing the perceptions and understanding the exceptions of key stakeholders build robust project relationships and improves the chances of project success; application of a methodology and visualisation tool such as the *Stakeholder Circle*^{TM} will contribute to the perception of these key stakeholders that the project is being well-managed. Finally, because a system such as this gathers data relating to stakeholder characteristics, beliefs and behaviours, it becomes a very valuable information-mining source to be used in ways similar to that of a CRM system.

Downstream stakeholder supply chain management value

We have discussed how project and (whole of project) team stakeholders can define and articulate their perception of what value the project represents to them and how they can contribute to that value generation. We also discussed how stakeholders can be engaged and their influence sought and managed as part of the procurement process. We also indicated that value propositions vary between stakeholders and, as is elaborated further in Chapters 4 and 6, project outcomes that lead to a sustainable economy add social value as well as potential economic advantage. This section focuses upon the downstream supply chain comprising sub-contractors and suppliers that are often small- or medium-sized enterprises (SMEs) that form a vital part of any economy.

A most interesting recent development in the management of procurement systems is the attempt to incorporate the supply chain into the system in order to ensure sustainability. In general, worldwide, this approach has been championed and led by the public sector. As indicated by Cheung (2006) public sector clients have the opportunity to create managerial, organisational and structural improvements and develop economic, social and environmental sustainability in various market sectors which support their asset portfolio. Supply chain sustainability (SCS) also supports the public sector's organisational, political industry development objectives through proactive supply chain management. This section reviews recent initiatives in both Australia and the United Kingdom.

Recent work in Australia by Rowlinson and Cheung in the pre-cast concrete and recycling sectors and reported in (London, 2005; Cheung, 2006) has investigated SCS in SMEs and (*inter alia*) sought to

- improve economic sustainability;
- stabilise employment levels and reduce high staff turnover;
- improve skill levels, occupational health and safety (OHS) performance and thus productivity as a result of a state-wide smoothing of market investment strategies;
- improve product quality;
- reduce remedial work and
- reduce waste in government resources in having to monitor poor performing sectors.

These issues have also been addressed in the UK, through framework agreements and other mechanisms, and this will be discussed later. As a general rule, the long-term goals of all such approaches is to improve competitive behaviour, improve market sector performance and thus improve both business-process efficiency and effectiveness of public sector procurement systems. This is undertaken by influencing policy development, changing organisational behaviour and culture and by implementing only those developments which lead to sustainability in the economic, social and environmental markets of the region. In order for such approaches to be successful a strategy is needed whereby regional and countrywide smoothing of demand in the various markets takes place through deliberate, planned government policy. Thus, investment strategy is driven with a view to long-term benefits accruing through stabilisation of employment levels, a commensurate reduction in employee turnover and the development and improvement of the industry-wide skills base. This will also have a knock-on effect in terms of OHS performance and the expectation is that there will be improved product quality as well as reductions in rework. All of these objectives, if achieved, should lead to a reduction in the resources needed by government to monitor performance, in that an assurance system is in place. Although improvements in OHS performance might be seen as something of a by-product of SCS, such improvements do add to competitiveness in an increasingly global market.

Competitiveness

Attitudes to OHS, quality and human resource development play a major part in how an organisation is judged by its shareholders, stakeholders and hosts. Whilst maintaining high levels of OHS performance, companies may bear heavy transaction costs in order to maintain this standard. Thus, there is a conflict between competitiveness and safety performance. However, given the global nature of the markets and the need to be seen to be acting responsibly, in terms of corporate governance, safety and health should be seen as essential elements in the goals and policies of companies. Thus, it is becoming increasingly difficult to separate the approach to safety and health and human resource development from other technical and financial issues without jeopardising a company's competitiveness. A strong safety and health management system and a clear policy on maintaining the highest standards of OHS is essential nowadays for the company's continued existence. Indeed, many companies derive competitive advantage from their impeccable safety and health records and this is an issue which can be enhanced through proactive supply chain management.

Key issues in SCS

A major issue to be addressed in maintaining a sustainable supply chain is the move towards more direct and long-term employment with commensurate skills and educational improvements within the work force. Outsourcing has been seen to attract substantial transaction costs in terms of setting up, maintaining and auditing the outsourcing system. It is also seen to be a social cost in the way that its implementation often leads to a downgrading in training and skill development, and so impoverishes the work force and puts at risk OHS performance. The objectives of SCS require some degree of stability of market demand and so its implementation is an area on which the public sector client has the opportunity, through management of demand, to have a significant impact. There are various mechanisms used to develop such an approach and two novel methods were reported by Khalfan and McDermott (2006) and Khalfan *et al.* (2006) in respect of framework agreements and a novel partnership undertaken in England.

Aggregation

They report that in the northwest of England a group of local authorities and, as described in English legislation, Registered Social Landlords (RSLs), formed an alliance to manage procurement for those involved in social housing renewal in their region. The innovative element of this approach was to bundle or aggregate present and future demand of different client organisations and put this bundle into the market to ensure a balanced pattern of demand, and also to negotiate with subcontractors and suppliers in return for resulting workflow certainty. This alliance undertook separate agreements with the contractors and installers to supply labour only and initiated another agreement with suppliers and manufacturers to supply products and materials only. In addition, the social agenda was addressed by ensuring skill development amongst the local labour force in order to improve economic and social sustainability as well as achieving hard environmental targets.

Benefits of aggregation

In discussing the benefits of aggregation the authors indicate that bundling of demand by local governments, social landlords and facilitation by initiatives of central government departments has enabled more companies, particularly SMEs, to collaborate and offer their own services as a package to these alliances. Thus there are both economic and social benefits accruing in the region from this particular type of initiative. Indeed, the authors go on to highlight some of the benefits of aggregation which are:

- direct and continuous employment, and subcontracting opportunities offered by the contractors to the local labour because of continuous work load for both their own employees and other subcontractors;
- skills development within the local community through apprenticeship and training schemes. Contractors have to take on board trainees and give them both on-the-job training and flexibility for attending training and education courses.

Community benefit

Platten *et al.* (2006) discuss the concept of community benefit in relation to regional alliances and procurement and elaborate community benefit as including:

- local employment;
- training provision;
- a commitment to diversity and representation in the workforce;
- a commitment to using the local supply chain;
- health and safety;
- sustainable construction and
- community awareness, consultation and profiling.

These are seen as being delivered through a project alliance which is encouraged to address these issues in the pre-qualification questionnaire (PQQ) and invitation to negotiate (ITN) thus indicating a shift in both procurement practice and key performance indicators. This change has, to a large extent, been facilitated by the Office of the Deputy Prime minister but implemented autonomously in the regions. These principles are incorporated in Framework Agreements, and these are described as having core values which are based on the partnering/alliancing concept which is agreed by the client body and all other participants, including the supply chain. These core values include trust, honesty, openness, commitment, cooperation and respect. In their case study, Khalfan *et al.* (2006) describe the Council's vision as a framework agreement that will deliver good-quality school buildings which will give

- better educational results;
- greater inclusion within the community;
- better safety and environmental performance and
- reduced demand on future school budgets by addressing whole life-cycle costing at the inception of the projects.

Benefits of framework agreements

The major benefits that the researchers found to be achieved by such agreements, in general, were:

- improved design;
- less waste and duplication;
- improved delivery;
- greater quality;
- greater certainty of cost;
- better whole life-cycle costing;
- building of trusting relationships and
- bringing of all 'project knowledge' together at the inception of a project.

Thus, it can be seen that society-based agreements for construction works are both possible and beneficial. This sea change in the paradigm of procurement systems in the UK has led to a situation where both community benefits and SCS go hand in hand. It shared some similarities with the concept of on-call contracting described in Walker and Hampson (2003b: 21–22) and by Jensen and Hall (1995) in relation to medical practitioners and others undertaking locum work in the medical system. Chan and Ibbs (1998) describe its use in a PM context where projects are split into a series of main contracts with the use of small-enterprise practitioners undertaking phased small chunks of work under work packages as min-contracts on an on-call basis. They tender on a services-agreement basis for an overall bulk of likely work with the on-call contracts providing the mechanism to formalise the contractual arrangements for each individual package.

Currently, the UK is probably the most advanced in adopting such approaches to new procurement paradigms but it should be borne in mind that this has been a politically driven agenda which has been in many ways forced upon the construction industry as a mechanism for restructuring and developing both communities and supply chains. Indeed benefits, as gleaned from the case studies, for the supply chain can be enumerated as follows:

1   Benefits are derived even for subcontractors (self-employed people hired for labour-only by main contractor).
2   Continuity of work is given with additional attractions that include prompt payment (one week).
3   Pay as You Earn (PAYE) paper work is done by the contractors.
4   Participating organisations do not have to incur tender costs because they get to know upcoming work around 12 months in advance.
5   People working on site (both direct employees and subcontractors) are trained in the underpinning concepts of the working arrangements. This includes the understanding about the partnership among the local authorities and RSLs and their initiatives to encourage apprenticeships. One of the supply chain partners describes the relationship in the following terms – 'The relationship of suppliers and contractors is changed because there is no money involved between them!' This is because the whole procurement is an open book. Another supply chain partner sheds light on the benefits as; 'Since RSLs are working together in one area, therefore, there are no conflicts and no problems in getting the material. If they were working against each other, then contractors would be fighting among and with suppliers for material supplies.'
6   Workforce smoothing is facilitated – a simple management concept is now being practised within the supply chains associated with the framework, for upcoming years.
7   The power to select the product and allocate the profit margin is shifted from contractors and moved to clients. But on the other hand, there is also a guaranteed profit to all the involved supply chain partners for a longer period of time.

Chapter summary

In this chapter we have used the stakeholder theory and applied it to concepts of trust and commitment to describe and illustrate the vital role that stakeholders can take in making a contribution to adding value to the procurement process. This is dependent upon the views and perceptions of value gained from stakeholders and how their potential contribution can be recognised. We described how a recent cutting-edge tool the *Stakeholder*

Circle™ can be applied and used to engage key identified stakeholders. This led us to one of the often-neglected stakeholder groups, the downstream supply chain. We discussed how their contribution to value adding could be built into a procurement process. We also introduced an emerging approach which has benefits that include developing the potential for local industry SMEs to participate in an alliance-type arrangement (Framework Agreement) that allows them to more fully participate in projects rather than being closed out of a relationship-based procurement system. This latter aspect not only potentially provides cost benefits but also facilitates building social capital and delivering value through other social benefits.

Vignette

ChangeByDesign Plc is a consultancy that specialises in helping organisations to restructure their organisations in a holistic way that incorporates brand and image changes, together with cultural change programs and training and development. It has been contracted to assist AutoCustomeyes, a company that has grown to 500 employees operating through its Oakville, Ontario, North America Regional centre, its European-based centre at Stuttgart, as well as its Melbourne Australia Asia-Pacific centre, to change focus from customising automotive cars to a new range of leisure water craft and light aircraft. It also has recently had a growing business (from a small base) with the US, Australian and NATO military commands for their 'badging' and identification tracking 'cradle-to-grave' system. ChangeByDesign Plc sees its automotive market sector in terminal decline whereas the leisure water craft sector is showing strong growth, particularly in the Asia-Pacific area. Its identification tracking systems that it developed as an 'expensive hobby' of one of the directors has been targeted as a key future market segment for strong future sales growth. Recent European legislation requires originators of consumer goods such as boats, marine leisure water craft and light aircraft, for example, to trace and demonstrate a recycling of at least 80% of the product's contents. Additionally, as valuable military assets are being seen to need unique identification, embedded technologies that allow these goods to be tracked by location and ownership, military facilities management groups have been searching for systems that track their assets on a 24/7 basis.

AutoCustomeyes realised, after undertaking a set of soul-searching strategic workshops, that it needs to transform its business from an auto-centric organisation to one that focuses on the dual business opportunities of the marine and air leisure craft business, and being part of the military supply chain for its badging and identification device business. ChangeByDesign Plc has the task of preparing a project plan for a five-year organisational transformation that will include shifting and consolidating its regional headquarters. The military business will be located closer to its military client base in Washington DC and the Supreme Headquarters, Allied Powers Europe (*SHAPE*), at Casteau, north of the Belgian city of Mons. Its consumer-badging business will be based in Chicago and London, with the Asia-Pacific offices in Brisbane. The first deliverable for the transformation plan is the stakeholder management plan.

Issues to ponder

1 What are the five key elements of this plan would you expect to see?
2 Identify five key stakeholder groups and explain some of the issues that they would be concerned about.
3 This business strategy envisages two quite separate business streams that could spawn a spin-off; if that were true, identify five stakeholder issues that this could introduce.
4 To what extent do you think that existing supply chain partners should be involved in the planning process?
5 With new market segments, and each segment's supply chain stakeholder group being largely unidentified at present, how do you think that ChangeByDesign Plc and AutoCustomeyes might identify and persuade these groups to gain their trust, support and commitment to engage with them – what might be their value proposition?

Web resources

See URL www.stakeholder-management.com for more details about the *Stakeholder Circle*^{TM} tool.

References

Bessant, J., Kaplinski, R. and Lamming, R. (2003). 'Putting Supply Chain Learning into Practice'. *International Journal of Operations and Production Management*. **23**(2): 167–184.
Bourne, L. (2005). Project Relationship Management and the Stakeholder Circle. Doctoral thesis. Doctor of Project Management, Graduate School of Business. Melbourne: RMIT University.
Bourne, L. and Walker, D. H. T. (2004). 'Advancing Project Management in Learning Organizations'. *The Learning Organization*, MCB University Press. **11**(3): 226–243.

Bourne, L. and Walker, D. H. T. (2005). 'Visualising and Mapping Stakeholder Influence'. *Management Decision.* **43**(5): 649–660.

Bourne, L. and Walker, D. H. T. (2006). 'Using a Visualising Tool to Study Stakeholder Influence – Two Australian Examples'. *Journal of Project Management.* **37**(1): 5–21.

Briner, W., Hastings, C. and Geddes, M. (1996). *Project Leadership.* 2nd edn. Aldershot, UK: Gower.

Carroll, A. B. and Buchholtz, A. K. (2000). *Business and Society: Ethics and Stakeholder Management.* Cincinnati, OH: South-Western College Publishing.

Cavaleri, S. (2004). 'Principles for Designing Pragmatic Knowledge Management Systems'. *The Learning Organization*, MCB University Press. **11**(4/5): 312–321.

Cavaleri, S. and Seivert, S. (2005). *Knowledge Leadership: The Art and Science of the Knowledge-Based Organization.* Oxford: Elsevier Butterworth Heinemann.

Chan, S.-T. A. and Ibbs, C. W. (1998). 'On-call Contracting Strategy and Management'. *Journal of Management in Engineering, ASCE.* **14**(4): 35–44.

Cheung, F. Y. K. (2006). *Supply Chain Sustainability: The Role of Trust and Relationship.* CIBW112, Sustainable Development Through Innovation and Culture, Dubai, UAE, 26–29 November, Dulaimi M.

Cleland, D. I. (1995). 'Leadership and the Project Management Body of Knowledge'. *International Journal of Project Management.* **13**(2): 82–88.

Cleland, D. I. (1999). *Project Management Strategic Design and Implementation.* 3rd edn. Singapore: McGraw-Hill.

DETR (1998). *Rethinking Construction*, Report. London: Department of the Environment, Transport and the Regions.

Dinsmore, P. C. (1999). *Winning in Business with Enterprise Project Management.* New York: American Management Association.

Donaldson, T. and Preston, L. E. (1995). 'The Stakeholder Theory of the Corporation: Concepts, Evidence, and Implications'. *Academy of Management Review.* **20**(1): 65–91.

Elkington, J. (1997). *Cannibals with Forks.* London: Capstone Publishing.

French, W. A. and Granrose, J. (1995). *Practical Business Ethics.* Englewood Cliffs, NJ: Prentice Hall.

Gibson, K. (2000). 'The Moral Basis of Stakeholder Theory'. *Journal of Business Ethics.* **26**: 245–257.

Greene, R. and Elfrers, J. (1999). *Power the 48 Laws.* London: Profile Books.

Hersey, P., Blanchard, K. and Johnson, D. E. (1996). *Management of Organizational Behaviour.* 7th edn. London: Prentice Hall International.

Inkpen, A. C. and Beamish, P. W. (1997). 'Knowledge, Bargaining Power, and the Instability of International Joint Ventures'. *The Academy of Management Review.* **22**(1): 177–203.

Jensen, P. and Hall, C. (1995). 'New Arrangements for Radiology at Sydney Hospital'. In *The Contracting Casebook – Competitive Tendering in Action.* Domberger S. and C. Hall (Eds). Canberra: Australian Government Publishing Service: 87–97.

Jones, T. M. and Wicks, A. C. (1999). 'Convergent Stakeholder Theory'. *Academy of Management Review.* **24**(2): 206–221.

Khalfan, M. M. A. and McDermott, P. (2006). 'Innovating for Supply Chain Integration within Construction'. *Construction Innovation: Information, Process, Management.* **6**(3): 143–157.

Khalfan, M. M. A., McDermott, P. and Kyng, E. (2006). *Procurement Impacts on Construction Supply Chains: UK Experiences*. Symposium on Sustainability and Value Through Construction Procurement, CIB W092 – Procurement Systems, CIB Revaluing Construction Theme, The Digital World Centre, Salford, UK, 29 November–2 December, Peter McDermott P. and M. M. A. Khalfan: 449–458.

Latham, M. (1994). *Constructing the Team*. Final Report of the Government/ Industry Review of Procurement and Contractual Arrangements in the UK Construction Industry. London: HMSO.

Leonard-Barton, D. (1992). 'The Factory as a Learning Laboratory'. *Sloan Management Review*. 34(1): 23–38.

Leonard-Barton, D. (1995). *Wellsprings of Knowledge – Building and Sustaining the Sources of Innovation*. Boston, MA: Harvard Business School Press.

Lewicki, R. J., McAllister, D. J. and Bies, R. J. (1998). 'Trust and Distrust: New Relationships and Realities'. *Academy of Management Review*. 23(3): 438–459.

London, K. (2005). *Supply Chain Sustainability*. CRC CI Report. Brisbane, 2004-016-A: 33.

Maqsood, T., Walker, D. H. T. and Finegan, A. D. (2007). 'Facilitating Knowledge Pull to Deliver Innovation through Knowledge Management: A Case Study'. *Engineering Construction and Architectural Management*. 14(1): 94–109.

Maslow, A. H. (1943). 'A Theory of Human Motivation'. *Psychology Review*. 50: 370–396.

Mayer, R. C., Davis, J. H. and Schoorman, F. D. (1995). 'An Integrated Model of Organizational Trust'. *Academy of Management Review*. 20(3): 709–735.

McAllister, D. J. (1995). 'Affect- and Cognition-based Trust as Foundations for Interpersonal Cooperation in Organizations'. *Academy of Management Journal*. 38(1): 24–59.

Meyer, J. P. and Allen, N. J. (1991). 'A Three-Component Conceptualization of Organizational Commitment'. *Human Resource Management Review*. 1(1): 61–89.

Mitchell, R. K., Agle, B. R. and Wood, D. J. (1997). 'Toward a Theory of Stakeholder Identification and Salience: Defining the Principle of Who and What Really Counts'. *Academy of Management Review*. 22(4): 853–886.

Morris, P. W. G. and Hough, G. H. (1993). *The Anatomy of Major Projects – A Study of the Reality of Project Management*. London: Wiley.

Platten, A., Dobrashian, T. and Dickenson, M. (2006). *Innovative Approaches to Developer Selection and Procurement for Housing Market Renewal*. Symposium on Sustainability and Value Through Construction Procurement, CIB W092 – Procurement Systems, CIB Revaluing Construction Theme, The Digital World Centre, Salford, UK, 29 November–2 December, Peter McDermott and M. M. A. Khalfan: 449–458.

Rousseau, D. M., Sitkin, S. B., Burt, R. S. and Camerer, C. (1998). 'Not So Different After All: A Cross-Discipline View of Trust'. *Academy of Management Review*. 23(3): 393–405.

Shelley, A. (2007). *The Organizational Zoo: A Survival Guide to Work Place*. Fairfield, CT: Aslan Publishing.

Sherer, S. A. (2005). 'From Supply-chain Management to Value Network Advocacy: Implications for e-Supply Chains'. *Supply Chain Management*. 10(2): 77–83.

Spekman, R. E., Spear, J. and Kamauff, J. (2002). 'Supply Chain Competency: Learning as a Key Component'. *Supply Chain Management.* 7(1): 41–55.

Von Hippel, E. (1988). *The Sources of Innovation.* New York: Oxford University Press.

Walker, D. H. T. (2003). 'Implications of Human Capital Issues'. In *Procurement Strategies: A Relationship Based Approach.* Walker D. H. T. and K. D. Hampson, (Eds). Oxford: Blackwell Publishing: 258–295.

Walker, D. H. T. and Hampson, K. D. (2003a). 'Developing Cross-Team Relationships'. In *Procurement Strategies: A Relationship Based Approach.* Walker D. H. T. and K. D. Hampson. Oxford: Blackwell Publishing: chapter 7, 169–203.

Walker, D. H. T. and Hampson, K. D. (2003b). 'Procurement Choices'. In *Procurement Strategies: A Relationship Based Approach.* Walker D. H. T. and K. D. Hampson, (Eds). Oxford: Blackwell Publishing: chapter 2, 13–29.

Ward, S. C. (1999). 'Assessing and Managing Important Risks'. *International Journal of Project Management.* 17(6): 331–336.

Ward, S. C. and Chapman, C. (2003). 'Transforming Project Risk Management into Project Uncertainty Management'. *International Journal of Project Management.* 21(2): 97–105.

Weaver, P. and Bourne, L. (2002). *Project Fact or Fiction – Will the Real Project Please Stand Up.* Maximising Project Value, Melbourne, 21 October, PMI Melbourne Chapter: On CD-Rom: 234.

Wenger, E. C., McDermott, R. and Snyder, W. M. (2002). *Cultivating Communities of Practice.* Boston, MA: Harvard Business School Press.

Yukl, G. A. (2002). *Leadership in Organizations.* 5th edn. International edn. Upper Saddle River, NJ: Prentice Hall.

Chapter 4

Business ethics and corporate citizenship

Derek H. T. Walker, Michael Segon and Steve Rowlinson

Chapter introduction

Some aspects of business ethics and corporate citizenship have an obvious relationship to project procurement. From an ethical point of view, the need to banish corruption and expunge exploitation is a given – or at least the espoused position of all reasonable people is that there is no place for these behaviours in a civilised world. So, there must be rules, standards and provisions that safeguard us from these extreme and unconscionable behaviours. This is where governance and cultural pressures from within organisations and society impose a way in which procurement of projects happens such that it does not solely satisfy the need for one particular group, organisation or individual to benefit unsustainably at the expense of others. The ontological perspective that we adopt in this chapter, and indeed in this book, is that the outcome of projects is a beneficial change that adds value and that this value should be sustainable and meets the needs of genuine project stakeholders. In this respect, this chapter links very closely with Chapter 3, Chapter 6 and Chapter 9 and somewhat to Chapter 8 and Chapter 10.

Sustainability has an important link to procurement. People influence and drive the process of initiating, designing and delivering the mechanisms that result in completed projects. People who influence that PM process have a range of motivations and needs, and different perceptions of what success and the project outcome should look like. This is why this chapter links to Chapter 3. People (the market) determine what they demand, and if business destroys their capacity to procure projects then the reason for these projects is removed. The definition of project success and what constitutes value (as discussed in Chapter 6) directly relates to ethics, corporate citizenship and governance issues that view value with a focus on sustainability in its broader sense. Much of ethics and the governance issues associated with corporate citizenship relate directly to culture, both organisationally and socially. This is closely anchored in notions of trust and commitment as discussed in Chapter 9. Projects require talented people to deliver projects and the expected benefits from those projects, so attracting talented people is a vital sustainability issue. Determining procurement governance structures and processes links with a project's capacity to deliver innovation and learning (discussed in Chapter 8) because rules, procedures and reward systems can be used to encourage and facilitate innovation. Similarly, ethics and governance are linked to providing an environment where supply chain partners and project team members choose organisations that they consider to be partners of choice (discussed in Chapter 10).

This chapter is presented in three broad sections, with a vignette at the end of the chapter which prompts relevant questions. The first section deals with theoretical underpinning relating to business and social ethics because this also forms the basis for corporate citizenship and corporate responsibility. It is relevant to project procurement choices because it helps us to understand the perspective of any given set of procurement arrangements that the project appears to endorse or be guided by. This section also explains the underpinning of what can be seen as a particular style of professional behaviour. Behaviour of course is cultural, value-laden and thus contestable so it makes an interesting point of debate and strategic advantage differentiation.

The second section relates to explaining and exploring the triple-bottom line (3BL) concept and its impact upon project procurement decisions because it considers the financial perspective as well as social and environmental dimensions. This has informed decisions about how to structure a procurement process for some industries, most notably the oil and gas industry. An interesting recent development is the increasing use of instruments that can be used for capitalising on green credits, and this is an area that most texts on procurement have failed to recognise or discuss. The emerging nature of this aspect means that it will receive an exploratory and somewhat cursory treatment that hopefully will develop in future editions.

The third section relates to the impact of governance on procurement options. This section briefly discusses corporate governance because projects fit into programs that are managed within portfolios to achieve corporate strategic goals – these days the strategic advantage is strongly influenced by competitive advantage based upon differentiation as well as cost efficiencies. Governance is both fundamental and practical for procurement because it is about the rules, regulations and measures that protect integrity and fair play, as well as protect project investors against muddled and mishandled approaches to managing projects. In some industries where the project outcome is highly tangible such as buildings, planes, ships, etc. there is a special focus on health, safety and the environment, and so we will devote a sub-section of governance to that which impact upon procurement choice and governance requirements.

Fundamental theoretical underpinnings of ethics

As managers, we are expected to make decisions that advance the interests of the organisation. In fact, we become agents of an organisation and in effect are legally bound to advance the interests of the organisation and its shareholders. Decision making is the essence of a manager's job and yet, as discussed in Chapter 5, Mintzberg (1990) identified most managers as being reactive decisions makers. Child (1986) also identified that managers tend to look to the first solution that appears to satisfy minimum criteria, rather than the best option. Such decision-making practice often leads to poor-quality outcomes which may expose organisations and managers to risk. This chapter is concerned with improving decision making by addressing the ethical dimensions. Including an ethics filter as part of the processes that enhances decision making by identifying the degree of risk associated with decisions, and the filter assists managers to develop defensible positions for their decisions and actions.

What is ethics?

In general, ethics is concerned with what is right, fair, just or good; about what we ought to do, not just what is the case or what is most acceptable or expedient (Preston, 1996: 16). Ethics is a multifaceted concept that includes the study of morality, the legitimacy of moral claims and the basis of justification of decisions. As we will see, organisations develop policies and procedures as part of corporate governance for individuals to follow – this is an example of normative ethics or the clarification of behaviour standards about what we ought to do. As ethics is directly concerned with behaviour it therefore has relevance to the way people behave in organisations.

Ethics can best be understood by considering a number of perspectives.

- philosophy and philosophical thinking about morality, moral problems and moral judgment (Beauchamp and Bowie, 1997);
- defined, as a study of what is good or right for people, what goals they ought to pursue and what actions they ought to perform (Velasquez, 1998);
- concerning itself with human conduct or activity that is done knowingly and consciously (Robbins and Mukerji, 1990) and
- actions and practices directed towards improving the welfare of people and attaining a good life (Buchholz, 1989).

An important question to answer is why should we consider ethics at all in business? There are some who argue that business transactions and business activity are limited activities, and as such managers are constrained to only consider actions that advance the interests of the organisation. This position advocates that managers are constrained in their decision making to only consider decisions that advance shareholder interests and it is wrong for them to engage in socially responsible activity, other than obeying the law – a view of the classical approach to socially responsible practice which was popularised by economist Milton Friedman in his significant article

'The social responsibility of business is to increase its profits' published in 1970 (Donaldson and Werhane, 1999).

However, Allinson (1998) has argued that ethics is fundamental to business conduct. He contends that a moral relationship lies at the basis of agreements and contracts between two people or two organisations, and that unless there is a basis of trust, business cannot proceed. This view suggests that the organisation is a moral player in society with duties and responsibilities, and in order to advance its interests it must engage in behaviour that is acknowledged to be at least the minimum moral standard. More recently, business ethics literature has focused on the business case for organisations developing ethical cultures. Related to the issue of trust between businesses and stakeholders as identified by Allinson (1998), the development of ethical cultures and business practices, as evidenced by the existence of codes of ethics, integrity systems and socially responsible strategies, strengthens the relationship between the organisation and the key stakeholders such as consumers (Freeman, 1999). This suggests that there is a direct relationship to business profitability, as evidenced by improved sales and sustained growth or a negative relationship to costs. This can be seen as another argument in support of organisational ethical frameworks.

Whilst research is clearly emerging that supports the long-held view that ethics is good for business (Carroll and Meeks, 1999), there is another reason why we should consider ethics at the micro level of managerial decision making. We all have slightly different understandings about what is ethical. Our moral values are shaped over time, influenced by a range of factors including our parents and upbringing, religion, formal and informal education and our cultural identity among other experiences (Buchholz, 1989). We should not presume that our understanding and application of ethics is necessarily correct or that others will agree with our interpretation. What is required is a considered and informed approach rather than an intuitive or gut feel to ethical decision making. We should incorporate a systematic approach to defining the ethical dimensions to our decisions and ensure that our business practices are ethical because it's the right and profitable thing to do. The study of business ethics does not just mean moralising about what should or should not be done in a particular situation. Rather it systematically links the concepts of morality, responsibility, and decision making in organisations (Ferrell and Fraedrich, 1997).

What is an ethical issue?

An ethical issue can be seen as a problem, situation or opportunity, which has several possible options, and where each must be evaluated as right or wrong, ethical or unethical. Aguilar (1994: 33) states 'Managers who are sensitive to the ethical content of business decisions and operational activities, who agree to behave ethically, and who hold compatible views on how to analyse ethical situations still must confront the inherently perplexing nature of many ethical problems.' Managers talking about workplace ethical problems often describe dilemmas, situations in which they were faced with a difficult choice and where no clear-cut right answers existed.

Managers face ethical issues on a daily basis, issues such as dealing with clients, marketing and advertising, product liability and safety, privacy of employee records, employee discipline and relations with competitors, the tricky topic of gifts, invitations from suppliers – all have ethical dimension. Ferrell, Fraedrich and Ferrell (2002) classified the ethical issues relevant to most organisations into four key areas: First, conflicts of interest; second, honesty and fairness; third, communications and fourth, relationships within the organisation.

A conflict of interest exists when an individual has the opportunity to take a decision which advances his or her own interest rather than that of the organisation. It is important to acknowledge that people often will make judgements about perceived, rather than actual, conflicts of interest. Separating private interests from business dealing helps employees avoid conflicts of interest. In project procurement terms, accepting bribes, personal payments and gifts are considered conflicts of interest in most 'Western' countries. However, in some countries, gift giving is commonly accepted as part of relationship building to establish trust and rapport with business partners. This concept is referred to in China as '*guanxi*' (Low and Leong, 2000) and should not be confused with bribery. We discuss this aspect in more detail in Chapter 9. Maintaining transparency is an effective strategy in minimising conflicts of interest.

Honesty and fairness can be seen as principles that most people would acknowledge as typifying ethical behaviour. As Ferrell *et al.* (2002) suggest, these relate to a decision maker's general characteristics or traits. It is expected that people, at the least, follow all applicable laws and regulations and should not knowingly harm customers, employees, clients or competitors through deception, misrepresentation or coercion.

Fraudulent behaviour[1] and corruption can be seen as a subset of conflict of interest; however, it can also be systemic and involve many individuals and groups. Ferrell *et al.* (2002: 33) define fraud as 'any purposeful communication that deceives, manipulates or conceals facts in order to create a false impression'. This would include financial records, advertising messages, tender selection, etc. where the organisation's message to customers and stakeholders needs to be truthful.

This last focus looks at areas such as relations with colleagues, meeting obligations and responsibilities, and not placing pressure on others to encourage them to take unethical actions. We should recognise that many laws exist that directly deal with interpersonal relationships and their ethical dimension. Anti-bribery, anti-collusive tendering practices, equal employment opportunity and anti-discrimination legislation are examples of laws which seek to ensure that people are not related unethically.

Explaining ethical behaviour

Behaviour is a function of many variables. We cannot simply say that people are born good or bad and thus their actions are always predictable. What we can do is examine how people make decisions and the factors which seem to impact on or influence their decision-making process.

One of the best explanations for how people make ethical decisions is probably that developed by Lawrence Kohlberg. He found that in order for people to behave morally they must, among other things, decide what course of action is morally right and then they must choose the morally right path over others. Kohlberg's theory primarily focuses on the first process, the process by which people *decide* what is morally right (French and Granrose, 1995).

Kohlberg proposed that individuals arrive at differing moral judgments about the same issues because they possess differing stages of cognitive moral development. He proposed a six-stage model of cognitive moral development, with each stage representing progressively more sophisticated moral evolution. These six stages can be grouped together in pairs to form three levels: pre-conventional, conventional and post-conventional. By implication, persons at higher stages of moral development tend to consider factors other then their own benefit, when identifying and resolving moral or ethical issues, however, they can comprehend all reasoning at stages below them, but cannot comprehend the reasoning of stages any more than one above there own (French and Granrose, 1995). This suggests that within organisations individuals may reach different decisions concerning the same action or circumstances if not governed by specific guidelines or decision criteria. Kohlberg suggests that when a dilemma presents itself, which cannot be resolved, explained or seems to contradict our own current level of reasoning, we gain maturity and move through these stages. This can occur through general interaction with others, peers, colleagues or through more formal methods such as education and training (French and Granrose, 1995).

Level 1 – pre-conventional level

The pre-conventional level is characterised by people making decisions which are based on self-interest, mitigated by rewards and punishment, and externally imposed rules such as law or organisational policies.

In *stage one* how people decide whether an action is right is largely determined by the rewards, punishments and favours associated with the action. This suggests people follow laws and rules automatically because they do not want the negative consequences, for example, reprimands, fines, etc. or because they are rewarded through prizes, salaries, etc.

At *stage two* people begin to recognise that they need to work with others so that they can attain their goals – they acknowledge others and a duty to them; however, they are still motivated largely by self-interest. This concept is known as reciprocity. The 'scratch my back and I'll scratch yours' approach. People engage in behaviour that they know will yield possible favours in return, or feel obliged to repay a debt.

Level 2 – conventional level

At this level people have recognised the importance of a group in society, usually family or some important clan/grouping. Generally, living up to the roles and expectations of others and fulfilling duties and obligations and following rules and laws within that group is determined as doing the 'right thing'. This is highly relevant to organisations as it highlights the importance of leaders and supervisors in influencing the behaviour of employees. According to business research most people operate at this level within organisations, highlighting the importance of leadership and culture, and at the following stage where the importance of organisational policies and procedures act as guiding principles, rendered as distinct from strict compliance (Ferrell *et al.*, 2002).

At *stage three* people value their belonging to a group and value the importance of group interaction. They tend to judge actions as right, based on the norms and standards of the group, or based on the belief that the action or outcome will be favoured by the group or the leader of the group. The important concepts are those of trust and loyalty. So people may do what is asked of them by a boss or managers because they want to please them or because others whom they respect say it's important to follow the directions of the bosses.

At *stage four* people's perspective broadens to consider the wider societal group. People tend to make decisions based on the agreed duties and following rules, which are designed to promote the common good. Whilst people have a broader perspective than that described at stage three, there is still a boundary that people use to define the wider societal group to the exclusion of others. This might be at a state or national level or some other basis of categorising the group. This could be used to explain why some state governments will take a decision that advances their own state but is to the detriment of another.

Level 3 – post-conventional or principled level

At this level people have gone beyond notions of self-interest or referent groups; rather they make their own decisions based on more principled notions consistent with justice and rights. This tends to be of greatest relevance to organisations when engaged in 'socially responsible' or corporate citizenship activities.

At *stage five* people still regard rules and laws are important because they maintain social cohesion (known as the concept of social contract); however, people are prepared to change laws for usual social purposes. In addition, people consider the concept of moral law, that which exists beyond the written law, as contributing toward societal well being.

At *stage six* people have moved to a higher level in which the notion of universal laws and principles are applied.

Kohlberg's approach is not about understanding the action. In fact, people operating at different levels may make the same decision or take the same action but for different reasons. This concept explains *how* they came to make their decision and helps explain how they see the world and critical factors influencing their decisions (French and Granrose, 1995).

An ethics bank

Recent research suggests that some people's approach to ethical decision making is not necessarily determined by levels of moral maturity, rather it is a conscious decision to follow an action, which they know to be wrong but argue that it is mitigated by other good deeds they have done in the past. This has been referred to as an *'ethics bank'* in which an ethics balance is established where good behaviour is deposited so as to allow the occasional withdrawal via bad behaviour.

This suggests that people generally follow a path which results in accepted morally right actions. Either consciously or unconsciously they keep track of the 'value' of these actions in effect building up credits of 'good'. Then, when confronted by a decision which they know to be ethically unacceptable, but which may be of sufficient importance or value to them, they may engage in a conscious and rational process of determining how many ethically right actions they could trade off in order to pursue the action.

Perhaps this type of decision making is most evident at an organisational level where companies engage in actions, which have a real or perceived detrimental impact on society, or pursue actions which are knowingly wrong, for example, pollution, or use or sell unsafe products etc. The organisation then does something which society will recognise as beneficial in order to restore balance. For example, a mining company may build roads, schools or hospitals for a community in compensation for ecological damage, which they may not redress for economic reasons.

Ethical reasoning

There are many ways of studying and analysing management decision making and action. Ethical theories describe general approaches to thinking that allows ethical principles to be drawn and applied to concrete actions and management practices.

What questions should we ask? What factors should we consider? Velasquez (1998) suggests that the first step in analysing moral issues is obvious but not always easy: get the facts – something we know from Mintzberg's (1975) study of what managers do and seem reluctant to do. Some decisions result in ethical controversies simply because we haven't collected enough information about the decision and its consequences. However, having the facts is not enough to ensure that the decision is not just profitable but also ethical. As Preston (1996) suggests, facts can be verified but they do not tell us what to do. We need to use facts as part of the decision-making process but according to Velasquez (1998), we also need to resolve the ethical issue through an appeal to values.

There are a number of ways that we can analyse the ethical dimension associated with a decision. Buchholz (1989) suggests that ethical theories are not just ways of analysing a decision but they may also enable us to justify it. He argues that a critical issue in decision making is being able to defend a decision using a variety of perspectives.

Determining ethics according to outcome

Often individuals and organisations determine whether a decision or action is ethical based predominantly on the actual or perceived value of the outcome that is generated. These approaches are also called consequentialist theories, and the two most often discussed are *utilitarianism* and *egoism*. They hold that the moral worth of an action or practice is determined solely by the consequences of the practice or action and not on the intent of the action. A decision or an act undertaken by an individual or a group would be considered right if the outcome generates benefits greater than the harm produced (the 'greater good'). This approach to decision making uses a decision-making system that parallels cost–benefit analysis and thus is very appealing to business. It assumes that everything can be valued and measured.

THE EGOTIST APPROACH (SELF-INTEREST)

As the name suggests, this approach focuses largely on benefits that can accrue to the individual. An act is considered morally right or wrong solely on the outcomes for one's self. Self-interest underpins ethical egoism and many argue that the pursuit of self-interest will lead to optimal allocation of resources in a free market system. Self-interest will provide the motivation for people to work hard, take risks and innovate, etc., thus guaranteeing economic growth.

Whilst this is a popular approach to defining ethical behaviour, many theorists suggest that it is a narrow interpretation. Under this concept, many actions that are clearly not beneficial to society could be argued as beneficial for the individual and therefore perhaps seen as acceptable. We can see egoism at work in organisations, in particular through reward systems such as commission-based salaries. Egoism can be also used to explain why people take actions that others would see as unethical, such as accepting gifts or matters that others argue as conflicts of interest. To the individual it could be seen as merely advancing their self-interest.

THE UTILITARIAN APPROACH (MAJORITY OR GROUP INTEREST)

Utilitarianism was developed by Jeremy Bentham and John Stuart Mill in the nineteenth-century Britain to help legislators determine which laws were morally best (Velasquez, 1998).

The objective of this approach is to generate the greatest possible benefit for the largest number of people, whilst minimising the damage or harm done to others. The ethical action is the one that provides the greatest good for the greatest number. As identified above, this is akin to a cost–benefit analysis. The issue here is not how these outcomes are achieved, but rather the size of the cost or benefit that is incurred. The primary way of determining the moral worth of an action is to evaluate all the associated social cost or benefit (Buchholz, 1989). Utilitarianism has not only an obvious appeal but it also has important limitations that are often overlooked by business.

First, the emphasis on measuring value often means that only those things that can be measured are included; issues such as emotions, feelings, sense of duty are often not part of a utilitarian calculus. Second, determination of value is somewhat subjective (see Chapter 6 for more on performance measures). One person's view of a decision might be very different to another's interpretation. Third, a significant problem with utilitarianism as it is applied in business is that it is not used to calculate the greatest benefit for the greatest number as originally intended; rather, a restricted utilitarianism is used whereby the benefits are calculated for only a certain group such as shareholders or customers. Fourth, utilitarians typically include only the direct costs, and fail to include the indirect and opportunity costs associated with a decision. This can often lead managers and business to fail to consider the impact of their decision on certain groups, as discussed later in this chapter with reference to 3BL issues raised by Flyvbjerg, Rothengatter and Bruzelius (2003). Perhaps the most important failing for us to consider is that utilitarianism emphasises outcomes and not how a decision is made – that is, the ends justify the means. Clearly it is possible to achieve what one could argue are very ethical outcomes, but in a manner which is unethical or illegal. The lack of attention to how decisions are made by managers exposes the organisation to substantial downstream risk, not to mention possible legal action, when making project procurement decisions.

Determining ethics by process and principle

Another way of determining the ethical nature of a decision is not to look at the outcome but rather the process or actions used in arriving at the decision or outcome. This branch of ethics, referred to as *deontology*, maintains that actions and practices have intrinsic value – some actions and practices may be morally wrong, no matter how good their outcomes might be (Buchholz, 1989). Deontology also suggests that it is almost impossible to accurately calculate benefits and harms because there are many unknown variables. As absolute certainty does not exist, as is the case in the business world, we can never really be sure of the full consequences of our actions, thus we need to use other means of determining the moral worth of an action.

In effect, this branch of ethics says that focusing on the decision itself is important, thus following duties, respecting rights and using principles to guide decisions such as fairness, justice and due process are advanced as the means by which ethical decisions can be justified. This has great relevance to organisations because they use policies and procedures as a means of guiding decision making. Professionals such as doctors, lawyers and engineers have codes of ethics and conduct which stipulate duties that need to be respected. Similarly, there are many laws such as Equal Employment Opportunity, Anti-Corruption and Fair Trade legislation, which are based on notions of justice and fairness.

THE RIGHTS APPROACH

The rights approach to determining ethics essentially proposes that people have a set of rights that must be respected. It emphasises that people should not be seen as objects, and must be afforded the right to freely choose a course of action. Velasquez (1998) provides a list of rights that can assist in guiding decision making. Whilst not a complete list they are considered to be indicative and have clear application to many business contexts.

- The right to the truth; we have a right to be told the truth and to be informed about matters that significantly affect our choices.
- The right of privacy; we have the right to do, believe and say whatever we choose in our personal lives so long as we do not violate the rights of others.
- The right not to be injured; we have the right not to be harmed or injured unless we freely and knowingly do something to deserve punishment, or we freely and knowingly choose to risk such injuries.
- The right to what is agreed; we have a right to what has been promised by those with whom we have freely entered into a contract or agreement.

THE DUE PROCESS AND PRINCIPLED APPROACH

This approach also focuses on the principles and processes that characterise the decision. Concepts such as fairness, justice and due process are identified by Buchholz (1989) as means by which the integrity of a decision can be demonstrated. Perhaps the most obvious application of these concepts is that of law. Law is often defined as the minimum moral standard in society. These standards are accepted and for the most part there is no need to undertake a cost–benefit analysis (a utilitarian calculus, or to analyse rights, etc.); we simply follow the law and thus the decision is seen as ethical. Fairness and justice is seen to be done when laws are followed and when due process is afforded. Its application to organisations and managerial decision making is very clear. Managers and organisations must abide by applicable laws, be they equal employment opportunity, health and safety, or advertising requirements. In addition, organisational policies and procedures can also be seen as part of this approach because they provide information about how to make a decision.

Clearly, determining ethics by process seeks to ensure the integrity of the decision-making process; however, there are limitations which should be acknowledged. Perhaps the most significant problem is the emphasis on how we make the decision without consideration of the outcome. Often courts will pass judgements on the basis that due process was not followed. This can occur in unfair dismissal cases in industrial relations, where an employee may have committed a serious offence, but because managers and organisations did not follow the required disciplinary procedures, the employee is seen to have been denied natural justice. The other danger with the due process approach is that it may encourage organisations to meet only their minimum legal obligation. This is identified by McAlister, Ferrell and Ferrell (2005) as a minimalist social-responsibility position.

DETERMINING ETHICS NATURE BY VIRTUE

The third major way in which people make ethical decisions is not as clear or calculating as the previous approaches; rather, the virtue approach has similarities to the controversial trait or character approaches to leadership.[2] This ethical approach proposes that there are certain ideals towards which we should strive – demonstrated by characteristics or virtues. It is concerned with people's motives or intention of their actions. In this way we need to recognise that it is a subjective analysis of the doer of the action. Preston (1996) suggests that this approach is not centred upon the question 'what ought we to do?' but rather 'what kind of person ought we to become?'.

The virtue approach focuses on personal disposition, attributes, traits/character[3] that enable one to become, and to act in ways that develop, what human beings have the potential to become. It enables the pursuit of adopted ideals ('virtues') such as honesty, courage, compassion, generosity, fidelity/loyalty, integrity, fairness, self-control and prudence. The goal is to encourage living properly so that good and right behaviour emanates from within the person. We can see the application to business, in that many organisations develop value or mission statements, as well as project charters for their projects, which are built on such virtues. The purpose of these is to establish the basis for organisational strategy, culture and to provide people with a series of principles that can assist in resolving an ethical conflict when one of the other approaches fails to deliver a clear or satisfactory resolution.

Virtues are developed through learning and through practice. This approach assumes highly developed cognitive processes; that is, the ability to rationally assess and understand complex environments and how they affect our decision-making. It also assumes that we have the time to exercise reasoning to the extent that virtues become habits, so that in practical life people act instinctively through habit. Consequently, virtue approaches suggest that a truly good person does not calculate or reason about truthfulness or lying; she/he is honest. However, despite a person's reasoning about what is right or good, that person may still not choose to do the right or good thing. Consequently, by becoming clearer about our identity (who we are), it does not follow that actions we must take are pre-determined.

The virtue approach's greatest strength, its generality, is also its greatest weakness. It does not provide specific guidance on how to resolve a dilemma – it assumes that the virtuous person will know the answer. Nor does it provide a framework for the resolution of conflicts of virtues. For example, how does one resolve a conflict between honesty and loyalty? Clearly some of these concepts are also open to interpretation. The vagueness about what constitutes a virtue like 'integrity' can also result in its devaluation – this has been a criticism of mission and value statements as being merely 'window dressing' because organisations and managers fail to demonstrate or live up to these expectations.

Applying the ethical theories as a decision-making filter

Whilst it is important to know the differences between the various ethical theories, a further challenge is how to integrate them into decision making – when should you use utilitarianism, or process and deontology or perhaps virtue ethics?

In this chapter we have identified that all the three major ethical approaches have advantages and disadvantages. To pick and choose a theory simply to justify a decision is consistent with what Benjamin identifies as the *moral opportunists*, constantly shifting their moral position so as to gain the most significant short-term advantage (French and Granrose, 1995: 166–169).

Velasquez (1998) and Francis (1999) suggest that decision making needs to reflect a holistic approach, which seeks to identify ethical implications from all three major approaches. Both agree that the first step is to collect the facts surrounding an issue and ensure that we have clarified the decision context. Managers need to be sure that the decision is theirs to take and that they have the knowledge and skills to make such a decision. It is a basic ethical duty to ensure that we act within our defined authority and capability.

Having ascertained the facts, we should ask ourselves a series of questions, which will assist us to identify the ethical dimensions associated with the decision.

1   What are the potential risks or harms that will result from the decision(s)?
2   Which alternative will lead to the best overall consequences?
3   What are the moral rights of the stakeholders (including those affected and those involved in the decision) and which course of action best upholds those rights?
4   Which decision is consistent with both the spirit and letter of applicable laws?
5   Which decision is consistent with policies and procedures of the organisation?
6   Which decision is consistent with principles of justice and fairness?
7   Which decision would be seen as having the characteristics of integrity, honesty, truthfulness and trustworthiness?

In effect, these seven questions ask managers to use all the three approaches as part of the decision-making process. Questions 1 and 2 reflect consequentialist perspectives, Questions 3–6 deontology and process and the final question reflects virtue approaches. If any one of the seven questions causes concern then the decision may have questionable ethics or not be justifiable. Rather than pursue such actions and possibly expose organisations and themselves to risk, managers need to revisit their decisions and seek alternatives that are consistent with the principles identified above.

This method of asking key questions does not necessarily guarantee that all decisions taken will be ethical, but it does increase the likelihood of ethical dimensions being uncovered and thus considered. As with any skill, the more attention given to developing ethical decision-making skills the more effective they will be.

Francis (1999) suggests that two further tests can be applied if there is still uncertainty, both dealing with your decision, and perceptions that are relevant to making project procurement decisions.

1   Imagine that you have made a decision, implemented it and for various reasons the matter comes before a public court: imagine further that

you are called to the witness box and are asked to defend your position. Would you be able to provide such a defence?

2   Consider a similar scenario; however, on this occasion it is your mother, father, spouse or children who ask you to explain your decisions. After the decision would they still be proud of you?

Discourse is fundamental to increasing one's understanding of ethics and the challenges of working in a morally complex world. Managers ought not to assume that a decision must be taken without input from others. A rich resource for many individuals are the numerous professional bodies which have codes of ethics and other forms of assistance such as forums that enable people to discuss and resolve challenges. Similarly, other managers and colleagues in your organisations have a wealth of experience and knowledge and may be in a position to provide valuable counsel; however, remember that their ethical perspective may differ substantially from your own. Also, stakeholder consultation can reveal pertinent and salient issues.

Project procurement and 3BL

The triple-bottom-line (3BL) concept came to the forefront of management thinking with Elkington (1997) arguing that organisations should report performance in more than purely *financial* terms because this does not fully adequately address the value proposition of significant stakeholders. The second two 'bottom lines' are social and environmental. The *social* bottom line relates to the social impact of any activity to enable us to evaluate the type and intensity of its impact upon society, because the aim of businesses is to serve society by delivering goods and services. If the social impact is ignored then the current and future market can be undermined. The *environment* also sustains business activity and so any activity that harms the environment can potentially harm existing and future markets. The reasoning behind Elkington's argument is that customers and clients should be provided with sufficient information to be able to judge the level of their support for a company's product (or service) so that they can make an informed choice. Those who feel that this is an intrusion into the prerogative of managers and leaders do so because they fear that these other two 'bottom lines' are highly politically charged. A number of global organisations are joining into support groups that implement 3BL initiatives. For example, the Australian Bank, Westpac, which operates globally, has signed up to the Equator Principles initiative and as a project financier this has important implications. An example of their 3BL reporting can be found on the web.[4] A full text of the principles, as well as frequently asked questions, can be found on the Equator Principles[5] website. In summary, the website explanation of the principles contains the following:

> Project financing, a method of funding in which the lender looks primarily to the revenues generated by a single project both as the source of repayment and as security for the exposure, plays an important role in financing development throughout the world. Project financiers may encounter social and environmental issues that are both complex and challenging, particularly with respect to projects in the emerging markets. The Equator Principles Financial Institutions (EPFIs) have consequently adopted these Principles in order to ensure that the projects we finance are developed in a manner that is socially responsible and reflect sound environmental management practices. By doing so, negative impacts on project-affected ecosystems and communities should be avoided where possible, and if these impacts are unavoidable, they should be reduced, mitigated and/or compensated for appropriately. We believe that adoption of and adherence to these Principles offers significant benefits to ourselves, our borrowers and local stakeholders through our borrowers' engagement with locally affected communities. We therefore recognise that our role as financiers affords us opportunities to promote responsible environmental stewardship and socially responsible development. As such, EPFIs will consider reviewing these Principles from time-to-time based on implementation experience, and in order to reflect ongoing learning and emerging good practice.

Project procurement and the financial bottom line

Procurement assumptions dominated by the financial bottom line present dangers in generating value for both customer procuring projects and those in the supply chain that help deliver projects. Customers who are highly price sensitive and seek to obtain the lowest cost may drive price-competitive oriented organisations into a competitive advantage cycle that is dominated by lowest price-seeking behaviour. A cost advantage is generally short-lived and unstable. Theory based upon a transaction cost economics (TCE) view assumes this to be driven by the 'efficiency' goal (Ghoshal and Moran, 1996). This cost efficiency view can be contrasted with a competitiveness view gained through having a differentiation competitive advantage that remains sustainable until that differentiation advantage can be replicated or improved upon (Porter, 1985). The broader view of resources includes knowledge, reflexivity and dynamic capabilities that are used to respond to competitive pressures (Teece, Pisano and Shuen, 1997; Eisenhardt and Martin, 2000). However, delivery effectiveness is also seen as crucial and this is expressed in project and PM terms through focussing on strategy (Morris and Jamieson, 2004). A strategy view of value relates to delivering a bundle of offerings within a project outcome that meets the often complex and diverse needs of numerous and diverse project stakeholders, rather than a dominant focus on gaining value through lowest cost. An unbalanced focus on projects, to surmount a return on investment hurdle, or designed and delivered to maximise profit, can be both distracting and ineffective. Thus, supporters of the 3BL concept believe that it is sub-optimal and often counter-productive to be blinded by the financial bottom line as the only valid criterion for procurement choice decisions.

For mega projects, where complex procurement decisions are made, there is justifiably an acute focus on the potential risk for large and embarrassing cost overruns. In Chapter 2 of *Megaprojects and risk: an anatomy of ambition*, Flyvbjerg *et al.* (2003: 14) cite cost overruns on major projects such as: the Boston artery/tunnel project, 196%; The Humber Bridge, UK, 175%; the Great Belt rail tunnel, Denmark, 110% and the France/UK Chunnel, 80%. They also provide chilling examples of revenue deficit problems with many mega projects, and cite actual traffic as a percentage of traffic forecast in the opening year of the Chunnel project at only 18%, and the UK Humber Bridge at 25% (Flyvbjerg *et al.*, 2003: 25). The cited figures are well substantiated from reliable evidence presented in a range of authoritative sources, so they can be accepted as representative for many large-scale projects. They also stress the risk impact of non-technical issues and claim that their data base, drawn from their own studies as well as many authoritative studies conducted by government auditing agencies and independent research bodies, suggests that

> A main cause of overruns is a lack of realism in initial cost estimates. The length and cost of delays are underestimated, contingencies are set too low, changes in project specifications and design are not sufficiently taken into account, changes in exchange rates between currencies are underestimated or ignored, so is geological risk, and quantity and price changes are undervalued as are expropriated costs and safety and environmental demands.
>
> (Flyvbjerg *et al.*, 2003: 13)

They conclude that we should not trust cost estimates and demand forecasts presented in business cases. This suggests unethical behaviour, through lying and misrepresentation of situations. It also suggests naivety and a lack of consideration of the real impact of ignoring social and environmental aspects of a project. In concentrating upon a dominating financial focus when preparing a business case for projects, the social and environmental legs of the 3BL stool often are ignored or at best poorly considered.

Throughout this book, we stress the need for looking beyond the financial dimensions of project procurement because society as a whole (including 'the market' and both 'customers/clients' as well as 'supply chain' members) demand a more holistic approach to procurement that considers a range of directly and indirectly influential stakeholder groups. Those who dismiss the importance of considering the 3BL as being of marginal relevance should review risk management, stakeholder management and change-management literature. A well-thought-through project business case should address adequate stakeholder analysis and issue mapping as a vital first planning step in managing risks that can jeopardise any project. Chapter 6 discusses a wider base of project performance measures than is often considered in project evaluation and monitoring and relates closely to the social and environmental aspects of the 3BL. This chapter's section on project governance stresses the need for clear accountability and transparency of expected standards which includes social and environmental aspects.

A financial bottom-line focus manifests itself in attempts to achieve efficiency through sourcing the least expensive project inputs (see Chapter 1 on outsourcing), effectiveness in terms of finding innovative ways to deliver projects and learn from past experience (see Chapter 8) and financial sustainability through ensuring that the project returns sufficient financial benefits that it justifies its existence.

Project procurement and the social bottom line

The concept of a social bottom line is at least partially anchored in the belief that the market is a broad set of immediate customers and potential customers that, given the financial means to access an opportunity, could and perhaps would become a consumer of goods and services on offer (Elkington, 1997). Thus, if business adopts a social responsibility agenda as part of its market obligations then it will benefit financially in the longer term. The concept of embracing social responsibility can also be justified from an intangible deliverables perspective (as is done so in Chapter 6 of this book) that provides the deliverer (of the project to a market or to the project delivery coordinator, being part of a supply chain) with a potential unique differentiation factor contributing to its competitive advantage.

In the value for money debate, the issue of publicly sponsored or funded projects having a critical need to deliver social benefits as their reason for being, as Donnaly (1999: 47) argues, the reason why the public sector conceived such projects as delivering clear water, educational facilities, medical facilities and the other trappings of a civilised society was because the private sector could not satisfactorily deliver these – at least during much of the twentieth century. This aspect of the public policy debate argues that to obtain best value in project procurement, the social bottom-line needs of projects designed to deliver social outcomes must be carefully considered (Staples and Dalrymple, 2008). This can translate in a number of ways relevant to the strategic intent of a given project. This notion accords with those who argue that project selection and outcomes must be aligned with the strategic intent (Archer and Ghasemzadeh, 1999; Morris and Jamieson, 2004; Norrie and Walker, 2004). Imagine, a highway infrastructure project is proposed in a rural setting, and a significant part of its timing decision for being promoted and supported was to maintain workflow in the local area to sustain local industry capacity and employment. It may be entirely appropriate, therefore, that the project be designed and delivered in a way that helps to maintain the sustainability of that community, together with its supporting industrial and workforce base. Making social bottom-line factors part of project procurement selection criteria is not new. In the National Museum of Australia (NMA) project for example, an indigenous workforce employment selection criterion was part of the evaluation process explicitly required by those submitting proposals to form an alliance to deliver that iconic project in 2001 in Canberra, Australia. Alliance partners had to demonstrate substantial acceptance of the draft alliance document for the project that included related codes of practice, proposals for support of local industry development and employment opportunities for Australian indigenous peoples. There was also a focus on demonstrating an outstanding record of ethical and socially responsible working with government and local communities and accepting broader responsibility (Walker and Hampson, 2003).

The incorporation of social bottom-line benefits to identified project stakeholders who may not have a critical and pressing lobbying capacity to demand social benefits, can deliver kudos as well as provide future brand *cache* value for organisations adopting this approach. Much of the philanthropic activity that openly takes place (as opposed to anonymous contributions) to deliver social benefit by corporations and individuals is partially justified as a brand-enhancing or kudos-building activity. Recognition of the value of strategically factoring in social bottom-line outcomes makes sense and can be achieved by linking these to tangible outputs more usually recognised as project outputs (Nogeste and Walker, 2005; Nogeste, 2006). In the Flyvbjerg *et al.* (2003) book they caution about promoting mega infrastructure projects on the basis of generating regional economic growth. Indeed, while these authors recognise and support the need to be transparent about the strategic intent for such projects they point to a number of examples where providing infrastructure to advantage one group has severely damaged others. One example they cite is the new transport fixed-link infrastructure that has been constructed between Denmark and Sweden. The intention was to synergistically stimulate economic activity between two centres and that there was an expectation that people might move from Copenhagen, Denmark to Malmö, Sweden but this appears not to have happened to the extent expected. In fact, they cite a report from the Danish Transport Council of predictions (subsequently realised in the two years following their report) that the Great Belt link (a mega infrastructure project) will have the affect on port cities of losing ferry operations, thus affecting that constituency, that regional economic effects will be small and that positive economic benefits that could be generated will occur in areas not targeted to be improved (Flyvbjerg *et al.*, 2003: 69). In considering corporate social responsibility more generally, Bansal (2006: 2) identifies several best practices that can be applied to business projects and other business activities:

- including social issues in risk assessment models;
- measuring the cost or revenue impact of social issues and using these as criteria in numbers-based models;
- lengthening decision-time horizons to make these fuller assessments;
- involving external stakeholders formally and informally and
- building social capital.

While considering the social bottom line may be seen as a generic ethical or 'socially responsible' thing to do, we stress that to be effective in a PM sense, these should be strategically targeted, and that performance measures be developed and used in monitoring 'success' of this particular project outcome aspect. It is clear that careful thought and planning is needed so that the unintended consequences of good intentions do not trigger 'collateral damage'.

Project procurement and the environmental bottom line

The environmental bottom line also sustains business activity. Any activity that harms the environment can potentially harm existing and future markets. Most projects have specific legal safety health and environment (SHE) provisions to protect people and the environment that they need to comply with. Indeed the NMA project alliance provides an example of stringent requirements (Walker and Hampson, 2003). Another example of a client that purposefully sought out green solutions to building design and delivery to present a showcase project that seeks a strong environmental bottom line is Melbourne City Council House 2.[6] This bold project provides a window on future projects that aspire to the values espoused by the website and provides an inkling of a possible future where environmental sustainability is considered a key element of any project. An example of a company designing and building a regional office for itself, incorporating 3BL principles for pragmatic reasons,[7] is the new Bovis Lend Lease P/L Sydney office. Here was the evidence of a concern for developing not only a building that had many 'green' credentials and features as well as promoting a workplace that added to people's enjoyment of their working day experience (Yates, 2006).

   While it is true that legislation has partially taken care of some of the above types of issues, there is much more that can be done by designing project performance to be more environmentally sustainable (Hart, 1997). He argues a three-stage environmental strategy that can inform procurement strategies: Stage one is a shift in emphasis from pollution control to *pollution prevention*. There are standards that move towards this – for example, ISO14000. Hart cites organisations such as the German chemicals giant BASF as helping China and other countries build less-polluting plants than have historically been provided. Their differentiation competitive advantage in responding to placing subsidiary or outsourced chemical plants in these countries comprises their advice, experience and development of improved pollution standards. This also enhances their 'brand' image.

Stage two is *product stewardship* where not only the pollution-generating product but other parts of the environmentally affected product cycle are addressed. This entails cradle-to-grave analysis of products (projects and/or portfolios of projects) where pollution may occur, and investigating innovative ways of overcoming these. The Rocky Mountain Institute's URL www.rmi.org/ provides an extensive link to materials and stimulating ideas for ways in which a proactive approach to specifying materials, methods, approaches and resources that can be more eco-friendly. For example, in a radio broadcast in October 2000, Amory Lovins explained how carpet-tile-floor finishes and air conditioning were provided as 'rental' services as part of a building project procurement approach (Lovins, 2000). That approach sought to prevent pollution by better conserving resources. Another design for the environment (DFE) approach, amongst a number of others cited by Hart, is an asset-recycling strategy adopted by Xerox for using leased recycling photocopier as low-cost high-quality Xerox parts to be reassembled into new machines (Hart, 1997: 72).

Hart's third identified stage is *clean technologies*. The fast-expanding area of nanotechnologies has the potential and promise to revolutionise production and therefore, procurement. Nanotechnology is a catch-all phrase for materials and devices that operate at the nanoscale. In the metric system of measurement, 'Nano' equals a billionth and therefore a nanometre is one-billionth of a meter.

Much of the literature (as at early 2007) suggests that this is a real revolution even though it is yet to have widespread impact upon us in the immediate future (i.e. up to the end of the first decade of this century). However, it is becoming obvious that industrial processes that use polluting by-products will be superseded by nanotech processes or products that not only provide superior environmental bottom-line impacts but are also less expensive. In the shorter term, nanobots (tiny machines) can be developed to work like enzymes by eating up waste products but before long they will also be programmed to actually build components. Under such circumstances, organisations that dismiss this potential differentiating competitive advantage technology do so at their own peril.

The above is a rapidly evolving field that will no doubt affect procurement practice. Before leaving this subject it is worth introducing a concept that is starting to creep into the procurement lexicon – emissions trading and more specifically, carbon credits.

This system has been in existence for many years; the sulphur dioxide trading system evolved out of 1990 Clean Air Act legislation in the USA designed to deal with acid-rain problems and has been more recently used to trade carbon credits as a way to encourage reduction in greenhouse gases and recognise that the environment (and all of us) bear the ultimate cost of global warming due to greenhouse gas emissions. A fast-growing trade in carbon dioxide (CO_2) credits has emerged with London becoming the carbon-credit trading capital of the world. In 2006, a major report was presented in the UK by Sir Nicholas Stern in which the financial implications of global warming and the need for a carbon-trading system was made abundantly clear.[8] This will have serious ramifications for procurement choices once the requirement for carbon offsets becomes law and part of the procurement culture. We can envisage that procurement choices in future will stipulate not only some form of global warming 'footprint' measure but also CO_2 reduction that footprint and carbon offset strategies.

On the Australian Broadcasting Corporation (ABC) programme Background Briefing, the programme noted an exchange between a reporter and a woman in a Sydney shopping centre accepting a free handout of energy-saving light bulbs. The woman was signing a document in accepting the gift that provided a named manufacturer with abatement certificate that it could subsequently use in its business activities to offset its pollution responsibilities as a carbon credit (ABC, 2006). This example is just one of many that are happening throughout the world. While this recent development may be seen to be of marginal impact upon most people's and organisations' current procurement choices, this will become more widespread and push the market to consider the environmental bottom line more closely.

Safety, health and environment (SHE) and project procurement

One can visualise safety for both the environment as well as individuals engaged in delivering projects as part of the 3BL concept. Ensuring a safe environmental outcome relates to the environment part of the 3BL, and occupational health and safety (OHS) relates to the social part of the 3BL. In order to address the issue of OHS it is necessary to look at the development of legislation and practice in these areas in order to give a perspective on the way ethics has shaped legislation and practice. We focus on a small part of this as it relates to project procurement because this is a broad subject area.

Prescriptive versus performance based

In the 1970s in the USA and the UK there was a move away from prescriptive legislation to performance-based legislation. This change in philosophy is discussed in Lingard and Rowlinson and this change has shaped the way organisations deal with OHS ever since. The basic premise behind this change in legislation was that organisations should be encouraged to develop their own OHS management systems which fitted their own particular circumstances. As a consequence, it was not necessary to provide detailed, prescriptive lists of how to organise safety: this was left to individual companies to devise their own safety management plan and implement this plan and management system accordingly.

> A review of the international literature reveals that the same type of work-related deaths, injuries and illnesses occur in construction industries all over the world. Many construction hazards are well-known and, in some cases, have been studied in great depth. What is apparent is that the construction industry fails to learn from its mistakes. We understand where deaths, injuries, and, to a lesser extent, illnesses occur in the construction industry but we still fail to prevent them. The same methods of working that have been used for generations are still being used, giving rise to the same hazards and ultimately resulting in the same incidence of death, injury and illness. Furthermore, the construction industry's organisation, structure and management methods militate against the identification and implementation of innovative solutions to the industry's OHS problems. Improvements in OHS will not occur unless new methods of working that reduce known OHS risks are developed. However, this is likely to require that the industry's structural and cultural barriers to the adoption of new methods of working be overcome.
>
> (Lingard and Rowlinson, 2005: 13)

Enforcement versus self-regulation

Unfortunately, the drive which brought about this change was predicated on the belief that existing, prescriptive safety management systems were adequately monitored, policed and enforced; the same can be said for environmental management systems. However, such an assumption was not true in many instances, as indicated in the preceding quote, and particularly so in project-based industries. Project-based industries are notoriously flexible and could be described as typical of Burns and Stalker's (1961) organic organisations. As such, these industries are difficult to monitor and control as participants, at both management and operational levels, are constantly moving and rosters are constantly changing as well. Thus, implementing a structured OHS system which is self-regulating is a much more difficult task in such an industry as compared to the much more mechanistic setting of a factory or an office. Consequently, the concept of self-regulation has been found to be far less effective in project-based industries than in more mechanistic organisations. However, it is important at this juncture to point out that one project-based industry, the petrochemical industry, has performed far better than many other project-based industries and this superior performance has stemmed from a particular attitude, organisational culture, in approaching safety and corporate governance in this sector.

In industries such as real estate development and construction there has been a tradition of lax management and this has been underlined by a notoriously poor OHS record in this sector. Hence, one might ask the question: Is the sector mature enough to cope with self-regulation or should a prescriptive and enforcement-based approach be adopted? After all, this would reflect a contingency viewpoint and there is absolutely no reason why all sectors should follow the same approach. If a sector as a whole can be considered immature, then there is a need to embark on a different approach to the implementation of a performance-based OHS system. Thus, an industry-wide education and training programme might be most appropriate in such circumstances. Once a level of industry-wide maturity is reached, then the leading performers can move towards a performance-based system. Having taken such a lead these high performers can then be encouraged to partner with their supply chain and allow the knowledge, expertise and experience they have developed to flow down the supply chain in terms of the implementation of plans, procedures and protocols which enhance the overall performance of the industry.

Worker engagement

An often-neglected aspect of a performance-based approach to OHS is the mechanism by which workers are engaged and enabled to manage and direct performance on their projects. Engagement is crucial if workers are to bring their specific knowledge of tasks and hazards to management's attention and might also be considered a basic human right. By forming, say, safety circles (and quality and sustainability circles) and engaging workers in discussions concerning working methods and precautions there is an excellent opportunity for companies to improve both performance and safety culture within an organisation. Hence, specifically focused engagement programs are a characteristic of high-performing project organisations. This is particularly important in a project-based industry where teams are forming, breaking up, reforming and breaking up once more. By engaging workers at all levels and across the whole of the project time line, it is possible to ensure that good ideas and specific hazards are identified and implemented or eliminated.

However, the nature of the industry does lead to problems of control, at all levels.

> The fact that construction is a project based industry is also an important contextual issue. When attempting to manage a dynamic, changing environment such as a construction site and, indeed, a construction

firm, it should be borne in mind that there needs to be an appropriate organisation structure to deal with the changing nature of the project and – as it moves from design to construction to in-use phases and as problems arise (such as late delivery of materials or labour shortages) on a day to day basis – there is a need for rapid, decentralised decision making, contingency planning and an appropriate, organic form of organisation. This engenders a free, independent spirit in construction site personnel and has, traditionally, led to a healthy disregard for authority and regulations. This has, in many instances, been taken too far and descended into unacceptable corruption and malpractice. The Housing Authority short piling case in Hong Kong (LEGCO, 2003)[9] and the Report of the Royal Commission into the building industry in Australia in 2003 are examples of how this can drastically affect the well-being of the industry and those working in it: the cherished characteristic of independence and initiative has also given the industry a bad name. In turn, these characteristics also make it difficult to implement programs across the whole of the industry and safety and health on construction sites is badly affected.

(Lingard and Rowlinson, 2005: 11)

Thus, the very nature of the industry inevitably leads to ethical issues being raised which must be dealt with if the industry is to become both effective and efficient and pass muster as addressing issues of corporate responsibility.

Safety culture

Safety culture, better referred to as health and safety culture, reflects the attitude of all in the organisations towards a continuous improvement of the health and safety situation within the organisation. A safety culture does not naturally occur in project-based industries, as workers come to projects with different experiences, attitudes, training skills and knowledge. Hence, the culture must be nurtured. Understanding this and risk perceptions within organisations is increasingly important for company survival.

Organizations...increasingly recognise the importance of managing such risks effectively. As well as becoming more internally sensitive to risk issues, organizations in high-risk sectors are coming under increasing pressure from external parties, particularly regulators and media, to manage the risks associated with their operations effectively. Organizations in high-risk sectors therefore increasingly seek greater understanding of their internal and external risk contexts in order to better manage their operations within a level of risk that they can control, including avoiding, minimising or mitigating effects of disasters.

(Glendon, 2003)

The effectiveness of OHS legislation

So, is OHS legislation effective? There is no straight answer to this question but, adopting a contingency viewpoint, it can be said that both prescriptive and performance-based legislation have their merits and one, or the other, or both are appropriate in different situations. Hence, the role of management is to address these difficult questions and provide a safety management system which is appropriate to the circumstances. This means having a good understanding of the competence of people within the organisation (technical, administrative and managerial competence) and providing a safety management system which fits both these competences and the nature of the tasks being undertaken. This is a decision which has to be taken after careful consideration. In addition, the context in which the decision is made needs to include the nature of enforcement and the consequences of non-compliance. However, this implies a trade-off and the issue arises: Is it ethical to trade off cost against safety when human life is at stake? This is a dilemma which safety practitioners, legislation enforcers and company directors must all address.

Project procurement and governance

The Wikipedia entry[10] for corporate governance reads:

> Corporate governance is the set of processes, customs, policies, laws and institutions affecting the way a corporation is directed, administered or controlled. Corporate governance also includes the relationships among the many players involved (the stakeholders) and the goals for which the corporation is governed. The principal players are the shareholders, management and the board of directors. Other stakeholders include employees, suppliers, customers, banks and other lenders, regulators, the environment and the community at large.

Rodney Turner offers the following premise relating to project governance:

> Project governance provides the structure, through which the objectives of the project are set, and the means of attaining those objectives are determined, and the means of monitoring performance are determined. Project governance involves a set of relationships between a project's management, its sponsor, its owner and other stakeholders.
>
> (Turner, 2006: 93)

The term corporate governance can sound very grand and intimidating but in essence it should be seen as setting reasonable, acceptable (ethical) rules and regulations that protect those who have a stake in a project including supply chain and team-member participants. It is about accountability and response – who does what, under what rules and for what rationale. Further, it revolves around structures being in place to facilitate accountability and second, it requires the will and capacity to ensure that governance structures are effective in ensuring ethical standards are in place. The section on theoretical underpinnings of ethics in this chapter has indicated the different ways in which ethics can be viewed, hence the beliefs and values that underpin what is seen and believed to be 'ethical behaviour'. Governance is mainly about ensuring that espoused (cultural and ethical) values are translated into matching behaviours. The ethical behaviour that most people would relate to as acceptable is

- a lack of criminal action;
- the adherence to the notion of fair play, that is, not taking unfair advantage over those in less-powerful positions – customers/clients, supply chain teams etc and
- maintaining a sustainable system so that the long term is not jeopardised by short-termism.

The above fits in well with building an organisational culture that supports positive behaviours that are discussed in Chapter 9 of this book. It also links in well with some aspects of risk management, notably taking due diligence to ensure that risks of this kind are mitigated against where possible and planned for.

A simple way of looking at governance as a means of getting things done is illustrated in Figure 4.1. This model is limited to the extent that it assumes that the majority of effort and engagement is involved in deterring and preventing people from taking opportunistic behaviour that inhibits or harms 3BL sustainability. This assumption (Williamson, 1975) draws upon TCE, which recognised that there is a cost of conducting a transaction that includes the cost of policing standards. This approach has not been universally accepted as necessarily pivotal because, it is argued, people are human and have a repertoire of behaviours and motivations and opportunities behaviour as viewed from a TCE perspective (Ghoshal and Moran, 1996). The Figure 4.1 model also illustrates ethical standards and cultural practices as well as contributing to governance.

The key to project delivery fulfilment is developing a means of project delivery that meets the project purpose, has supportive structures to help make this happen using a system and people with the capacity to do so that it is designed with accountability frameworks in place that have visible, credible and reliable performance measures that are transparent. Transparent means that there is a tangible and understandable link between what is being done, the way it is being done and the project outcome. One may view this model as the synthesis of this book – finding

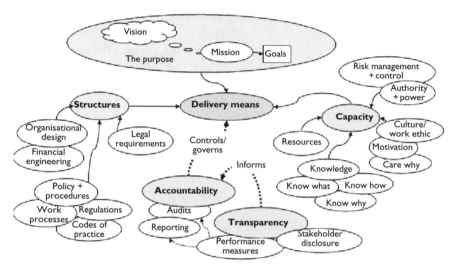

Figure 4.1 Simple model of governance.

procurement choices and systems that deliver positive intended results (outcome rather that just outputs) in a way that we can learn from and improve upon.

This section will discuss aspects of this model in more depth but first let us get a grasp of the essentials. The first element in the figure is the purpose. Governance does not just happen by itself. Projects by definition are a change process (Turner, 1999). Effective change first requires those who must be involved to appreciate and support the need for the change so that they 'own' the change and are committed rather than compliant with regard to that change (Kotter, 1995). Thus, the first step is in developing a vision that people are likely to embed in their psyche. Some projects that have maintained a strong sense of vision can overcome poor PM performance to facilitate project success (Christenson and Walker, 2004). A strong vision should inform the project mission that identifies what tangible ways in which the vision can be realised and this cascades into a set of objectives.

To allow this to happen there needs to be a structural framework in place that represents considerable intangible value to the organisation, delivering projects and providing a real differentiated competitive advantage, simply because each organisation is unique and has a unique set of initial conditions to set that organisation upon a trajectory of project delivery. This requires an organisational design that provides certain functions and people (a project sponsor to help shape, support and champion a project's vision and transform that into a systemic entity that can deliver the project vision outcome) to deliver the intended results. The financial structure – the way that the project will be financially engineered – also needs to be considered. The extent of risk capital can influence how a project's feasibility will be approved, and how it may be monitored and controlled. For large-scale infrastructure projects, a government raising only sovereign debt (that which is backed by an ability to raise taxes to cover costs) can introduce a degree of laxity in monitoring and control, and so some experts in this topic suggest that ensuring some market participation with capital injections that are at risk enforces a higher degree of rigour in transparency and governance than might not otherwise occur (Flyvbjerg *et al.*, 2003: 122). There will also be administrative components to the project delivery support structure such as policies and procedures, regulations and codes of practice. This is where explicit ethical codes can make a difference to the flavour of the administrative practices and influence the organisational culture. There will be legal requirements that force certain actions such as workplace relations laws, cultural practices and observations (feast days etc.), environmental laws, statutory obligations and a host of other legal requirements. Often a tacit 'cultural' or 'traditional' overlay of practices can also impact upon this element.

The capacity for 'the system' developed for the project is also impacted upon by capacity sub-elements. Resources are a natural and obvious part of this. Resources include funding, people, equipment and knowledge to name just a few. In terms of a collective and individual team knowledge capacity, there needs to be 'know how' or technical competence to deliver the vision, knowledge relation to how to achieve goals set, as well as the rationale of why this is important. We will see in Chapter 8 that the notion of 'sticky knowledge' or degree of difficulty in knowledge transfer, is affected by people's understanding of the cause and effect link that exists between action and outcome (Szulanski, 1996). Authority and power are also highlighted in the model as part of the capacity to deliver, because without delegated authority and the power to enforce the design of the project delivery strategy, the project vision and its validity will be undermined. This includes the capacity to use knowledge to recognise and evaluate risk, manage that risk or have the authority and control to be able to pass that risk to others (with due recompense for doing so) who are in a better position to accept and deal with that risk. Additionally there is the capacity input of the cultural work ethic that influences behaviour. Motivation drives this in part. This can include the 'care-why' element by shifting people from being compliant to the achievement of a set of specific project objectives, to being committed to achieving the project vision, even though the mission and specific objective may be upgraded beyond the stated (and possibly outdated) specified requirements. This is where the spirit of the vision is seen as being more valid that temporarily fixed mission and objectives that lead to a richer sense of success.

To ensure that the project delivery design operates within the scope of its intended outcome, a number of checks and balances need to be put in place. These are the more obvious governance aspects that gain much attention in the literature but cannot be of use without effective configuration of the above elements of a delivery strategy. This is often understood in terms of accountability and transparency.

The use of appropriate performance measures, informed by and consistent with the project vision and goals is necessary to be able to measure and then monitor what is happening as opposed to what was expected to happen during the project life cycle. These measures support monitoring taking place to enable accountabilities to be specified and ingrained (within the culture of the organisation, so unspecified required aspects that need to happen 'just do' happen because that is the way that the organisation culture drives tacit or unspecified required actions to be undertaken) to ensure that required actions do happen. Accountability allows people and functions to have attributed to them responsibility for outcomes so that it is clear what actions these people and functions are meant to take, when and how those actions will be taken and how project stakeholders with a clear and mandated influence can be assured that those accountable perform as required. Accountability is informed by transparency. This means that all concerned with a project should have a clear indication of who the stakeholders are on a project, the nature of their interest and their influence upon decision making and disclosure of project information. Stakeholder analysis and mapping as well as developing stakeholder communication strategies is also part of a robust stakeholder management risk plan, and can use tools such as the Stakeholder[11] Circle™ (Bourne and Walker, 2005) that are discussed further in Chapter 3. Disclosure is an important part of governance and ethical and professional behaviour as it seeks clarity and communication.

Having mapped out the scope of what project governance can mean we will now justify the above model's elements from expert evidence that has appeared in the literature.

From a corporate governance perspective, there is an argument that an organisation's board of directors, when effectively subject to sound governance principles, can lead to superior strategic decision making and hence improved organisational performance (Young, 2003). The rationale behind this assertion is that well-qualified independent-thinking directors who intelligently question board proposals and reports on performance strengthen accountability within the organisation. This quality can prevent organisations from failing to deal with conflict of interest issues that have been earlier cited as important ethical issues. In project governance terms this can translate to a project advisory board, and an example citing experience of such a company board that had input into product development as 'An effective advisory board, properly composed and structured, can provide non-binding but informed guidance and serve as a tremendous ally in the quest for superior corporate governance' (Reiter, 2003: 1). However Reiter noted that such boards must have a clear mandated purpose, scope and focus and be appropriately constituted to encourage open discussion and debate about relevant issues that introduces valuable review and wisdom rather than being a closed policing agency; such a board should open up possibilities as well as review existing proposals. In terms of the composition of corporate boards and their capacity to provide valuable review and make sound decisions, Richard Leblanc's work has had a profound impact in Ontario, Canada and elsewhere in North America. His study of corporate boards and their impact upon company performance indicates that a balance of skills and abilities is often lacking in many organisations because of bias in board member recruitment that blinds the board to certain entrenched views (Leblanc and Gillies, 2003; Leblanc, 2004). He has advocated (and this has been accepted into legislation in Ontario) that effectiveness of board decision-making can be enhanced by transparency of board composition in terms of not just their identity but the skills and capacities that they bring to the board, so that investors can make an informed choice about investing in that organisation. Relating this back to project governance issues, transparency of the composition and workings of any project review, oversight or overall governing group that initially approves, monitors and permits a project to continue (particularly if it becomes a distressed project), is vital so that those on the project team know that their outcomes are being rigorously assessed during the project delivery phase. The lesson that we should learn from corporate governance literature is that for many projects, particularly those of large scope or high complexity, a system of advisory and oversight-monitoring boards should be established with suitable expertise and capacity to initiate, review and approve continuance of a project – already part of the Prince 2 methodology (Bentley, 1997).

Recent PM thinking relating to project governance has centred on the role of project sponsor, broker and steward. A project *broker* is responsible for the relationship between the client and the project team. A client with a requirement approaches an appropriate person who knows where the resources to meet that requirement can be sourced and mobilised. A project *steward* marshals and puts together the resources necessary to deliver the project that the project manager will be responsible for leading (Turner and Keegan, 2001). The steward role can be seen as similar or the same as what Peter Morris describes as the project *champion*. He says that a champion obtains the resources needed to accomplish the project with much of the work being externally and internally focused (Morris, 1994: 257). In this way, it describes a combined broker/steward role. Perhaps to add to any confusion of terms is the term 'sponsor' whose role is to make resources available, and a 'champion' who is a senior representative who convinces the sponsor that the project should have priority and supporting resources (Turner, 1999: 51).

If we try to simplify the concept in order to be clear about where these roles fit in with project governance, we can conclude that there is a role for an entity to be accountable to the client or owner of a project. This entity could be an individual such as a broker or sponsor who provides a single interface between the project and the entity that will receive the realised project deliverables, or it could be a board that reports to the project sponsor. This entity should have a clear accountability to deliver and be able to effectively communicate progress towards that end.

Within the board or in terms of an individual being accountable to deliver a project, there should be an effective champion/steward role which is to be accountable for effective support of that project. This leadership role ensures that the structures and capacity are in place – described briefly in Figure 4.1. If it becomes clear that resources are insufficient, that authority constraints inhibits performance, or that there is insufficient knowledge available to deliver, or structural impediments such as organisational design or policy and procedures, then the champion/steward should transparently advise the board (or sponsor) of the crux of problems encountered, with potential solutions to resolve any difficulties, and be able to either gain support to provide the means to overcome highlighted distress or to abandon the project (Crawford, 2006).

Dinsmore and Cooke-Davis (2006) identify five roles that a board-level sponsor champion/steward would fulfil: being governor of the project and avoiding either abdicating responsibility to the project team or micromanaging the project; being owner of the business case, that is being committed to it and understanding its nuances intimately; harvesting benefits, that is ensuring that both tangible and intangible benefits delivered are optimally absorbed; typical of a harvested benefit is the learning gained through delivering the project; being a friend in high places; often the project manager does not have the political clout to make an impact when seeking priority over resources when heading for distress or to influence the board to gain due recognition for project achievements; finally, being a visible champion of the project, both within and outside the board, so that the project's profile and vision is fully supported.

In a study of project success and project governance involving a survey data set of 168 valid records recently reported upon (Crawford, 2006), results suggest that projects are not generally good at delivering the benefits for which they were undertaken. Analysis presented about that research suggests that about 26% of the variance in success can be attributed to lack of governance effectiveness of project sponsors. This supports the general corporate governance meta-analysis of studies reported upon by Young (2003). Crawford (2006) argues from her study results that in terms of effectiveness, the project sponsor is in the best position to

1   ensure that all strategic options are considered before commencing a project;
2   ensure that the project receives needed resources;
3   ensure that the project team has sufficient authority to complete the project to realise expected benefits;
4   influence the quality of the business case presented to sanction a project;
5   ensure that responsibility and accountability levels rest with the right people in the organisation and
6   ensure that, if technical performance requirements of the project can be achieved, the business case will also be achievable.

From the top-down perspective (i.e. the board level) it becomes clear that project governance is enhanced or limited by the quality of the structure, capacity and accountability of a board member who can effectively champion or sponsor a project. However, the governance system must be designed and implemented to provide the detailed knowledge and information about the project to the champion/sponsor at a board level.

Underpinning and supporting the project manager's ability to deliver a clear picture of a project's status to the project champion/steward is a series of iron triangle (cost, time and quality) monitoring and control mechanisms. Also, the actual procurement process (to reduce the likelihood of opportunism and exploitation and to manage uncertainty relating to the ability to deliver the required outcomes) underpins this ability, and features procedures which help or hinder the project manager, such as contractual means to deal with changes in scope and modifications to the existing specification of details of work to be undertaken (Winch, 2001). There would be organisation-wide and legal requirements that require transparency being provided through audits and reporting mechanisms that should support accountability and transparency.

Chapter summary

In this chapter we have explored a number of project procurement issues that relate to how project procurement policies can, and should, take account of long-term sustainability – of not only the organisation making decisions but the wider community. The ethics section helps provide us with frameworks to make decisions and choices that facilitate us to 'do the right thing'. In Chapter 3 we argued that trust is an essential ingredient in making realisable and valuable commitments and so ethical behaviour plays an important role in demonstrating and facilitating trust between parties engaged in projects. Similarly, the section on 3BL in this chapter relates to another practical aspect of 'doing the right thing' as it helps us understand the balance needed between creating financial, social and environmental value through our projects so that we build 'brand' image and loyalty while addressing value to stakeholders in a more holistic manner. A critical part of that is OHS, as it helps shape procurement choices and actions to conform to an ethical path to deliver value, as well as addresses our duty of care to our stakeholders. Emerging issues of CO_2 emissions and potential impact of carbon trading are raised. The section on governance shows how rules, regulations and protocols of delivering projects can provide transparency and accountability to govern the way that value is delivered.

Vignette

CanWheat, a major Canadian, Saskatoon-based exporter of wheat products had decided to build a set of wheat storage facilities in Urqatistan, a former Soviet republic now independent, with untapped large deposits of natural gas and oil and located on the route of a proposed hydrocarbons pipeline grid connecting Siberian fields to Europe. Urqatistan is ruled by an ex-Soviet regional military commander and highly placed *party* official. CanWheat also is a joint venture (JV) partner with several food-processing businesses that are subsidiaries of Global-Nutrition+ plc that, while basing its headquarters in Chicago, has most of its operations and logistics group based in St Petersburg, Russia and its marketing operations based in The Hague. The final JV partner in the holding company that will own and operate the facilities is the charity TasDentAID an association formed by a group of Tasmanian dentists and periodontists that had raised AUD$50 million for the project.

The storage facilities project has 60% of the necessary US$1.5 billion funding, financed through the JV and the rest as loans by a consortium of banks. The project comprises 10 grain-handling silos, associated railway-access spurs, a series of 30 small local bakeries spread out across Urqatistan to be leased back to local operators and a data centre for operational and marketing IT systems with Global-Nutrition+ plc providing all IT services through its be-spoke system developed and enhanced over the past 5 years. The advantages of the project were stated to: 'provide much needed employment in an economically depressed part of the world; to provide a best-in-class central distribution logistics point for wheat distribution in a region critical to global hydrocarbons during the next few decades; and to help kick-start skills development opportunities'. Urqatistan had served as a base in the Soviet Union days for regional expertise in logistics and engineering science through its highly regarded University of Urqatistan Engineering School and Science School.

Recent media reports raised disturbing allegations involving bribery, money laundering and labour exploitation. These were dismissed as misunderstanding the nature of consulting commissions, permits and administrative costs incurred and transfer costing arrangements that are perfectly legal in a global economy. A Russian reporter had broken a story published in the Times in London alleging that Global-Nutrition+ plc had recently opened up a number of clinics in central Africa, ostensibly as part of its 3BL initiatives, that had been implicated in highly suspect drug-testing practices for an unnamed major pharmaceutical company. The reporter alleged that a small number of villages (considered to be of sub-standard housing) would be removed to make way for the rail spurs and that the Urqatistan government had pledged to relocate and re-house people from those villages in a new suburb of the capital that had been reported in National Geographic magazine as a model new 'Silk Route City'. Additionally, villages had agreed to compensation of US$2,000 each for relocation of three community burial sites dating back 3,000 years.

A subsidiary of Global-Nutrition+ plc is the principal distributor of genetically modified wheat. Much of the financial dealings have been undertaken through their St Petersburg office. CanWheat provided the engineering expertise for design briefing of the silo and rail-spur logistics engineering in cooperation with local engineering consultants. TasDentAID consultants, directed by one of its key administrative staff, was involved in much of the preparatory work in negotiating real estate deals and logistics involved with transport of equipment and materials into Urqatistan as the TasDentAID staff member had intimate local knowledge of the country and key government and business stakeholders through family connections. The project had strong government support from Australia, Russia and USA.

Issues to ponder

1   Identify and discuss three ethical issues raised and, using Kohlberg's stages model, analyse these issues and any paradoxes that may be raised to identify procurement requirements that could address these issues.
2   Identify and discuss an example of a 3BL initiative being pursued by JV participants and evaluate, with justification, your considered 3BL maturity level.

3 What other 3BL maturity level that you can assess from information provided is indicated, and what initiatives do you think could be incorporated into this project?
4 What information in this vignette is missing about the project JV members that you would like to know more about to better understand the ethical and SHE issues raised here?
5 In terms of project governance, identify and discuss three formal procedures for addressing accountability and transparency issues do you consider appropriate for this project.

Useful links

The 2004 KPMG Forensic Fraud Survey see www.cgs.co.nz/files/Fraud-Survey-2004.pdf

The Stern Report see www.hm-treasury.gov.uk/independent_reviews/stern_review_economics_climate_change/sternreview_index.cfm

The better practice guide for governance is at the Australian Auditor General's office site www.anao.gov.au/WebSite.nsf/ViewPubs!ReadForm&View = LatestBetterPracticeGuidesByTitle&Title = Latest%20Better%20Practice%20Guides&SubTitle = (Released%20within%20the%20last%2030%20days)&Cat = &Start = 1&Count = 10

Other links are indicated in footnotes in this chapter.

Notes

1 See KPMG's bi-annual Forensic Fraud survey www.cgs.co.nz/files/Fraud-Survey-2004.pdf
2 For an interesting ABC broadcast on this subject readers can download an MP3 file of the program Bryson, G. (2006). The Ethics of Virtue. Encounter. Australia, ABC, http://www.abc.net.au/rn/encounter/stories/2006/1759630.htm
3 For more on the four virtues: temperance, prudence, courage and justice see the Wikipedea entry on http://en.wikipedia.org/wiki/Virtue#The_four_virtues
4 See URL www.westpac.com.au/internet/publish.nsf/Content/WICREV+Equator+principles
5 See URL www.equator-principles.com/
6 Their website www.melbourne.vic.gov.au/info.cfm?top=171&pg=1933 can be accessed for further details.
7 For an example of the BLL 2005 sustainability report go to www.bovislendlease.com/llweb/bll/main.nsf/images/pdf_sustainability_report_05.pdf/$file/pdf_sustainability_report_05.pdf this report clearly states their values and 3BL actions taken.
8 See www.hm-treasury.gov.uk/independent_reviews/stern_review_economics_climate_change/sternreview_index.cfm for discussion from the Stern Report relating to carbon trading see Part IV: Policy responses for mitigation (Chapters 14–17) on www.hm-treasury.gov.uk/media/986/E1/sternreview_report_part4.pdf

9 See HKLEGCO (Hong Kong Legislative Council) 2003, www.legco. gov.hk/yr02–03/english/sc/sc_bldg/reports/rpt_1.htm
10 Reproduced from the Wikipedia article http://en.wikipedia.org/wiki/Corporate_ governance for which the use is licensed under the GNU Free Documentation License
11 www.mosaicprojects.com.au/Techniques.html#Stakeholder%20Circle

References

ABC (2006). The Rise of the Carbon Traders. Carlisle W. Sydney, ABC National Radio, Background Briefing.

Aguilar, F. J. (1994). *Managing corporate ethics: Learning from America's ethical companies how to supercharge business performance*, New York, Oxford University Press.

Allison, R. E. (1998). Ethical values as part of definition of business enterprise and part of the internal structure of the business enterprise. *Journal of Business Ethics.* **17**(9/10): 1015–1028.

Archer, N. P. and Ghasemzadeh, F. (1999). An Integrated Framework for Project Portfolio Selection. *International Journal of Project Management.* **17**(4): 207–216.

Bansal, T. (2006). Ivey on . . . best practices in corporate social responsibility. *Ivey Business Journal.* **70**(4): 1–3.

Beauchamp, T. L. and Bowie, N. E. (1997). *Ethical theory and business*, Upper Saddle River, NJ, Prentice Hall.

Bentley, C. (1997). *PRINCE 2: A practical handbook*, Oxford, UK, Butterworth-Heinemann.

Bourne, L. and Walker, D. H. T. (2005). Visualising and mapping stakeholder influence. *Management Decision.* **43**(5): 649–660.

Bryson, G. (2006). The Ethics of Virtue. *Encounter.* Australia, ABC, http://www. abc.net.au/rn/encounter/stories/2006/1759630.htm (accessed on 28 July 2007).

Buchholz, R. A. (1989). *Fundamental concepts and problems in business ethics*, Englewood Cliffs, NJ, Prentice Hall.

Burns, T. and Stalker, G. M. (1961). *The management of innovation*, London, Tavistock Publications.

Carroll, A. B. and Meeks, M. D. (1999). Models of management morality: European applications and implications. *Business Ethics: A European Review.* **8**(2): 108–116.

Child, J. (1986). *Organization: A guide to problems and practice*, London, Harper & Row.

Christenson, D. and Walker, D. H. T. (2004). Understanding the role of 'vision' in project success. *Project Management Journal.* **35**(3): 39–52.

Crawford, L. (2006). *Project Governance – The Role and Capabilities of the Executive Sponsor*. PMOZ – Achieving Excellence, Melbourne, Australia, 8–11 August, PMI – Melbourne, Chapter: 1–11 CD-ROM paper.

Dinsmore, P. C. and Cooke-Davies, T. (2006). *The right projects done right!: From business strategy to successful project implementation*, San Francisco, CA, Jossey-Bass.

Donaldson, T. and Werhane, P. (Eds) (1999). *Ethical issues in business: A philosophical approach*. Series Ethical issues in business: A philosophical approach. Upper Saddle River, NJ, Prentice-Hall.

Donnaly, M. (1999). Making the difference: Quality strategy in the public sector. *Managing Service Quality.* 9(1): 47–52.

Eisenhardt, K. M. and Martin, J. A. (2000). Dynamic Capabilities: What are They? *Strategic Management Journal.* 21(10/11): 1105–1121.

Elkington, J. (1997). *Cannibals with forks,* London, Capstone Publishing.

Ferrell, O. C. and Fraedrich, J. (1997). *Business ethics: Ethical decision making and cases,* Boston, Houghton Mifflin Co.

Ferrell, O. C., Fraedrich, J. and Ferrell, L. (2002). *Business ethics: Ethical decision making and cases,* Boston, Houghton Mifflin.

Flyvbjerg, B., Rothengatter, W. and Bruzelius, N. (2003). *Megaprojects and risk: An anatomy of ambition,* New York, Cambridge University Press.

Francis, R. D. (1999). *Ethics and corporate governance: An Australian handbook,* Sydney UNSW Press.

Freeman, R. E. (1999). Divergent stakeholder theory. *Academy of Management Review.* 24(2): 233–236.

French, W. A. and Granrose, J. (1995). *Practical Business Ethics,* Englewood Cliffs, NJ, Prentice Hall.

Ghoshal, S. and Moran, P. (1996). Bad for practice: A critique of the transaction cost theory. *Academy of Management Review.* 21(1): 13–37.

Glendon, A. I. (2003). *Understanding and managing risk perceptions.* Quarantine and Market Access Conference: Maximising Trade – Minimising Risk, Canberra, 24–25 September, pp. 36–43.

Hart, S. L. (1997). Beyond greening: Strategies for a sustainable world. *Harvard Business Review.* 75(1): 67–76.

Kotter, J. P. (1995). Leading change – Why transformation efforts fail. *Harvard Business Review.* 73(2): 59–67.

Leblanc, R. (2004). Preventing future Hollingers. *Ivey Business Journal.* 69(1): 1–9.

Leblanc, R. and Gillies, J. (2003). The coming revolution in corporate governance. *Ivey Business Journal.* 68(1): 1–11.

Lingard, H. C. and Rowlinson, S. (2005). *Occupational health and safety in construction project management,* London, United Kingdom, Taylor & Francis.

Lovins, A. (2000, 8 October). Natural Capitalism – A Lecture by Amory Lovins. *Background Briefing.* Sydney, Australia, ABC, www.abc.net.org/rn/talks/bbing/stories/s196391.htm

Low, S. P. and Leong, C. H. Y. (2000). Cross-cultural project management for international construction in China. *International Journal of Project Management.* 18(5): 307–316.

McAlister, D. T., Ferrell, O. C. and Ferrell, L. (2005). *Business and society: A strategic approach to social responsibility,* Boston, MA, Houghton Mifflin.

Mintzberg, H. (1975). The manager's job – Folklore and fact. *Harvard Business Review.* 53(4): 49–62.

Mintzberg, H. (1990). The design school: Reconsidering the basic premises of strategic management. *Strategic Management Journal.* 11(3): 171–195.

Morris, P. W. G. (1994). *The management of projects a new model,* London, Thomas Telford.

Morris, P. W. G. and Jamieson, A. (2004). *Translating corporate strategy into project strategy,* Newtown Square, PA, PMI.

Nogeste, K. (2006). Development of a Method to Improve the Definition and Alignment of Intangible Project Outcomes with Tangible Project Outputs. Doctor of Project Management, DPM, *Graduate School of Business*. Melbourne, RMIT.

Nogeste, K. and Walker, D. H. T. (2005). Project outcomes and outputs – Making the intangible tangible. *Measuring Business Excellence*. 9(4): 55–68.

Norrie, J. and Walker, D. H. T. (2004). A balanced scorecard approach to project management leadership. *Project Management Journal, PMI*, 35(4): 47–56.

Porter, M. E. (1985). *Competitive advantage: Creating and sustaining superior performance*, New York, The Free Press.

Preston, N. (1996). *Understanding ethics*, Leichhardt, NSW, Federation Press.

Reiter, B. J. (2003). Corporate governance and firm performance: Is there a relationship? *Ivey Business Journal*. 68(1): 1–7.

Robbins, S. P. and Mukerji, D. (1990). *Managing organisations: New challenges & perspectives*, New York, Sydney, Prentice Hall.

Staples, W. and Dalrymple, J. (2008). Best Value Public Sector Construction Procurement. *International Journal of Public Sector Management*. (forthcoming).

Szulanski, G. (1996). Exploring internal stickiness: Impediments to the transfer of best practice within the firm. *Strategic Management Journal*. 17(Winter special Issue): 27–43.

Teece, D., Pisano, G. and Shuen, A. (1997). Dynamic capabilities and strategic management. *Strategic Management Journal*. 18(7): 509–533.

Turner, J. R. (1999). *The handbook of project-based management: Improving the processes for achieving strategic objectives*, London, UK, McGraw-Hill.

Turner, J. R. (2006). Towards a theory of project management: The nature of the project governance and project management. *International Journal of Project Management*. 24(2): 93–95.

Turner, J. R. and Keegan, A. (2001). Mechanisms of governance in the project-based organization: Roles of the broker and steward. *European Management Journal*. 19(3): 254–267.

Velasquez, M. G. (1998). *Business ethics concepts and cases*, Upper Saddle River, NJ, Prentice Hall.

Walker, D. H. T. and Hampson, K. D. (2003). Project alliance member organisation selection. In D. H. T. Walker and K. D. Hampson (Eds), *Procurement strategies: A relationship based approach* (Chapter 4). Oxford, Blackwell Publishing: 74–102.

Williamson, O. E. (1975). *Markets and hierarchies, analysis and antitrust implications: A study in the economics of internal organization*, New York, Free Press.

Winch, G. M. (2001). Governing the project process: A conceptual framework. *Construction Management and Economics*. 19(8): 799–808.

Yates, M. (2006). *The Building as an Organism for Knowledge and Cultural Change in Organisations*. The 6th International Conference on Knowledge, Culture and Change in Organisations, Prato, Italy, 11–14 July, Kalantzis M., Common Ground: 1–8 Electronic.

Young, B. (2003). Corporate governance and firm performance: Is there a relationship? *Ivey Business Journal*. 68(1): 1–5.

Chapter 5

The role of business strategy in PM procurement

Derek H. T. Walker, Mario Arlt and James Norrie

Chapter introduction

We have stated throughout this book that the purpose of procuring a project is part of a process to derive benefit from a planned and executed transformation. Moving from state X to Y, development of a new building or product, or moving from a stand-alone IT system to an integrated solution; all start with a need that is expressed as being part of a strategy to generate a benefit from an identified opportunity. So any one project is probably part of a series of projects that together, in the context of a programme or portfolio sense, deliver an outcome that contributes to strategic goals of the organisation. The value proposition is not so much the success of the project but the success of the strategy. This is why some failed individual projects can lead to strategic success because, as explained in Chapter 6 and Chapter 8, the project outcome may have led to accelerated learning and improvement on subsequent projects that would not have happened without the initial apparent failure. If we use the original sense of the word strategy being associated with warfare, a project may be considered as a tactical battle, in a long war. This may be of little consolation to a project manager who has been haunted by the spectre of failure – so this chapter is important in that it may bring solace to such individuals, it places projects in a meaningful organisational context, and supports the focus on the need for experimentation, innovation and learning to be encouraged in project management, as highlighted in Chapter 8.

In addition, we must not forget that some strategies emerge and evolve. They often take altered paths from that originally envisaged perhaps because of environmental turbulence or conscious change. In this context a failed project may well emerge as being seen as playing a pivotal role in overall success. An Australian example of this that many readers may relate to is the case of the Sydney Opera House. This iconic project was, during its construction, described in the press and professional journals of the time as an unmitigated disaster because of both significant cost and time overruns (Murray, 2004). As an opera building it also appeared to cater for a tiny minority of potential consumers and would not be thought of as a tourist destination. Naturally the design brief and intent was to produce a spectacular monument that might embrace the image of a vibrant and emerging lively place where people would like to visit; however, the strategy for providing an icon was somewhat masked at the time by more functional performance criteria. To judge this example's success in terms of delivering an opera house on budget, on time and in a way that facilitates superb music being played and listened to there is to miss the point (Heneghan, 2004; Murray, 2004). This icon can be seen to have played a significant part of almost literally placing Sydney on the map.[1] The 'success' of this building as an icon may well have been a catalyst to change the image of Sydney, capturing an image of itself as a centre of sophistication and culture and its perceived place as the major Gateway to Australia for many international tourists. This resulted in much of the finance industry moving their headquarters and global operations in the 1980's from Melbourne to Sydney.

The purpose of this illustration is to stress the importance of linking project success to strategy through procurement choices based on strategic thinking rather than delivering a project limited to the iron triangle performance criteria. The interface between business strategy and project/ programme management needs to be explored for appropriate project strategies to be developed.

The purpose of this chapter is to help make explicit the link between strategic intent and decision making related to making project procurement choices. This chapter is closely linked to Chapter 6, in which we discuss project performance measures and procurement. Naturally, performance is established in context with performance objectives and aims. Our central argument is that the strategic aim of performance is paramount and individual project performance, while very important and not to be underrated, is after all just a strategic performance contributor. It is for this reason that Chapter 6 includes discussion of intangible performance aspects because these aspects can provide a greater strategic impact than more tangible output-specified quality measures – such as on-time, on-budget. To achieve the objects set for this chapter, it is segregated into four sections together with a vignette that presents a hypothetical situation together with relevant questions that if answered, could reinforce learning from this chapter.

The first section deals with theories of the firm to identify what is important to organisations that will frame strategies to remain sustainable and competitive to deliver value to their constituents. The second section discusses strategy and associated value propositions that frame the chapter's discussion. The third section focuses upon programme and project management. The fourth section then concentrates on ways in which project approval and initiation decision can be made to be aligned with strategic intent.

Theories of the firm-influencing strategy and project procurement

In Chapter 1, The Make-or-Buy Decision Section, we introduced several views or theories of the firm. We will expand upon those ideas in this section. We argue that organisations delivering projects need to have dynamic capabilities and high levels of knowledge competence to be able to deliver a known product/service or to be adaptive and innovative to address dynamically changing needs or potential problems to deliver solutions even when they begin the project process not quite knowing how that solution will emerge. It is interesting to think about what kind of organisation could best do this. This prompts some interesting questions.

Fundamental terms and definitions

What is a firm – an organisation, an enterprise? Why does this kind of entity exist? These are sound questions to ask and probe in this book because an organisation's *raison d'être* is critical to the theme of this book. An organisation can be seen as an entity that creates value by transforming resource inputs to outputs as products, services and ideas, using a variety of resources that it sources, as and when required, and manages to facilitate that transformation. It facilitates 'projects' happening and it undertakes routine operational activities. Simplistically, we can see that many organisations do the same things time and again in a kind of factory-line way, or they do things differently by transforming a set of resources as inputs into a product, service or idea. The transformational view can either assume that essentially the same things are produced (as in a factory assembly line or looking at business processes as ways of doing things in a coherent and structured manner) or that different things are produced because the operating environment is likely to vary (subtly or radically) with each iteration. This means that firms make choices, take strategic directions and deploy specific tactics either proactively or reactively. The subtext behind their choices can be said to be influenced by their dominant philosophy, and this philosophy can shape procurement choices and their impact on the creation of value to stakeholders.

Organisations have a vision of what their role in society is and how they can best deliver value to their constituents. This vision translates to a mission that defines what they will do – 'our mission is to (verb)'. There must be a point to this activity and that is spelled out in terms of objectives, and the way that these will be carried out is defined by the strategy to be employed. Some parts of that mission translate to somewhat continuous ongoing operational activities and others more clearly have a defined start and end. The project management profession has developed over the past 50 or so years through leading some activities (that were seen as best undertaken using a world-view that these activities will get done as part of the organisation's day-to-day operations) to be undertaken using a PM perspective (that they should be seen as having an independent and defined life span with defined life phases).

The shift from seeing many activities that were undertaken with an operational work view to a 'management by projects' view is gaining ascendancy. This view sees activities as being undertaken by temporary organisations assembling these as a project, delivered within a defined start and completion boundary (Turner, 2006). We do not suggest that everything is a project, or could be twisted in some way to be seen as a project. We argue that many operational activities that were understood to be stable routine processes (such as building cars) that could be viewed as projects (because even car models have life cycles and are also subject to mass customisation with short production runs) should be treated and managed as projects. Case management, in terms of government agencies and even such things as customer relationships (and student experience management), is managed as a project with a focus on the case and case subject. So, much of the current business context is dominated by the need for a project management approach to deliver value, be that a product, service and more commonly some kind of transformation.

This brings us to the interesting question of what is an organisation – be that a firm, business unit, project team or an enterprise consisting of many divisions, sub-ordinate legal entities and a conglomerate of commercial or social groupings. An organisation or firm is often attributed to have a physical structure, and when incorporated (literally made human in the flesh) is legally treated as an eternal entity. However, there are a variety of views of the firm (or organisation) and each view influences the way that strategy might be developed. Fundamental to this discussion, we need to clarify PM terms related to defining a project, programme and portfolio to understand how a project organisation can deliver beneficial outcomes through a project framework.

The PMI has recently turned its attention to explaining how operations and a project live side-by-side in organisations; Figure 5.1 illustrates this (PMI, 2006: 7). The PMI (2004: 5) defines a project as 'a temporary endeavour undertaken to create a unique product, service or result'. It is thus defined in scope by time duration (rather than be ongoing) and is unique (a customised offering). Therefore, viewed from a procurement perspective, a project requires an organisation or assembly of resources that need to be sourced either internally or externally to deliver the specified outcome. By inference, if not by definition, choices need to be made, and so an overall strategy is required with cascading tactical plans that need to be managed – coordinated, monitored and controlled (PMI, 2004: 7). This can be further explained as experiencing pressures of uncertainty, integration and transience using flexible goal-oriented staged processes (Turner and Müller, 2003: 2).

Figure 5.1 An organisational context of managing portfolios of projects.
Source: PMI, 2006: 7.

A project may be a small part of the overall strategy. A series of connected projects that strategically link in such a way can be referred to as a 'programme' (Turner and Müller, 2003; Morris and Pinto, 2004: xx). There will most likely be a range of programmes underway grouped into strategic interests such as (taking the automotive industry example) cars, trucks, even perhaps a hybrid fuel source vehicle. Operational activities, such as equipment or software maintenance for example, could be seen as being separable. Routine maintenance, where the maintenance crews respond to faults, and regular scheduled maintenance activities may work in a day-to-day fashion where the scope may be readily identifiable or even unknown until the problem is investigated in depth – delivery expectation is 'within a reasonable time'. Emergencies and scheduled maintenance can be viewed as a project. With emergency maintenance, completion time and reliability is a key performance measure and with scheduled maintenance, cost and reliability are key considerations within an envelope of expected time delivery. There are other operational activities such as legal services, many general administration activities, etc. that could also be viewed in the same way, as illustrated by the maintenance example. The main differentiator between operational and project activities lies with scope definition and other boundary considerations (time, cost, quality, for example).

The PMI has defined a programme as 'a group of related projects, managed in a coordinated way to obtain benefits and control not available from managing them individually' (PMI, 2004: 368). Thus a sense of logic or strategy is associated with a programme – it can just be a set of projects undertaken by a specific sub-unit within an organisation. One advantage of looking at program delivery in this way is that a highly complex infrastructure facility such as, say, a tollway, can be looked at as comprising a programme of projects, some linked in similar resource bases (roads, bridges, tunnels), others as linked by the outcome's logic (e-tag systems to read and process transponder data for electronically charging and collecting tolls, development of spin-off business enterprises, marketing projects, community consultation projects, arranging finance, etc.). Such a very large infrastructure programme may even be viewed as a portfolio of projects. The PMI offers a useful definition to help us understand what a portfolio may be. A portfolio is 'understood as a collection of projects or programs, and other work that are grouped together to facilitate effective management of that work to meet the strategic need of the business' (PMI, 2004: 367). The terms programme management and portfolio management are similar and the distinction is somewhat confusing. The term portfolio management is becoming dominant, referring to the management of a series of related and connected projects that are purposefully grouped together to obtain synergies and further the strategic direction of the organisation served by those constituent projects. This is the essence of the term as currently expressed by researchers who have studied this area in great depth, for example, by Elonen and Artto (2003).

The PMI defines a portfolio as a collection of projects and/or programs to meet strategic benefit (PMI, 2006: 4). Morris and Pinto define portfolio management (2004: xx) as a strategic alignment activity 'concerned with decisions on choosing and prioritizing programs and projects' – the PMI includes authorising projects and programmes (PMI, 2006: 5).

In highlighting these definitions it becomes clear that the dominating ontological assumption is that projects are real things, that they need management and that bundles of resources are assembled and deployed to deliver intended outcomes, and that there must have been some strategy or at least *raison d'être* for the project, programme and portfolio. This perspective must be stated because many projects are ephemeral and envisage intangible outcomes such as a cultural change management project. Even for these difficult-to-physically-see projects, there is the same ontological perspective of these having a strategy that an organisation uses to underpin the introduction and germination of projects and programmes.

How do firms and organisations deliver projects?

In Chapter 2, in the Make-or-Buy-Decision Section, we cited (Barney, 1991) as one of the resource based view (RBV) proponents of explaining the nature of a firm. We also introduced other sub-elements of that RBV that look at the set of competencies that a firm may have to offer its clients (Prahalad and Hamel, 1990), and of those competencies we noted that agility and flexibility (dynamic capabilities) to respond to turbulence were key competences (Teece, Pisano and Shuen, 1997; Eisenhardt and Martin, 2000). Within the dynamic capabilities view of the firm is the learning organisation view in which organisations are seen as an organic metaphor in which they need to be able to learn and effectively use knowledge to function flexibly in a turbulent environment. Further, projects, programmes and portfolios exist within an institutional context. Thus, part of an organisation's competences and resources is its capacity to respond within a political and institutional constraints and opportunities context so that an organisation with the resources to be able to combine its knowledge and assets to address institutional barriers, as well as having the resources to do so, can provide the precise resources to deliver the desired results (Oliver, 1997). The political and dynamic capabilities view of organisational competency again highlights the need for organisations to be seen as being knowledge based (Grant, 1996; Spender, 1996; Spender and Grant, 1996). Nelson (1991: 70) and raises an interesting competency – its dynamic capabilities – that of an organisation's ability to destroy as well as create production infrastructure and technologies to enable nimble adaptation of new approaches and purposeful destruction of outdated approaches to make way for the development of any new capabilities. In procurement terms this means not being rooted to a specific dogmatic approach despite prevailing political pressure (i.e. the lowest-cost versus best-value or public sector versus private sector dichotomies). Often there is an underlying logic to a particular approach that fits a situation, and organisations must adapt to deliver this logic. The requisite capability of an organisation/firm is to be sufficiently adaptable to respond to a project's needs.

Strategy and strategic management schools of thought

This brings us to the next two terms that need to be made explicit. The term strategy is widely accepted as a form of overarching plan. The way that strategy is developed and deployed varies. The second important term to define is the organisation, its purpose, role and how it operates. Without exploring these terms and understanding how meaning can influence PM and procurement perspectives, we would be in danger of not recognising how various project types (as discussed in Chapter 2) can need different organisational foci and strategic management styles.

The dominant view of strategic management in much of the PM literature is one of conscious planning at a high level that is cascaded down the hierarchies that fall in line with the prevailing wisdom from on high – that is, that everyone is singing from the same hymn sheet. Much of Section II of the Morris and Pinto (2004) edited book,[2] for example, falls into this paradigm and this reflects the managerialism and thoughtful control-seeking behaviour that much of the PM profession seems to identify with.

One weakness that plagues many professions is that those writing and commenting on practice tend to avoid looking outside their immediate theoretical comfort zone. This is understandable because we all have limited time to study, read and think and so we tend to seek the opinion of those we have professional respect for in our fields of endeavour. Many PM commentators have for many years accepted that there are very different types of projects that range from the highly concrete to the highly intangible (Turner and Cochrane, 1993; Shenhar, Dvir and Shulman, 1995; Shenhar and Dvir, 1996). Different project types may and will require different strategic management approaches. It is for this reason that we will provide a summary of the various strategic management schools drawing heavily from the text by Mintzberg, Ahlstrand and Lampel (1998). Mintzberg *et al.* (1998) list 10 schools of strategy that fall into three groups prescriptive, descriptive and configurative. We will summarise these with a PM perspective so that it enriches our understanding of how a given strategic approach may make more or less sense for a given project type.

Prescriptive strategic schools

This set of three strategic schools is more concerned with how strategy should take place than with how it does take place.

The Design School sees strategy formation as a process of conception in which strategy being created from an analysis of external factors is represented by threats and opportunities and internal appraisal of the organisation's strengths and weaknesses. It was developed initially in the late 1950s and drew upon classical military theory for designing battle conditions that could minimise weakness and exploit strengths through detailed knowledge of the environmental factors in which the battle would take place. This strengths, weakness, opportunities and threats (SWOT) analysis, as it has been widely known, helps us to explore options, prioritise them and select what appears to be the most likely one to successfully implement. It depends upon consistency in not presenting mutually inconsistent goals and policies, and needs to be adaptive to an external environment and be reinforced by deploying the organisation's competitive advantage. It is a very deliberate process and presupposes that the SWOT analysis can be undertaken accurately. In terms of project types discussed in Chapter 2 it seems appropriate where objects and rules are reasonably well known, the environment is reasonably stable and the organisation undertaking the project is also stable so that its competitive edge does not evaporate or is easily hindered.

This represents 'Type 1' projects as defined by Turner and Cochrane (1993). However, this school seems to be at odds with the characteristics of many projects that are highly volatile in terms of their operating environment, goals and priorities, or where there are high levels of uncertainty. Type 1 projects typically are suited to the range of techniques espoused in the PMI's body of knowledge (PMI, 2004), such as use of detailed work-breakdown structures, detailed scheduling tools such as PERT, high levels of monitoring and control using methodologies such as PRINCE2 (Bentley, 1997) and generally manage projects in a way that assumes that it is possible to plan in great detail. The approach assumes that while plans are likely to need to be changed to reflect changed circumstances, it is indeed worth the effort to invest time and effort to develop detailed plans and to monitor these, and that the output from that monitoring process leads to a worthwhile degree of control over the progress of the project. Responsibility for the designed control resides with senior management at the organisational apex (for portfolio and to a large extent programmes of projects). At the project level the project manager and the PM team will respond to the governance arrangements as discussed in Chapter 4.

According to Mintzberg *et al.* (1998) the design school's view of strategy formation requires that it should be kept simple and informal, made explicit in simple terms and also that it should be one of a kind and customised after consideration of the SWOT process. Perhaps the most telling signature of the design school's strategic development process is that the plan is developed first and followed like clockwork in a disciplined way through its realisation during the implementation phase. The SWOT analysis remains its most recognised feature. This approach thus tends to rely on a plan being formulated that should work, the procurement system adopted, and processes and techniques used all aim to control the project according to the plan.

Some of the features of this approach are salutary when considering PM outcomes for projects other than a Type 1 project, see Chapter 2, and (Turner and Cochrane, 1993). First, this school makes strategy explicit and discipline demands an inflexible adherence (Mintzberg *et al.*, 1998: 36) to 'following the dots'. Second, the strategy design is separated from implementation (Mintzberg *et al.*, 1998: 36–37); clearly this requires the trajectory set by the strategy designer to be unwavering, with assumptions made by the strategy designer in the SWOT phase to remain stable and be largely correct. We know from experience that the environment experienced during most projects, even a Type 1 project, matches this requirement.

The Planning School views strategy as a formal process where the ultimate objective of 'the project' is made explicit in its strategic and operational terms. This is closer to how we see many projects being planned. It tends to use the strengths, weakness, opportunities and threats (SWOT) model but rather than try to explain strategy in a hierarchal 'must adhere to' mode, it introduces procedures, checklists and bureaucratically correct procedures that achieve the objective in a planned way. It is also a highly specialised field with experts setting the plans that are delivered 'according to plan' (Mintzberg *et al.*, 1998: 58).

Adherents to this school also make extensive use of scenario planning and or simulation and may mistakenly believe that scenario plans are predictions (Flyvbjerg, Rothengatter and Bruzelius, 2003). The key to the planning school approach is a belief that it is a mechanical system that can be understood and solved. In terms of project types it assumes that there are valid project types with different aims, that these conform to rules and that rules can lead to optimal solutions. This translates into a belief that project types have some kind of inbred solutions 'out there' ready to be deployed and that a perfect solution or methodology is ready to successfully deploy.

Projects are fraught with uncertainty. The planning school concentrates on objectives that must be achieved to meet the needs of stakeholders. In PM fields this may be viewed as more important that in operational management because project success is appreciated in the eye of the beholder – the stakeholder. The planning school is pragmatic and thus looks towards benefits derived from a project that is targeted at the aim of the project concerned; it is highly logical. It starts out with a premise that the project has a vision and mission that will make a difference to creating or maintaining value. This vision can be operationalised into a mission and objectives statement that can inform strategic planning. Plans can then be parcelled up into short- and long-term plans. This approach has many similarities with the strategic design school, except that it assumes that there is a way of foreseeing the future and hence preparing for that expected future.

Courtney, Kirkland and Viguerie (1997) argue that there are a number of different types of uncertainty which shape our view of appropriate strategies to pursue. They discuss a 'clear-enough future' where the likely future is reasonably predictable, traditional strategy development tools are appropriate, and a single forecast and plan to meet that forecast is probably appropriate. They see another category as exploring alternative futures, perhaps using game theory. A more complex 'future' involves scenario planning to explore a number of possible futures with full analysis of consequences. There is also a further state that embodies a somewhat chaotic and unknown condition in which true ambiguity prevails. Clearly, developing project delivery strategies matched by procurement choices is complex and suggests that there are few simple stock-standard prescriptive solutions.

The Positioning School sees strategy as an analytical process. Mintzberg *et al.* (1998: 83) argue that 'only a few key strategies – as positions in the economic marketplace – are desirable in any given industry: ones that can be defended against existing and future competitors. Ease of defence means that firms which occupy these positions enjoy higher profits than other firms in that industry.' They go on to describe the development of this school in three waves. The origin of this school was described as being linked to military maxims relating to defensive or offensive positions in a battlefield such as 'surround and destroy', and so on. The second wave was headed by groups of boutique consultants that applied various military maxims to commercial strategy problems as market imperatives. These consultants from the 1960s through to the end of the 1980s, sold packaged or mass-customised niche strategic-planning solutions to organisations to help them develop a position strategy for them, using building blocks of expertise that they gained from their practice. Later, these consultancies became dominated by a smaller number of larger consulting firms where global consultants, with experience of working with many clients on many problems in many countries and environments, build up a formidable data base (and knowledge base) to tap into.

The third wave, gaining momentum in the 1980s, was highly influenced by Porter's competitive strategy (Porter, 1985; 1990). He identified five forces that shaped competitive advantage. These are: the threat of new entrants; the bargaining power of suppliers; threat of substitute products; the bargaining power of buyers and intensity of rivalry among competitors. Through analysing the nature and trajectory of these forces, a firm could adopt a competitive advantage through focusing on a narrow or broad target with the aim of positioning itself with a cost advantage (being the lowest-cost provider), or a differentiation strategy (having a unique offering or position in terms of brand, reputation and gaining loyalty). This led to an analysis of where value is best derived – as is stressed in this book. This strategy school assumes that firms have the time, resources and means to plan and analyse in detail and carefully position themselves appropriately. However, this approach may not be quite so successful in highly turbulent situations where scope, scale and opportunity are rapidly changing. As Mintzberg et al. (1998: 121–122) argue, a host of soft factors need to be considered alongside hard ones in turbulent conditions, and while these three schools have strengths in measurement, data gathering and analysis, they do lack responsiveness to disruptive change. While some elements or phases of projects are relatively stable others are highly turbulent.

Descriptive strategic schools

This set of six strategic schools is more concerned with describing how strategy does in fact take place.

The *Entrepreneurial School* responds to market chaos and extreme turbulence and it focuses exclusively and intensely on a top-management perspective. This school sees strategy as being centred upon vision through the innate intuition, judgement, wisdom, experience and insight of the entrepreneur who formulates the strategy to meet the envisioned challenge. The fundamental basis of this school is rooted in innovation.

Schumpeter (1934) coined the terms (translated from German) as gales of creative destruction to describe the role of market turbulence as triggering change in both what customers demand and what is on offer. This assumes that first-mover innovators, who can position their product or service to take advantage of rapid change, will prosper. While undertaking analysis and design (and other features of the prescriptive group of strategy schools) is useful in developing an offering, there is frequently little time for first-mover innovators to perfect their strategy and product/service. This school of strategy may be appropriate to specific project types discussed in Chapter 2 and described by (Turner and Cochrane, 1993) – notably those with few methods known and low clarity of goals/objectives, that is, some 'water' Type 2 Projects and many Type 4 'air' Projects. Type 1 'earth' engineering projects tend to be stable in known methods and goals/objectives and often experience what Graham Winch (1998) describes as 'zephyrs of creative destruction' in a witty parody of Schumpeter's gales of destruction term.

Mintzberg *et al.* (1998: 133–136) describe the characteristics of the approach summarised as follows. First, it is characterised by an active and aggressive search for new opportunities, often not thought of or conceived before. For example, consider providing a portable music provision facility. This has moved radically from the courtly patronage that supported the likes of Mozart in his career where such artists were attracted to a particular location. Centuries later, gramophones that could be located in one's home, provided that service. Later still the transistor radio and walkman and currently the iPod and MP3 player evolved at several decade intervals. Each of these innovations has been radical and they have not only changed the way that music services are procured, but also the whole supply chain that delivered those innovations as complex portfolios of projects. It may have included a series of new innovations in how music is produced and stored as well as delivered. The procurement strategies had to adjust to suit these driving innovations; for example, procuring iTunes instead of CDs.

Second, in the entrepreneurial organisation, power is centralised in the owner's hands – in fact it may be more accurate to say the mind of the entrepreneur rather than hands because it is the perspective, vision and power of persuasive argument based on a great idea that drives *ad hoc* strategy. This style resonates with some projects where vision is the all-important guide that leads to success when techniques are poorly applied. The vision becomes the strategic plan and as long as it is clearly understood, compelling and motivational, credible and doable, and challenging to force smart new ways of delivering the vision and thus benefits from the project (Christenson and Walker, 2004). Following on from the last point, the entrepreneurial school of strategy is strongly characterised by dramatic leaps forward in the face of uncertainty. These strategies are aimed to take first-move advantages, often using unconventional approaches that extract high returns from a cost advantage before competitors and others in the Porter 5-force equation can catch up with them. It represents true differentiation competitive advantage because by the time other competitors have found a substitute to challenge the entrepreneur, they have moved on to another great idea. For those wishing to procure services of this kind of innovator they must take a radically different approach (procurement strategy) to tune into this kind of project deliverer.

Finally, the motivation of the entrepreneur, according to this school of strategy, is achievement in terms of any innovation delivered as well as growth in business derived from that innovation that functions as a proxy for validating the strategy. This strategy is highly risky, and those involved in procurement relationships with adherents of this kind of school are in for a turbulent ride. A highly bureaucratic client who seeks to work with a firm adopting an entrepreneurial school of strategic thought might demand a reporting or governance system (refer to Chapter 4) that is totally inappropriate and therefore might kill any potential benefits to be derived, by stifling an innovative entrepreneur's capacity or motivation to deliver an outcome.

The *Cognitive Strategy School* sees strategy as a mental process and takes an interest in the way that a strategist's mind works by using cognitive psychology to do this – how they make decisions, how they judge evidence and how they communicate what they believe to be sound strategic decisions. They do this within two wings of this school taking quite a different perspective of how to see a strategy. Mintzberg *et al.* identify these two wings (1998: 150–151): One wing is highly positivistic and objective, taking a view of the world as if seeing a video/DVD being played. The other wing sees the strategy as being subjective, and so they see the strategy being developed as a process of observation and interpretation of the world.

Both approaches are rational and subject to a range of judgemental biases. One of the main flaws noted by Mintzberg *et al.* (1998: 153) is that proponents of this school may falsely believe that a cause and effect relationship exists between things that are merely correlated. Also, they may suffer under the delusion that their past actions have led to a specific outcome when that may have been merely closely correlated to a random event. Blind belief in so-called facts can seriously cloud judgements; so can inappropriate application of statistical tools or when supported by poorly designed methodologies to gather data for analysis. This danger is flagged by a number of thinkers on project management and procurement, notably much of the work of Flyvbjerg (2006) and his colleagues (Flyvbjerg *et al.*, 2003). They argue that bias includes not just poorly structured governance systems allowing poor strategic decision making based on exhaustive data analysis, but also deliberate lying and manipulation by special-interest groups who paint a convincing picture of the way that they would like others to accept their position as reasonable facts. A metaphor may be used by cognitive strategy school disciples to encourage divergent ways of thinking to counter some of the bias found in being rigidly anchored to a 'factual' way of thinking and analysing situations. While this can provide a powerful and useful tool for creating innovative creative visions and meaning, it can also exaggerate inaccuracy because the context of the situation can be misrepresented to fit the metaphor or analogy.

If a strategist thinks that the correct way to develop a strategy is by rigorously gathering (quantitative/ qualitative) data and that facts (or theory) inevitably lead to good strategy, then they may be deluded. Mintzberg *et al.* (1998: 154–155) also draws to our attention the impact of different personality types referring to the Myer-Briggs-Type Indicator – (see Myers, 1962). The strength of this school is that it concerns itself with understanding the frames of reference, assumptions, mental models and problem-solving strategies that lie behind the strategist's mind. This approach to planning can be useful when trying to understand what happened during a project's initiation stage and its likely impact upon its capacity to deliver value.

This school is useful as it provides insights into the way that strategists think and approach strategy development – it naturally alerts us to strengths and weaknesses of human implications upon strategy formulation. This can be important on many projects that take years to deliver, when so many changes affect the project in so many ways so that a retrospective evaluation to gather lessons learned may be very difficult without considering this school of strategy.

The *Learning School* sees strategy as an emergent process. We touched upon strategy being seen as a dynamic ability to be flexible and based upon knowledge in the knowledge-based view of the firm (Teece *et al.*, 1997; Eisenhardt and Martin, 2000). In this view of strategy there is a departure from the CEO (or the highest echelon of management) being the strategy formulator; rather, it implies a flow of influence stemming across the organisation from top to bottom. Strategy and changes emerge from the organisation as a whole through its evolved culture, and how the interplay and actions between leaders and followers affects procedures and governance (discussed in Chapter 4). We will discuss the link between project procurement, innovation and organisational learning in more depth in Chapter 8. It is worth, however, highlighting in this section key aspects relating to strategy formulation and deployment.

The design school sees wisdom and knowledge flowing from on high, trickling to those below who make the strategy happen in a disciplined and mechanistic manner. The learning school sees a very different picture of what is happening. Knowledge is seen as residing at a range of levels throughout organisations as specific expertise that needs the perspective and practical application of pragmatic knowledge (Cavaleri and Seivert, 2005). Quinn (1992) brought to our attention the importance of ideas bubbling up from below, from which strategy later developed through a conversation between those with great ideas and those in the hierarchy who could support these with resources and plans of how to commercialise and gain value from these. Quinn (1992) cites companies such as 3M, see also (Shaw, Brown and Bromiley, 1998), and explains how many of these firms have tapped into specific technical expertise that is used to inform strategic decision making.

Bontis, Crossan and Hulland (2002) discuss managing an organisation's learning in terms of stocks and flows of knowledge. They see learning being available at the individual level as intuition being interpreted through socialisation and culture of an organisation, then flowing to the group level where it is integrated and reframed according to the circumstances facing that group. This is then institutionalised by the organisation in routines, rules, rites and regulations or codes of practice. This happens as a flow forward and feedback process (Crossan, Lane and White, 1999) so that each of the three levels, individual, group and organisation is enmeshed in a communal effort. Crossan *et al.* (1999) offer the 4 Is concept of organisation learning – individual *intuition* informing *interpretation* of ideas through reflection as an individual and through dialogue with colleagues. *Integration* of the reframed intuition becomes codified within groups and this leads to *institutionalisation* by the organisation. This is also mediated by power and influence at different levels (Lawrence, Mauws, Dyck and Kleysen, 2005) that link to governance issues discussed in Chapter 4. Individuals influence groups who through cultural adaptation or changed practices force the organisation's leadership to respond formally. This results in changed approaches (including strategy) and procedures (tactics and governance arrangements) and this imposes a discipline that reinforces the knowledge sharing and learning within the group and by an individual. Certainly the 3M approach to innovation that fits with this feed forward and feedback concept appears to have influenced 3M's organisational culture by allowing people their pet research and development projects that can then be considered by 3M senior management for further development by the organisation.

The learning school helps to explain what is going on within organisations and how new ideas, products or proposals for forming alliances with others shapes strategy. It also provides additional insights into what has been termed internal corporate venturing (Burgelman, 1983) and entrepreneurship where an internal project sponsor or champion strategically influences the selection of particular projects.

This theory of how strategy emerges radically differs from the top-down design school and requires different procurement approaches and a different leadership style. For ICT innovation diffusion, and the support and roll out of ICT innovation and change projects, 2 of the 11 factors determined through factor analysis that were shown to affect case study organisations in the Australian Construction industry among a select group of already IT-sophisticated organisations, included supervisor support and a culture of open communication about problems and solutions (Peansupap and Walker, 2005). This research fits well with the above view of strategy as learning emergence and the 12 prescriptions for logical incrementalism offered by Mintzberg *et al.* (1998: 183–184) that also matches with much of the change-management literature (Kotter, 1996; Wind and Main, 1998; Kanter, 1999) so well.

The natural development of the organisational learning strategy school has been the focus on how strategy appears to emerge. Here strategy may appear to randomly surface from complex processes and interactions occurring in organisations that permit and encourage diverse thought and experimentation; however, there is an underpinning rationale. There will still be clandestine players who find it effective to 'suggest' ideas to their organisational bosses (or others with more perceived influence) who then set in train a championing and lobbying process for the idea. As Mintzberg *et al.* (1998: 190) point out while individual(s) may press the idea forward in a deliberate and strategic way, the organisation may see the strategy as just emerging. The emergent strategy school adapts learning-school ideas and further develops these to introduce elements of power, influence, change management and diffusion to the concept of ideas and strategies just bubbling up. It also draws upon 'sensemaking' concepts. Karl Weick (1989; 1995) explains 'sensemaking' as a process where people retrospectively explain (or make sense) of actions from a chain of events and influences.

It is important to remember that emergence can occur through ignorance as well as learning, through brute power as well as sophisticatedly crafted and subtle influence. Strategies can be forced through even the most politically open and liberal organisations, and also they may be forced through by an external environment that imposes a particular logic. Government policy in the Australian higher-education system (as elsewhere) has provided a pervasive influence in (government) forcing desired strategy to 'emerge' within universities. This has nothing to do with a learning-organisation approach, other than 'sensemaking' imposing a particular logic as either inevitable or an easier path in some *quid pro quo* struggle for survival and sustainability of an organisation. Strategy can emerge from administrative fiat, collective choice, self-interest or professional judgement (Mintzberg *et al.*, 1998: 192). A broad umbrella set of constraints and opportunities can impose a logic with details that when analysed, leads to more detailed strategies emerging.

The literature relating to Honda's rise in motorcycle manufacturing and distribution within the UK and the USA in the late 1960s provides interesting insights into how Honda's strategy emerged (Pascale, 1984; Goold, 1996; Mintzberg, 1996; Mintzberg *et al.*, 1996). While that story triggered a great deal of debate about emergent strategy it also serves as a useful adjunct to developing the learning school of strategic thought. This strategy school has contributed to and drawn from the literature on core competencies (Prahalad and Hamel, 1990) and how that shapes strategic intent (Hamel and Prahalad, 1989) and adds to an organisation's dynamic capabilities (Teece *et al.*, 1997). It also draws upon chaos theory in that it influences organisational strategy as being seen to need to adapt to and be flexible in turbulent situations where dramatic change can rapidly occur (Strogatz, 2003;

Backström, 2004; van Eijnatten and Putnik, 2004; Snowden, 2006). This stresses the need when making procurement choices to be more cognisant of today's turbulent environment and that one size does not fit all.

The *Power School* sees strategy as a process of negotiation linked to influence and politics. Some aspects are subject to further discussion in Chapter 3, on how stakeholders influence a procurement strategy and so we will limit our discussion in this chapter. Influence can be legitimate, being designated by the governance provision such as the power to initiate, support or approve. It can be illegitimate by subverting legitimate power through hidden (non-transparent) influence or pressure – perhaps as part of mutual adjustment or more sinister motives. Some of these influences are discussed in Chapter 4 relating to ethics and professionalism.

Being aware of the political landscape can be an advantage when influencing others. Bourne and Walker (2004: 228) refer to project manager competencies as needing to include three dimensions of wisdom and these are further explained in the discussion on stakeholders in Chapter 3. Dimension 1 relates to knowledge of how to look forwards and backwards to apply correct PM techniques – in the case of this chapter, this would include the appropriate school of strategy to draw upon when making decisions. Dimension 2 relates to knowledge of relationships of how to look inwards, outwards and downwards, which is also relevant to the power school of strategy as it includes the ability to manage relationships with key influencing stakeholders. Dimension 3 skills relate to considering and ensuring that political influence and lobbying is addressed by looking sideways and upwards. This is what Bourne (2005), and Bourne and Walker (2004: 228) refer to as 'tapping into the power lines' of project stakeholder influence (see Chapter 3 for further discussion).

The politics school also believe that the ability to influence and negotiate is also reflected by the sense of commitment or compliance experienced by those taking part in negotiations. This will be discussed further, along with notions of trust in Chapter 9, Cultural Dimension of Project Procurement. To summarise the relevance of those concepts in this context, we can envisage compliance as being the lowest level of commitment (Meyer and Allen, 1991) where perhaps legitimate power is enforced to push through strategy, and may explain why the strategy may then be given only lip service. Other forms of persuasion may be employed to encourage high-level commitment that can lead to sustainable support for strategy. Similarly, trust is an important element of political persuasion because personal integrity, behaviour and the dynamics of power interact in negotiation (Lewicki, Litterere, Minton and Saunders, 1994). Trust can effectively cohabit with distrust where processes are put in place to ensure that transparency enables trust to build by being seen to be verified (Lewicki, McAllister and Bies, 1998). Mintzberg (1989: 238–240). This highlights a number of power-politic games that are played out in organisations that affect strategy development and adoption. Some of these games are overt, such as using authority to coerce people through intimidation or bribery by budget approval, by using expertise in an exclusionary way. Others are covert and might occur through such ploys as alliance building, empire-building, selective sponsorship based on reciprocity. These are a few of those described by Mintzberg (1989) but they give a flavour of the assumptions underpinning the power-school strategy that seek to understand how strategy evolves from this negotiation-dynamics perspective. Strategy becomes a bargaining process and its results can be explained not so much by rationality or even the fiat of the leader or any particular group; rather, it evolves as a series of compromises temporarily mapped by waves of influence that map out boundaries.

A related school to this is the *cultural school*, that sees strategy as a collective process. Culture can be studied from either the outside in, or inside out. Mintzberg *et al.* (1998: 267–268) outline five premises of the cultural school. First is that strategy is based upon beliefs and understandings shared by an organisation's members; this has profound implications for misinterpretations in cross-cultural teams because of the wide range of meaning that can be assigned by different groups to what seem to be common cultural artefacts (Grisham, 2006). Second, individuals acquire their cultural beliefs through socialisation which is highly tacit and value laden. Third, because much of culture pertains to the unspoken and assumed, artefacts at the superficial level may appear to represent culture coherence, while at deeper levels assumptions may vary. Thus, a strategy document may be interpreted quite differently by different individuals and groups. Fourth, strategy takes the form of perspectives rooted in collective intentions. Finally, Mintzberg *et al.* (1998: 268) argue that culture, and especially ideology, does not encourage strategic change so much as perpetuate the *status quo*.

The cultural school is seen by Mintzberg *et al.* (1998: 280) as being too conceptually vague to be useful for working out what strategy can be achieved, and how; rather, it is helpful for explaining what might be happening at any given time. This is because so much of the literature principally describes what is going on in people's heads and hearts rather than their physical observable behaviour; it all requires expert interpretation. That said, it introduces us to important aspects of change management and as projects are change programmes of one kind or another and this school offers us a useful way of understanding the dynamics of, for example, (internal team and external) stakeholder influence. Considerations of culture should be made when making procurement choices so as to minimise unpleasant unexpected outcomes happening.

The *environmental school* sees strategy as a reactive process. This school sees the environment as the defining actor, together with leadership and the organisation, shaping strategy through the organisation's reaction to forces external to it. Mintzberg *et al.* (1998: 288) argue that this school evolved out of contingency theory with ideas such as 'the more stable the environment, the more formalised the organisation's internal structure'. They present four premises of this school: First, the environment presents itself to the organisation as a set of general forces at the centre of the strategy-making process. Second, in a kind of Darwinian sense, organisations must adapt, react or be selected out of existence. Third, leadership becomes a passive element in reading the environment and proposing organisational adaptive action. Finally, the environment metaphor leads to a belief that organisations cluster into distinct ecology-style niches. Its Darwinian population-ecology focus leads this school to observe strategy from a population perspective. A group of organisations within a population niche compete; as one innovates it advances within that pack, provided that it has sufficient resources to sustain it, however, the environment has a limited carrying capacity, so in this Darwinian struggle less-efficient organisations die.

Another stream of the environmental school uses institutional theory which posits that there is an institutional pressure to conform, and so strategy becomes ways in which resources are converted into symbolic ones such as 'brand reputation', being a market leader (of that niche within the pack). This compels competing organisations to mimic each other's perceived successes. Being recognised as successful is a strategy for attracting more resources (share-market recognition, being a partner or employer of first choice, etc.). Benchmarking and 'best practice' can be viewed as expressing this strategy. Christine Oliver (1997: 708) adapts institutional theory and argues that far from being passive, organisations that survive are evolving internally by adaptation and change management to respond to external environmental factors. However, large institutional organisations are hindered by sunk costs, not in financial terms, but sunk costs in culture (organisational inertia through habit) and ways of valuing skills (in public-works government departments valuing the opinion and skills of, say, architects over engineers or vice versa). This can affect their type and rate of response because they are captives of their own history, and what appeared to work best in the past – being blinded rather than being informed – by history. Nevertheless, she states that 'Firms seek out competency blueprints or recipes in several ways, including direct imitation of a successful competitor (e.g., late-mover imitation of a rival's technology), indirect mimicry of role models (e.g., benchmarking), and the use of outside consultants to develop expertise employed competitors.' This tends to force organisations within a niche to adopt similar strategies until one completely breaks the mould and pole vaults into a separate niche, revealing a competency trap of others in their niche. Institutional forces, therefore, entrap some firms into pursuing limited strategic choices. In project procurement terms, we can observe the difficulty in government agencies and corporates moving away from a lowest-price to the best-value approach. No matter what the rhetoric is, the tender and evaluation process always seem to favour and slip towards trying to get a low tender cost.

Mintzberg *et al.* (1998: 294) cite a number of critics who argue that environment explains neither short- nor longer-term survival and decay because other internal forces, for example, irrational competitor actions, and other factors may have as much if not more influence upon organisational sustainability. However, they do value this perspective as being useful in trying to understand how organisational populations form and adopt strategy, and how that helps us think about strategy related to seeing how strategic choices may be perceived to be limited in organisations at a mature stage in their life cycle.

Configuration strategic schools

This last strategic school is concerned with strategy as a process of configuration and transformation. PM is described as creating a transformation, so this school may be of particular interest to readers of this book. Mintzberg *et al.* (1998: 302) identify two main labels for this school, strategy as *configuration* and as *transformation*. Some of the preceding schools can be seen as representing different phases or stages of an organisation's approach to strategy as it evolves and responds to the challenges it senses. Start-up organisations tend to have a charismatic entrepreneurial leader with a focus on visionary strategies while mature organisations that have been around a long time have built up not only cultures and habits but also have professional managers who conservatively approach strategy. Likewise, different types of project and different types of sponsor or clients will elicit different strategic approaches to project procurement that match what they are comfortable with.

Mintzberg *et al.* (1998: 305–306) states that *Configuration School* proponents believe that most of the time an organisation can be described in terms of a kind of stable configuration of its characteristics for a distinguishable period of time, adopting a particular structure by adopting its strategies to meet the challenges it faces from its external environment. There are periods of stability and periods of extreme turbulence when disruptive change requires a quantum leap to another organisational form. These successive states of configuration and transformation occur over time similar to the stages of organisational transformation triggered by turbulent shocks described by Greiner in (Greiner and Schein, 1988; Greiner, 1998). The key to strategic management, therefore, is to sustain stability most of the time when that maintains competitive advantage and to be able to shift gear to being transformational to cope with disruptive change to position the organisation to be ready for the next fairly stable period. This is where the other strategy schools have their usefulness because the process of strategy making can be seen, as Mintzberg *et al.* (1998: 305–306) argue as:

> 'one of conceptual designing or formal planning, systematic analyzing or leadership visioning, cooperative learning or competitive politicking, focusing on individual cognition, collective socialization, or simple response to the environmental forces: but each must be found at its own time and in its own context' – 'the resulting strategies take the form of plans or patterns, positions or perspectives, or else ploys, but again for its own time and matched to its own situation'.

Finally in this section it is worth looking at transformational change again because this is in essence a PM issue. Change-management proponents stress that in order to achieve change, people need to be persuaded that change is required. They need to see that the current state needs reform because it is either currently undesirable or will lead to imminent undesirable consequences. Further, a specific plan to make that change needs to be made. This plan should be credible, achievable and reasonable in terms of the balance between the efforts to be made (or sacrifices) and the benefits to be derived from that change (Kotter, 1995). This way of looking at strategy fits well with the PM concept of identifying a need and developing a brief (the business case), deciding upon a plan of action to achieve the strategy that includes working out how to assemble the required resources (the procurement plan), and then taking action to deliver the desired outcomes (the implementation stage) and finally, to deliver the outcome and learn from the experience (the closeout stage). Understanding the various strategic schools of thought is helpful for us to argue for change or to emphasise choosing an appropriate procurement strategy.

Linking business strategy to procurement

This section will explore the area of project portfolio management (PPM) – how projects may be strategically selected. It will outline three levels of activity involving project execution at the project, programme and portfolio level. We will then discuss current insights gained from study of how projects can and should contribute to creating and delivering value to their sponsors and stakeholders.

Project managers tend to see a project as a given task that must be carried out. This view is fast changing as the importance of managing complexity and risk associated with large projects is better understood and the trend develops towards managing businesses as a series of projects where an increasing amount of attention is directed towards this view of PM (Archer and Ghasemzadeh, 1999; KPMG, 2003; Turner and Müller, 2003; Morris and Pinto, 2004; Blomquist and Müller, 2006b).

Recent work funded in part by the PMI has led to some useful clarification around how PM and programme and portfolio management fit together

(Morris and Jamieson, 2004; Blomquist and Müller, 2006a,b). In simple terms, the whole process is about improving PM (doing the project right) and improving programme or portfolio management (doing the right projects). Modern portfolio theory was pioneered by a Nobel-prize-winning economist (Markowitz, 1991) and it has also been further developed by academics from marketing and product development backgrounds to propose models of selecting the strategically appropriate project from a range of choices (Cooper, Edgett and Kleinschmidt, 1997a,b; Cooper, 2005).

Elonen and Artto (2003) identified from an in-depth study of two portfolios that revealed common problems not only make sense but have also been used to map problems of roles and practices that indicate poor portfolio management performance identified in a comprehensive and detailed study undertaken by (Blomquist and Müller, 2006a: 63). The Blomquist and Müller (2006b) study comprised 11 semi-structured interviews with senior managers from large globally operating organisations representing 5 different industries. Their findings indicated that managers with a focus on portfolio management aim for maximising organisational results.

They also conducted a quantitative study analysing organisations' governance practices, roles and responsibilities of 242 participating programme/portfolio managers. They devised an index (Blomquist and Müller, 2006b) that measured benefit to the organisation. All participants indicated some level of programme and portfolio practice, so they divided the sample into one group that performed above the mean and another group at or below the mean. They then analysed the characteristics of roles and practices of those that performed below the mean-index results (designated as low-performing organisations) using the analysis of variance (ANOVA) statistical technique. Table 5.1 illustrates their findings (Blomquist and Müller, 2006a: 63). They discovered that, as expected, portfolio management practices are affected by the complexity of the environment. Higher complexity demanded more rigorous application of a programme/portfolio governance structure to meet the needs of their environment and project types. The organisations studied were primarily involved in for-profit business activity but an interesting question is raised by this, what can we learn from this research that can be applied to another very important sector, not-for-profits (to society in general rather than commercial organisations)?

First, it is instructive to understand what programme and portfolio management generally offers. It is instructive to consider two basic requirements that need to be satisfied by programme/portfolio management to improve value generation for project clients/sponsors/champions and other stakeholders. These are to first, select the 'right' project to realise and second, to design a delivery governance system to ensure that expected benefits are realised. The first part is related to developing a system to choose the 'right' project, the second relates in part to Chapter 4 in relation to governance and in part to performance measures, which is discussed in

Table 5.1 Mapping strategy research against known problem areas

			Elonen and Artto (2003) defined six problem areas					
			Inadequate project-level activities	Lack of resources, competencies and methods	Lack of commitment, unclear roles + responsibilities	Inadequate portfolio-level activities	Inadequate information management	Inadequate management of project-oriented business
		The middle manager is …						
	Roles	Involved in resource procurement	√	√		√		
		Involved in identification of bad projects	√		√			√
		Working in steering groups	√	√	√			
		Involved in project-group reviews	√		√			
		Involved in handling issues related to projects						
		The organisation …						
	Practice	Prioritises projects				√		
		Selects project on the basis of the organisation's strategy						√
		Communicates which projects are important					√	
		Uses a tool to collect and disseminate information about the status of all high-priority projects					√	
		All reporting to the Steering committee is done using the same template					√	
		Uses similar metrics for reporting similar projects					√	
		Decisions about groups of projects are made as joint management decisions						√
		Decisions about groups of projects are made in the best interest of the company						√
		Facilitates effective management of the work in order to meet strategic annual-business objectives (uses portfolio management)						√
		Portfolios are managed by a single manager or a group manager				√		
		Group projects together to obtain benefits and control not available from managing them individually						√

Areas low performing organisations scored significantly lower

Chapter 6 but will also be expanded here with a discussion of the role of the project management office (PMO) and its variants.

Choosing the 'right' project

In the private sector, practitioners and academics have begun to partially address this challenge by first adopting and then translating portfolio theory, as originally defined within the Finance discipline, into a project management context (Martino, 1995; Cooper *et al.*, 1997a). This approach is generally known as PPM. In effect, this extends the boundaries of traditional project management by attempting to include how projects are selected, prioritised and approved to proceed, rather than a simple focus on project execution.

An important underpinning of the original theory is the normal priority placed on maximising the financial return while minimising the risk of the selected portfolio. This point cannot be ignored when the theory is applied to project management. This is known as portfolio optimisation. This approach leads to the development of decision support tools (normally in the form of some kind of consistent scoring model) to assist in portfolio selection. Most of these scoring models emphasise selecting projects that offer higher financial returns as measured by traditional means, that is, return on investment (ROI), internal rate of return (IRR) or project payback. Considerations might, however, include amounts of available capital or resources, thus forcing the selection of a portfolio of 'affordable' or 'doable' projects based on cash-flow considerations. The emphasis and underpinnings of any and all of these more or less complex approaches remains primarily financial. While this may seem like an acceptable assumption in the private sector where the singular purpose of most corporations is to make profits at almost any cost (Bakan, 2004), it should be clear that this same assumption cannot and should not hold true in the public sector. In fact, others have even questioned this approach within the private sector (Elkington, 1997), suggesting its focus is too narrow and propagates concepts such as the 3BL and social responsibility as other points of reference for corporate perfor- mance in the private sector. This aspect is discussed in detail in Chapter 4 of this book. If managing a portfolio of projects for financial gain is not appropriate in the public sector and perhaps not even in the private sector, what are the right criteria then?

Since achieving alignment between strategy and projects is one of the oft-cited benefits of programme management (Morris and Jamieson, 2004), we must therefore conclude that one critically important aspect in any revised PPM methodology is the assurance of a firmer connection between the organisational strategy-and project outcomes. This suggests a need for project management professionals to be both deeply involved and to completely understand the strategy-making process and to properly address communicating eventual links within both single-project management and multiple-project management settings to those on their project teams.

This notion has been sufficiently explored. The literature rightly calls for a clearly defined and measurable business strategy and an accompanying vision and mission for every project. This leaves the question, how does one theoretically and practically link this to PPM methodology and practices in a way that will enhance the understanding of strategy among project team members?

It is also relevant to question why many PM professionals naturally assume a profit-centric model could be universally applied in a project management context. It does not seem as if this fundamental assumption has been challenged to any great degree in the existing literature. While this assumption may make sense in the original context of selecting a financial portfolio where investors, like corporations, are focused on maximising their rates of return, in the public sector strategic objectives are normally more complex to define and measure than a pure profit motive (for instance societal outcomes, regulatory compliance, life and death issues, community health, educational achievement, etc.). A more rational and effective approach to project selection would be to create conditions in the scoring model and accompanying decision-support tools where one could be satisfied that

- a sufficient portfolio of strategic projects actually existed and, if properly selected and managed, would lead to the organisation accomplishing its intended strategy and
- given a clear definition of strategy, it would be possible to select projects that were most likely to have the maximum strategic contribution balanced across all aspects of the stated strategy and not just its financial performance.

One approach offered here is an adaptation of the balanced score card (BSC) idea that will be discussed in more detail in Chapter 6. A BSC attempts to balance competing notions of how value can be represented and we suggest that all projects should be prioritised using a balanced assessment. Rather than discuss the underlying theory, we refer readers to Chapter 6 or to read about the underlying concepts, (see, for example, Kaplan and Norton, 1992; 1993; 1996; 1998a,b). Norrie (2006) developed and tested a system of evaluating projects for using this evaluation framework to guide approval of projects offered for development and realisation as part of a portfolio of opportunities for enacting change. He argues that while traditional scoring models simply only compared financial return to risk, the more sophisticated proposed scoring model must now compare strategic contribution across four individual dimensions to risk. The outcome of this change is a more complex, but more strategic and dynamic scoring model as shown in Figure 5.2. The four elements are based upon the Kaplan and Norton criteria modified for a PM and portfolio management purpose.

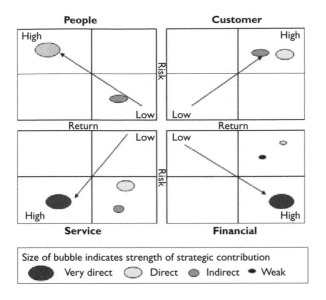

Figure 5.2 Norrie's proposed balanced PPM scoring model.

The model recognises outcomes being related to people, customers, service and financial outcomes that are strategic elements in an organisation's sustainable development. Each project proposal can be assessed according to criteria related to an organisation-specific template. Most organisations have a strategy that helps them develop specific measures and where this is not present they are advised to access expert advice to help them develop a model that fits their unique circumstances. As noted earlier in the theory of the firm section, organisations should aim to mould a unique and difficult-to-imitate set of characteristics of the value that they offer. If an organisation is confused over how to structure their equivalent to Figure 5.2 then it should commission a consultant to help it do so.

Each dimension of the graph represents its own traditional risk/ reward matrix, with reward on one axis and risk on the other. The centre of the graph represents (0,0) suggesting that as projects move outward from the centre they increase in risk and return – a normal assumption in most instances since rarely is their return without risk. Each project is analysed for its contribution (measured in terms from a very weak or indirect supporting contribution all the way through to a direct or very direct measurable impact) and the size of the circle represents the degree of contribution of the project in each domain. This is determined by comparing the contribution of the individual project outcomes (always in their most measurable terms) to the previously established gap identified during the planning process between current and future performance levels for that domain at the overall enterprise level. Similarly, projects which impact more than one quadrant concurrently (i.e. contribute to building strategic capability in several areas rather than only in one) are more valuable to the organisation and should be executed in order of highest priority. The more each project can establish a significant ability to contribute to a lessening of that strategic gap in measurable terms, the more reliable the project is to that quadrant attaining its intended performance level overall.

In relation to risk, the circles are then positioned to determine those projects as having relatively higher or lower assessed risk (depicted by moving outward on the graph from the centre as having more assessed risk). This combination allows for trade-off decision making in each quadrant on a project-by-project basis between risk and reward. The importance of this change in practice cannot be underestimated for its possible contribution to the private sector certainly; but, it is even more significant in the public sector. The complex multi-stakeholder strategy present in the public sector makes traditional, financially driven project scoring models quite limited in their applicability. To address this imbalance, we must use a method which allows us to measure, for example, outcomes related to the greater social good or achievement of social policy objectives, not just financial terms (e.g. improvements in overall health levels, decreases in waiting lists or treatment lapses, increases in educational success rates, access to cost-effective day care or improvements in the number of commercial patents issued to business, etc.). The BSC would allow one to do that. Furthermore, once the BSC is incorporated into a public sector setting, we have demonstrated how it can be used to balance strategic outcomes with the costs of programme implementation, enforcement and other factors (Norrie, 2006).

This balanced view of strategy is imperative in the public sector where responsibilities cannot be abandoned simply because they are costly. However, where applicable, financial efficiency is still equally valued, creating the right balance for application of the methodology in this sector. In addition, it retains the approach of mitigating potential project benefits with possible risks thus retaining an important theoretical foundation of effective PPM methodologies. By combining these methodologies, we find a way to acknowledge this complexity and we can learn to account for the true value of a project beyond its pure financial impact. By adopting the BSC as a method of measuring any particular part of the public sector's mandate across its four dimensions and attaching relevant measures to each, it finally becomes possible to score the relative strategic value of any project by measuring its potential individual contribution to the overall measures in each of the four dimensions. But there is another inherent value in this proposed change which is subtle but critically important. If an organisation's focus is on strategy execution, and if projects form part of the infrastructure required to execute strategy once it has been formulated, it becomes imperative that the organisation ensures that it has a sufficient number of projects in each dimension to actually accomplish its stated goals. This transfers the notion of 'balance' from the scorecard methodology into a project setting demanding that the project portfolio also be similarly balanced. This helps build expected performance levels into the selection of the project portfolio from the outset.

This approach complements the notion of stage-gates proposed by Cooper (Cooper, 2005) for selecting winning product developments where a series of go or no-go decisions are made at logical junctures of a product development cycle (read project phase for PM). This portfolio practice is particularly common in some industries, for example, in the pharmaceutical industry (Lehtoned, 2001; Morris and Jamieson, 2004: chapter 4) and in research and development (R&D) in telecommunications (Poskela, Korpi-Fippula, Mattila and Salkari, 2001).

The PMO enacting portfolio strategy

There has been much discussion regarding the PMO concept and where it is currently heading. Most construction projects are managed on site through a project office where specialists and much of the project team congregate in physical proximity to work together, to share ideas and to facilitate communication through rich personal contact. This has been happening for many decades if not centuries. The main difference that has more recently occurred that is relevant to this chapter is that the PMO has been evolving as a way to move beyond just gathering people together and better coordinating their actions and resources towards sharing standards (Rad, 2001; Ingebretsen, 2003; Blomquist and Müller, 2006b; Crawford, 2006). This facilitates programme or portfolio management (Turner, 1999: chapter 14; Artto, 2001) in a way that adds coherence.

PMOs at its lowest level merely provide a location where project team members gather to do their PM work and this is often restricted to those involved in the 'iron triangle' processes of time, cost and quality control. This level of PMO supports decision making, coordination and control and is a way of 'housing' project managers. Once they are linked either physically or through a form of communities of practice (COP) as for example illustrated by a construction industry example (Jewell and Walker, 2005) that the knowledge sharing which takes place allows standards to be set that make established PM processes more repeatable for that organisation. Thus the PMO facilitates organisation-wide impact of an incremental improvement and increased PM maturity. The PMI standard OPM3 (PMI, 2003) specifies a five-level maturity model: *Initial*, where *ad hoc* PM approaches are deployed and the project is characterised by chaos and confusion with the same mistakes being repeated for each project; *Repeatable*, where basic PM tools, techniques and processes are deployed and formalised often with a start in developing a procedures manual or system that allows success to be repeatable or failures not to be repeated; *Defined*, the procedures manual or system is formalised and tested to a point where it is accepted as a policy and guides PM activities; *Managed*, at this stage performance is recorded and data collected to demonstrate the effectiveness of the systems in place in the PMO, and it is possible to justify resources and initiatives and gauge the value of the PMO and ensure the systems are well understood and accepted; and finally, *the optimising level* may be reached; this is where it is continuously improved and is embedded in the organisational culture (Crawford, 2002; Rad and Levin, 2002: 282). The impact and degree of involvement of the PMI in projects can be mapped. Kendall and Rollins (2003: 284) illustrate a simple example where business units, function units or other organisational sub-units can be established as columns in a table and each split into columns for type of project, for example, development, integration, maintenance, etc. The rows can then represent a classification of type of projects undertaken that makes sense to the organisation, for example, large, medium and small. The cells of the table can then be populated by project references so that the scope, scale and type of projects undertaken by the organisation can be visualised. This can be useful for senior managers to understand what PM is happening, or it can be used as an input into a project filtering and screening process described in the previous section of this chapter.

We can now start to get a sense of where the PMO could be strategically heading. Chris Cartwright from Ericsson Australia presented at a PMI chapter seminar a detailed account of Ericsson's development of a strategic PMO for the Asia Pacific region that has won awards for excellence in business (Cartwright, 2006); this discussion is extended in Chapter 11. He outlined how PMOs evolve or inevitably are doomed to be restructured out of existence. The Ericsson Melbourne PMO started life as many others as an attempt to raise PM skills and capabilities, and a large part of its original remit concerned creating standards, promoting learning exchange and sharing experience and gathering data on success or otherwise of their projects. This evolved to being involved in decision making about which projects to focus on, and so its role crept into the portfolio management arena and further evolved to being one where it has a strategic focus in not only advising on the strategic selection of projects but also in helping to develop business cases, coach PM staff as well as advise sponsors and champions on how to optimally enhance their role. The PMO in its latest evolution is proactive and has been made more accountable. If projects now start to drift, the focus is not on auditing and crisis management; rather, it is on needs analysis, facilitating better quality resource decision making, coaching and mentoring, developing COPs and facilitating people-networking and supporting projects from the start so that they do not become distressed. The result has been that the PM and program management activities has moved from a back-of-house small revenue generator to being a revenue and social capital star that is recognised now at board level. This trend concurs with efforts by the PMI, for example, to get greater board-level recognition of PM and its value; in addition, boards and high-level advisory groups would have more PM skills to draw upon when making strategic decisions.

Chapter summary

This chapter starts with a general discussion of the nature of the organisation and various perceptions of its reason for being. The purpose for this section was to question how organisations create and generate value through projects. This led to a summarised discussion about the various strategic schools that currently attempt to describe why organisations behave in the way they appear to do. The rationale of this was that if organisations have to choose between a range of potential projects then they should choose those that deliver greatest benefit for stakeholders, given the purpose that organisations see themselves as fulfilling. A highly commercial organisation will have a different vision, priorities and objectives to a not-for-profit one, for example, and so it is likely that their strategic propensity will also differ. The section on strategic choosing the 'best' fit project naturally flows from understanding the organisation's strategic rationale. The BSC tool described provides a useful way of making strategic choices. We then introduced the concept of the PMO as a strategy-enabling device as it facilitates taking strategy into action by guiding projects in a strategic and formal way. Chapter 11 takes both the theory and practice of the PMO further so that readers may find more substance and content relating to that aspect in Chapter 11.

Vignette

XYZ Corp's strategy is defined as being geared towards a leading market position for private yachts in its respective geographic market. Rather than being a trendsetter, the company typically responds quickly to market trends and is known for being a reliable integrator of proven technologies and features. XYZ Corp's reputation in the market is characterised by modern, yet somewhat standardised, designs with limited customisation options. Compared to its competitors the company is best described as a 'fast follower', but it leads the competition in delivering a quality product in short time frames at a very reasonable price point. However, due to stiff competition, margins are very slim and the corporation is struggling to make the debt payments for substantial equipment purchases of the past.

The PMO, which is responsible for both PPM and project execution, has the following potential projects to choose from:

1   upgrade manufacturing facility with new kitchen and dining area, a project long promised by the CEO, benefiting staff morale. Cost: $200k over two years;
2   PMP training and Certification of the 10 project managers, who oversee large-scale yacht projects. Cost: $50k, Benefit: non-conformance cost savings of 120k over three years;
3   purchase of a new tool machine for manufacturing more custom accessories. Cost: $250k, Benefit: additional sales of approximately $2m p.a.;
4   hiring of an additional industrial designer: Cost: $120k p.a., Benefit: approximately $1m in sales through more customised design offerings;
5   R&D for new hybrid propulsion projects to offer a product which complies with more stringent noise and pollution restrictions for inter-coastal waterways. Cost: $5m over five years;
6   new construction of an R&D facility to allow projects like the hybrid propulsion and other basic research efforts. Cost: $400k and
7   replacement of the 'broken' order entry system and other related software. Cost: $100k, Benefit: time savings 50k p.a. and order generation of at least $1.5m per year. Update financial reporting system, required to be completed in the new fiscal year in order to sustain regulatory compliance. Cost: $75k.

Issues to ponder

The CFO assigns a project budget of $500k, which cannot be exceeded. The outcome of the first selection process is to 'shortlist' projects (2), (3), (4), (7) and (8).

1   What strategy model or combination of models best characterises XYZ Corp?
2   Is the provided information sufficient to make an optimal decision? If not, what information is missing?

3   If the financial benefit of the R&D investment would achieve a
     quick payback, should the company assume more debt funding to
     go ahead with them?
4   Provided projects (2), (3), (4), (7) and (8) could be funded, could
     this choice set be justified as the optimal portfolio?
5   What would be the decision steps after the 'short listing' exercise?

Notes

1 For readers interested in finding out more about this icon visit the
   Australian Government Culture and Recreation portal web site www. culture
   andrecreation.gov.au/articles/sydneyoperahouse/
2 The authors recommend this text as being a high valuable reference for project
   managers and PM students as it is a comprehensive and current collection of
   PM theory and practice.

References

Archer, N. P. and Ghasemzadeh, F. (1999). 'An Integrated Framework for Project
    Portfolio Selection'. *International Journal of Project Management.* 17(4): 207–216.
Artto, K. A. (2001). Management of Project-oriented Organization – Conceptual
    Analysis. In *Project Portfolio Management – Strategic Management Through
    Projects.* Karlos Artto M. M. and Taru Aalto. Helsinki: Project Management
    Association of Finland: 5–22.
Backström, T. (2004). 'Collective Learning: A Way Over the Ridge to a New
    Organizational Attractor'. *The Learning Organization, MCB University Press.*
    11(6): 466–477.
Bakan, J. (2004). *The Corporation: The Pathological Pursuit of Profit and Power.*
    New York: Free Press.
Barney, J. (1991). 'Firm Resources and Sustained Competitive Advantage'. *Journal
    of Management.* 17(1): 99.
Bentley, C. (1997). *PRINCE 2 A Practical Handbook.* Oxford: Butterworth-
    Heinemann.
Blomquist, T. and Müller, R. (2006a). 'Middle Managers in Program and Project
    Portfolio Management: Practices, Roles and Responsibilities'. *Project
    Management Journal.* 37(1): 52–66.
Blomquist, T. and Müller, R. (2006b). *Middle Managers in Program and Project
    Portfolio Management: Practices, Roles and Responsibilities.* Newtown Square,
    PA: Project Management Institute.
Bontis, N., Crossan, M. M. and Hulland, J. (2002). 'Managing an Organizational
    Learning System by Aligning Stocks and Flows'. *Journal of Management Studies.*
    39(4): 437.

Bourne, L. (2005). Stakeholder Relationship Management and the Stakeholder Circle. Doctor of Project Management, *Graduate School of Business*. Melbourne: RMIT University.

Bourne, L. and Walker, D. H. T. (2004). 'Advancing Project Management in Learning Organizations'. *The Learning Organization, MCB University Press.* **11**(3): 226–243.

Burgelman, R. A. (1983). 'A Model of the Interaction of Strategic Behavior, Corporate Context, and the Concept of Strategy'. *Academy of Management Review.* 8(1): 61–70.

Cartwright, C. (2006). The Journey to a 'Best in Class' – Corporate Project Office. members P. M. C. Melbourne, PMI Melbourne Chapter.

Cavaleri, S. and Seivert, S. (2005). *Knowldege Leadership the Art and Science of the Knowledge-based Organization.* Oxford: Elsevier Butterworth Heinemann.

Christenson, D. and Walker, D. H. T. (2004). 'Understanding the Role of "Vision" in Project Success'. *Project Management Journal.* 35(3): 39–52.

Cooper, R. G. (2005). *Product Leadership: Pathways to Profitable Innovation.* New York: Basic Books.

Cooper, R. G., Edgett, S. J. and Kleinschmidt, E. J. (1997a). 'Portfolio Management in New Product Development: Lessons from the Leaders-2'. *Research Technology Management.* **40**(5): 43–52.

Cooper, R. G., Edgett, S. J. and Kleinschmidt, E. J. (1997b). 'Portfolio Management in New Product Development: Lessons from the Leaders-1'. *Research Technology Management.* **40**(5): 16–28.

Courtney, H., Kirkland, J. and Viguerie, P. (1997). 'Strategy Under Uncertainty'. *Harvard Business Review.* 75(6): 67–81.

Crawford, J. K. (2002). *The Strategic Project Office.* New York: Marcel Dekker AG.

Crawford, L. (2006). *Project Governance – The Role and Capabilities of the Executive Sponsor.* PMOZ – Achieving Excellence, Melbourne, Australia, 8–11 August, PMI – Melbourne Chapter: 1–11 CD-ROM paper.

Crossan, M. M., Lane, H. W. and White, R. E. (1999). 'An Organizational Learning Framework: From Intuition to Institution'. *Academy of Management Review.* **24**(3): 522–537.

Eisenhardt, K. M. and Martin, J. A. (2000). 'Dynamic Capabilities: What are They?' *Strategic Management Journal.* **21** (10/11): 1105–1121.

Elkington, J. (1997). *Cannibals with Forks.* London: Capstone Publishing.

Elonen, S. and Artto, K. A. (2003). 'Problems in Managing Internal Development Projects in Multi-project Environments'. *International Journal of Project Management.* **21**(6): 395–402.

Flyvbjerg, B. (2006). *From Nobel Prize to Project Management: Getting Risks Right.* Research Conference 2006: New Directions in Project Management, Montreal, Canada, 16–19 July, Project Management Institute, CD-Rom: flyvbjerg_bent 1–17.

Flyvbjerg, B., Rothengatter, W. and Bruzelius, N. (2003). *Megaprojects and Risk: An Anatomy of Ambition.* New York: Cambridge University Press.

Goold, M. (1996). 'Design, Learning and Planning: A Further Observation on the Design School Debate'. *California Management Review.* 38(4): 94.

Grant, R. M. (1996). 'Toward a Knowledge-based Theory of the Firm'. *Strategic Management Journal.* **17**(Winter Issue): 109–122.

Greiner, L. E. (1998). 'Evolution and Revolution as Organizations Grow'. *Harvard Business Review.* **76**(3): 55–68.

Greiner, L. E. and Schein, V. (1988). *Power and Organisation Development.* New York: Addison-Wesley Management.

Grisham, T. (2006). Cross Cultural Leadership. Doctor of Project Management, *School of Property, Construction and Project Management.* Melbourne: RMIT.

Hamel, G. and Prahalad, C. K. (1989). 'Strategic Intent'. *Harvard Business Review.* **67**(3): 63–76.

Heneghan, T. (2004). The Architectural Review Opera Becomes Phantom. (The Saga of Sydney Opera House) (Book Review). *EMAP Architecture.* **215**(1): 95.

Ingebretsen, M. (2003). 'Win Project Battles Managers Enlist Project War Rooms'. *PMnetwork.* **17**: 26–30.

Jewell, M. and Walker, D. H. T. (2005). Community of Practice Perspective Software Management Tools: A UK Construction Company Case Study. In *Knowledge Management in the Construction Industry: A Socio-Technical Perspective.* Kazi A. S, Ed. Hershey, PA: Idea Group Publishing: 111–127.

Kanter, R. M. (1999). 'From Spare Change to Real Change'. *Harvard Business Review.* **77**(3): 122–132.

Kaplan, R. S. and Norton, D. P. (1992). 'The Balanced Scorecard – Measures that Drive Performance'. *Harvard Business Review.* **70**(1): 171–179.

Kaplan, R. S. and Norton, D. P. (1993). 'Putting the Balanced Scorecard to Work'. *Harvard Business Review.* **71**(5): 134–142.

Kaplan, R. S. and Norton, D. P. (1996). 'Using the Balanced Scorecard as a Strategic Management System'. *Harvard Business Review.* **74**(1): 75–85.

Kaplan, R. S. and Norton, D. P. (1998a). Putting the Balanced Scorecard to Work. In *Harvard Business Review on Measuring Corporate Performance.* Boston, MA: Harvard Business School Publishing: 147–181.

Kaplan, R. S. and Norton, D. P. (1998b). Using the Balanced Scorecard as a Strategic Management System. In *Harvard Business Review on Measuring Corporate Performance.* Boston, MA: Harvard Business School Publishing: 183–211.

Kendall, G. I. and Rollins, S. C. (2003). Advanced Project Portfolio Management and the PMO – Multiplying ROI at Warp Speed, Boca Raton, FL: J Ross Publishing.

Kotter, J. P. (1995). 'Leading Change – Why Transformation Efforts Fail'. *Harvard Business Review.* **73**(2): 59–67.

Kotter, J. P. (1996). *Leading Change.* Boston, MA: Harvard Business School Press.

KPMG (2003). Programme Management Survey – Why Keep Punishing Your Bottom Line?, General Report. Singapore: KPMG International Asia-Pacific: 20.

Lawrence, T. B., Mauws, M. K., Dyck, B. and Kleysen, R. F. (2005). 'The Politics of Organizational Learning: Integrating Power into the 4I Framework'. *Academy of Management Review.* **30**(1): 180–191.

Lehtoned, M. (2001). Resource Allocation and Project Portfolio Management in Pharmaceutical R&D. In *Project Portfolio Management – Strategic Management Through Projects.* Karlos Artto M. M. and Taru Aalto, Eds. Helsinki: Project Management Association of Finland: 107–140.

Lewicki, R. J., McAllister, D. J. and Bies, R. J. (1998). 'Trust and Distrust: New Relationships and Realities'. *Academy of Management Review.* **23**(3): 438–459.

Lewicki, R. J., Litterere, J. A., Minton, J. W. and Saunders, D. M. (1994). *Negotiation.* Sydney: Irwin.

Markowitz, H. M. (1991). *Portfolio Selection: Efficient Diversification of Investments.* Cambridge, MA: Blackwell.

Martino, J. P. (1995). *Research and Development Project Selection.* New York: Wiley.

Meyer, J. P. and Allen, N. J. (1991). 'A Three-Component Conceptualization of Organizational Commitment'. *Human Resource Management Review.* 1(1): 61–89.

Mintzberg, H. (1989). *Mintzberg on Management: Inside Our Strange World of Organisations.* New York: MacMillan.

Mintzberg, H. (1996). 'Reply to Michael Goold'. *California Management Review.* 38(4): 96.

Mintzberg, H., Ahlstrand, B. W. and Lampel, J. (1998). *Strategy Safari: The Complete Guide Through the Wilds of Strategic Management.* London: Financial Times/Prentice Hall.

Mintzberg, H., Pascale, R. T., Rumelt, R. P. and Goold, M. (1996). 'The Honda Effect'. *California Management Review.* 38(4): 80.

Morris, P. W. G. and Jamieson, A. (2004). *Translating Corporate Strategy into Project Strategy.* Newtown Square, PA: PMI.

Morris, P. W. G. and Pinto, J. K., Eds. (2004). *The Wiley Guide to Managing Projects.* Series The Wiley Guide to Managing Projects. New York: Wiley.

Murray, P. (2004). *The Saga of Sydney Opera House: The Dramatic Story of the Design and Construction of the Icon of Modern Australia.* New York: Spon Press.

Myers, I. B. (1962). *Introduction to Type: A Description and Application of the Theory of the Myer-Briggs Type Indicator.* Palo Alto, CA: Consulting Psychologists Press.

Nelson, R. R. (1991). 'Why Do Firms Differ, and How Does It Matter?' *Strategic Management Journal.* 12(Winter): 61–75.

Norrie, J. L. (2006). Improving Results of Project Portfolio Management in the Public Sector Using a Balanced Scorecard Approach. Doctor of Project Management, *School of Property, Construction and Project Management.* Melbourne: RMIT University.

Oliver, C. (1997). 'Sustainable Competitive Advantage: Combining Institutional and Resource-based Views.' *Strategic Management Journal.* 18(9): 697–713.

Pascale, R. T. (1984). 'Perspectives on Strategy: The Real Story Behind Honda's Success'. *California Management Review.* 26(3): 47.

Peansupap, V. and Walker, D. H. T. (2005). 'Factors Affecting ICT Diffusion: A Case Study of Three Large Australian Construction Contractors'. *Engineering Construction and Architectural Management.* 12(1): 21–37.

PMI (2003). *Organizational Project Management Maturity Model (OPM3) Knowledge Foundation.* Newtown Square, PA: PMI.

PMI (2004). *A Guide to the Project Management Body of Knowledge.* Sylva, NC: Project Management Institute.

PMI (2006). *The Standard for Portfolio Management.* Newtown Square, PA: Project Management Institute.

Porter, M. E. (1985). *Competitive Advantage: Creating and Sustaining Superior Performance.* New York: The Free Press.

Porter, M. E. (1990). *The Competitive Advantage of Nations.* New York: Free Press.

Poskela, J., Korpi-Fippula, M., Mattila, V. and Salkari, I. (2001). Project Portfolio Management Practices of a Global Telecommunications Operator. In *Project Portfolio Management – Strategic Management Through Projects*. Karlos Artto M. M., & Taru Aalto, Eds. Helsinki: Project Management Association of Finland: 81–102.

Prahalad, C. K. and Hamel, G. (1990). 'The Core Competence of the Corporation'. *Harvard Business Review*. 68(3): 79–91.

Quinn, J. B. (1992). 'The Intelligent Enterprise a New Paradigm'. *Academy of Management Executive*. 6(4): 48–64.

Rad, P. (2001). 'Is Your Organization a Candidate for Project Management Office (PMO)?' *AACE International Transactions*. PM.07.01–04.

Rad, P. F. and Levin, G. (2002). *The Advanced Project Management Office: A Comprehensive Look at Function and Implementation*. Boca Raton, FL: St Lucie Press imprint of CRC Press.

Schumpeter, J. A. (1934). *The Theory of Economic Development: An Inquiry Into Profits, Capital, Credit, Interest and the Business Cycle*. Cambridge, MA: Harvard University Press.

Shaw, G., Brown, R. and Bromiley, P. (1998). 'Strategic Stories: How 3M is Rewriting Business Planning'. *Harvard Business Review*. 76(3): 41–50.

Shenhar, A. J. and Dvir, D. (1996). 'Toward a Typological Theory of Project Management'. *Research Policy*. 25(4): 607–632.

Shenhar, A. J., Dvir, D. and Shulman, Y. (1995). 'A Two-Dimensional Taxonomy of Products and Innovations'. *Journal of Engineering and Technology Management*. 12(3): 175–200.

Snowden, D. J. (2006). 'Stories from the Frontier'. *E:CO*. 8(1): 85–88.

Spender, J.-C. (1996). 'Making Knowledge the Basis of a Dynamic Theory of the Firm'. *Strategic Management Journal*. 17(Winter Issue): 45–62.

Spender, J.-C. and Grant, R. M. (1996). 'Knowledge and the Firm'. *Strategic Management Journal*. 17(Winter Issue): 5–9.

Strogatz, S. H. (2003). *Sync The Emerging Science of Spontaneous Order*. New York: THEIA Hyperion Books.

Teece, D., Pisano, G. and Shuen, A. (1997). 'Dynamic Capabilities and Strategic Management'. *Strategic Management Journal*. 18(7): 509–533.

Turner, J. R. (1999). *The Handbook of Project-based Management: Improving the Processes for Achieving Strategic Objectives*. London: McGraw-Hill.

Turner, J. R. (2006). 'Towards a Theory of Project Management: The Nature of the Project Governance and Project Management'. *International Journal of Project Management*. 24(2): 93–95.

Turner, J. R. and Cochrane, R. A. (1993). 'The Goals and Methods Matrix: Coping with Projects With Ill-defined Goals and/or Methods of Achieving Them'. *International Journal of Project Management*. 11(2): 93–102.

Turner, J. R. and Müller, R. (2003). 'On the Nature of the Project as a Temporary Organization'. *International Journal of Project Management*. 21(3): 1–8.

van Eijnatten, F. M. and Putnik, G. D. (2004). 'Chaos, Complexity, Learning, and the Learning Organization: Towards a Chaordic Enterprise'. *The Learning Organization, MCB University Press*. 11(6): 418–429.

Weick, K. E. (1989). 'Theory Construction as Disciplined Imagination'. *Academy of Management Review*. **14**(4): 516–531.

Weick, K. E. (1995). *Sensemaking in Organizations*. Thousand Oaks, CA: Sage.

Winch, G. M. (1998). 'Zephyrs of Creative Destruction: Understanding the Management of Innovation in Construction'. *Building Research & Information*. **26**(5): 268–279.

Wind, J. Y. and Main, J. (1998). *Driving Change – How the Best Companies are Preparing for the 21st Century*. London: Kogan Page.

Chapter 6

Performance measures and project procurement

Derek H. T. Walker and Kersti Nogeste

Chapter introduction

Projects are procured to fulfil one or more project outcomes. Chapter 2 referred to a range of project types and their characteristics and a discussion of project vision and how it can be used to determine what is expected to be done and how to do it. Chapter 3 discussed the importance of stakeholders and identifying their expected project outcomes, and also how developing techniques and processes to address these should shape performance measures and provide an effective mechanism to engage with them. In Chapter 4 we discussed triple-bottom-line (TBL) concepts, ethics and governance, and the governance model presented in Figure 4.1 of that chapter illustrates how performance measures can be linked to accountability and transparency and the way that disclosure to stakeholders of how the project is to be delivered was to be performed. Chapter 5 discussed how strategy is the starting point in working out what gaps exist between a current value-generating situation and maintaining sustainable existing value or generating new value. Chapter 8 discusses how learning and innovation are dependent on feedback and consequently how project performance measures can provide crucial (feedback) knowledge to enhance an organisation's continued success.

Figure 6.1 illustrates the basis for this chapter. Project performance measures should be defined to reflect the full complement of project stakeholders' expected project outcomes – both tangible and intangible. Well-defined performance measures will comprise a combination of lag and lead indicators which satisfy audit and control requirements and also provide the basis for continuous quality improvement.

This chapter deliberately avoids detailed discussion on traditional project performance techniques relating to time, cost and quality, planning, monitoring and control because they are adequately addressed in many other texts on project control (Cleland and King, 1988; Harris and McCaffer, 1995; Turner, 1999; Cleland and Ireland, 2002). Instead, we take a more holistic view of how project performance measurement can be

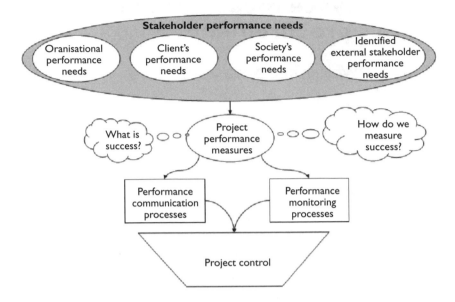

Figure 6.1 Basis of the chapter.

defined to align with stakeholders' expected project outcomes. This chapter is therefore structured as follows.

The following section discusses the concept of project success and how it may be measured. This involves a brief discussion of several tools that have emerged over the past decade or two, especially those related to a balanced scorecard approach that recognises 'hard' and 'soft' measures. The 'A more sophisticated view of success' section discusses 'intangible' outcomes in particular. The last section discusses how capability maturity models are related to project performance measures.

The prime objective of this chapter is to demonstrate how stakeholder-based definitions of project performance can be used to define project procurement performance specifications for both tangible and intangible outcomes. Outcomes include the full complement of benefits or value which is often delivered, but poorly defined, monitored and reported upon. In addition, this chapter will help readers to understand how procurement choices might incorporate performance measures that drive improved internal and supply chain team quality and innovation cultures. Encouraging and facilitating these cultures through shaping project procurement choices may then lead to improved effectiveness and efficiency in delivering value.

The concept of project success

There has been a lot of interest in the concept of project and PM success. Anton de Wit (1988) distinguishes between project success and

PM success. He states that project success is judged by the degree to which project objectives have been met. We can extend this distinction to argue that a successful project also delivers outcomes which provide the potential for benefits and that successful PM sustains or enhances the value of a project's objectives.

PM success refers to the extent to which efficiency-focused PM processes are applied. It is important to note that PM success cannot compensate for an organisation choosing the wrong problem to solve or for poor project definition and design (as indicated in Figure 1.2). The choice of the wrong problem to solve may be due to a number of factors, including the project sponsor's own personal motivations. We discussed the cognitive school of strategy in Chapter 5 which included examples provided by Flyvbjerg, Rothengatter and Bruzelius (2003) of project strategies based on poorly defined or suspect project sponsor motivations.

The concept of PM success has evolved over time. Early literature equated outputs with success, primarily in terms of time, cost and quality standards (de Wit, 1988). This then evolved to a broader definition of PM success including scope management, stakeholder management, communication management, the linking of causes and effects, and the proven ability to learn from experience (Cooke-Davies, 2002). PM success has also been defined through a process lens focused on project leadership and coordination spanning the project phases of conception, planning, production and handover (Munns and Bjeirmi, 1996). More recently the concept of PM success has evolved to include the definition and alignment of expected project outcomes and outputs, including intangible outcomes and tangible outputs (Nogeste, 2004; Nogeste and Walker, 2005). The definition and alignment of expected project outcomes and outputs will be discussed in more detail, later in this chapter. PM success has also been defined by relating an organisation's PM maturity to capability maturity models (CMMs) (Ibbs and Kwak, 2000; PMI, 2003; Cooke-Davies, 2004).

In a widely cited PM paper, Baccarini (1999) outlines how the Logical Framework Method (LFM), an approach developed in the 1970s by the American Aid Agency, may be applied to defining project success and how it can lead to the development of a balanced approach to measuring project performance. The LFM traces the links between *goal, purpose, outputs and inputs*. The project goal is understood to be the overall orientation and alignment with the organisation's strategic direction. The project purpose describes the near-term effects upon stakeholders; outputs are specific tangible results and deliverables, and inputs are the resources and activities used to deliver the outputs. Baccarini describes the LFM as a how–why logic chain where the relationship between the project objectives are transparently presented – the 'how' describing the means and the 'why', the ends. Project success is defined by starting with the project goals and asking how they can be achieved – thus generating a description of the project purpose.

Asking in turn how the purpose will be achieved generates a description of the project outputs, and again asking how the outputs will be achieved generates a description of the necessary project inputs. Working in reverse order, from project inputs through project outputs, purpose and goals, the logical basis for a project can be defined by asking the question 'why?' at each step in the logic chain.

A more sophisticated view of success

Projects may meet one or more of the following needs as outlined by de Wit (1988: 166):

1   A functional *reason d'être* such as responding to a commercial opportunity or delivering necessary infrastructure. In these cases, the project

    a   will work and should pay for itself;

    b   is justified on the basis of engineering and economic rationale and

    c   performance measures will include economic and technical, such as fitness for purpose.

2   A prestige need based on the overriding logic that the project is to boost the owner's brand image (examples being the Eiffel Tower or the Sydney Opera House – both of these may create large amounts of tangible and intangible value, simply through their existence). In these cases, the project

    a   Will be judged on political criteria.

    b   Is justified on the basis of ephemeral rationale such as pride, spiritual uplifting and the expected long-term intangible benefits from the project generating culturally transformational icons.

    c   Performance measures will most probably include political and perception-oriented criteria such as popularity, increased standing, enhanced reputation or even the increased power of sponsors and

3   A research need that is based on pure or applied research projects, such as medical research or experimental projects. In these cases, the project

    a   Will be judged according to how well it provides a platform for future pure or applied research initiatives.

    b   Is justified on the basis of reaching a solution to a complex problem that satisfies key stakeholders (even though goals and targets may be somewhat unclear at the outset of the project).

    c   Performance measures will most likely focus on the development of absorptive capacity and enhanced agility to react to new opportunities (see Chapter 8 for further discussion on this aspect).

Other categories of projects could also be added. Nevertheless, the point is that project success is closely tied to motive and need, and as Flyvbjerg *et al.* (2003) stress, motivation should be expressed as transparently as possible to help resolve any conflicts of interest at the pre-feasibility and feasibility stages of a project. Standard project performance measures of time, cost and quality should not automatically be given highest priority. Instead, project performance measures need to accurately reflect the 'true' priority of stakeholders' expected project outcomes; both tangible and intangible.

As discussed in Chapter 2 of this book, Shenhar *et al.* have undertaken much work to define different types of projects. Of particular interest is their relatively sophisticated association of the four dimensions of success with the time frame of expected results, which is based on their argument that 'project success planning should be an integrated portion of the organisation's strategic thinking and strategic management' (Shenhar *et al.*, 2001b: 719). Table 6.1 illustrates and comments on several emergent success dimensions, the way that they may be measured, and comments upon these measures. This table illustrates the presence and the degree of attainment of the highlighted emergent success dimensions together with comments on these measures (Shenhar *et al.*, 2001b: 712).

Figure 6.2, adapted from Shenhar *et al.* (2001b: 717), illustrates the association of the four (4) dimensions of success with the time frame of expected results. Dimension 1 has the short-term goal of project efficiency, Dimension 2 has the medium-term goal of customer success, Dimension 3 has the long-term goal of business success and Dimension 4 has the very long-term goal of preparing for the future. Figure 6.2 indicates a relatively high concern for project efficiency and customer impact, whilst only moderate importance is placed upon business success, with virtually no importance placed upon actively preparing for the future. Shenhar *et al.* suggest that this framework can be used as the basis for defining project performance success measures for different types of projects.

The concept of success would not be complete without understanding the role of a project vision that translates into a mission statement and explicit objectives. There are numerous well-publicised examples of failed projects. The Standish Group have reported numerous examples of failed IT projects (1994; 2003). In a recent book, the founder and chairman of the Standish Group presents 10 lessons learnt on the basis of the Group's experience and many case studies (Johnson, 2006). Two of these lessons learnt are of direct relevance to this chapter. First, that of having a clear vision and, second, that of having a well-primed project champion. One of the more spectacular project failures is the London Stock Exchange Taurus project where poor project vision led to massive scope creep and subsequent classical casebook failure, resulting in the project being cancelled after spending £500 million (Drummond, 1998). The failure of Project Taurus is attributed to a number of reasons including poor stakeholder management, and the

Table 6.1 Emergent success dimensions

Success dimension	Measures	Comments
1 Project efficiency	• Meeting schedule goal • Meeting cost goal	Goals need to be realistic. If 'stretch goals' are to be used then they need to be defined clearly along with the consequences of meeting the goals only part way
2 Impact on the customer/stakeholder	• Meeting technical specifications • Meeting functional performance • Solving a customer's problem • Fulfilling customer needs • Customer's use of the project product • Customer satisfaction • Meeting intangible needs • Meeting unarticulated needs	Many of these relate to orthodox quality measures and can be addressed using total quality management (TQM) philosophical approaches Shenhar et al. (2001b) refer to customers defining project success. However, as noted in Chapter 3, a variety of stakeholder groups can influence the perception of project success. Therefore, the relevant stakeholder groups need to be clearly identified, along with their respective success criteria Consideration also needs to be given to clearly defining expected intangible project outcomes Effort must be expended to clearly articulate as many expected outcomes as possible. Otherwise, unarticulated expectations may not be addressed
3 Business success	• Commercial success • Gaining increased market share	While commercial success is important, authors such as Shenhar et al. (2001b) ignore the situation where some organisations may choose to limit themselves to a particular niche market
4 Preparing for the future	• Developing a new technology • Creating a new product • Creating a new market	Shenhar et al. (2001b) identified that these types of measures have been relatively poorly represented in the relevant body of literature

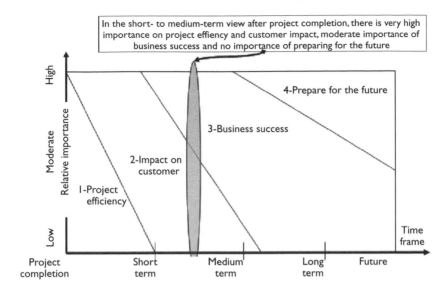

In the short- to medium-term view after project completion, there is very high importance on project effiency and customer impact, moderate importance of business success and no importance of preparing for the future

Figure 6.2 Relative importance dimensions of project success: Adapted from Shenhar *et al.* (2001b: 717) with permission from Elsevier.

inability to effectively and accurately gauge how well project objectives were being met.

A project first needs to have a clear vision of the transformation that it is expected to achieve and how it will add value. Development of a vision by interacting with stakeholders is discussed in depth in Chapter 3. What is relevant here is that a clear and unambiguous vision be developed and communicated to all project participants so that the image of success is clear in the minds of all concerned with the project. On at least one reported project, a clear image of success kept the project team and stake-holders focused upon achieving success – even when PM processes were being poorly applied (Christenson and Walker, 2003). A project vision statement can be assessed according to how easy it is to understand, whether it is motivational, inspirational, credible, promotes working smarter and specifies stretch goals (Christenson and Walker, 2003: 50). Project vision statements should be translated into specific mission statement/s that explain *how* the vision will be transformed into reality. The mission statement/s should clearly define project goals/objectives. Cascading from the vision, mission and objectives will be the detailed specification of the project composition and the methods and processes that will be used to realise the project. These methods and processes can include the definition and alignment of expected project outcomes with project outputs (Nogeste and Walker, 2005; Nogeste, 2006a).

A balanced scorecard (BSC) approach

For about 50 years it has been known that there is a need to clarify the links between a broad range of performance indicators and productivity (Likert, 1958). The value of a balanced approach to highlighting performance measures and linking these to the project's objectives and organisational vision has been more recently demonstrated on commercial projects (Norrie and Walker, 2004) and not-for-profit projects (Norrie, 2006). The principal advantage shown by using this approach was that a clear and significant cause-and-effect link between the PM methodology adopted and the goals and vision is achieved. This achieves the same level of clarity between a goal and its rationale as was noted earlier with the LFM. During the early 1990s in particular, the literature began to acknowledge the short-sightedness and lack of cause-and-effect clarity between inputs and performance. As Eccles (1991: 134) argues, 'developing a coherent, companywide grammar is particularly important in the light of an ever-more stringent competitive environment'. Eccles also highlighted how short-term (often lagging) indicators represented by financial performance measures can skew perceptions of performance. Companies need to report on leading performance indicators that illuminate the cause-and-effect links between inputs and outcomes.

Robert Kaplan and David Norton introduced the ideas of a BSC in the early 1990s. They recognised the value of leading indicators and successfully developed and operationalised these into a BSC tool that became widely known and used. The BSC comprises the following four elements (Kaplan and Norton, 1992: 72):

1   the financial perspective that poses the question 'how do we look to shareholders?';
2   the internal business perspective that poses the question 'what must we excel at?';
3   the innovation and learning perspective that poses the question 'can we continue to improve and create value?' and
4   the customer perspective that poses the question 'how do customers see us?'

A small number of critical goals are defined for each of these perspectives, along with associated measures. For example, Kaplan and Norton describe the financial perspective of a case study example as having three goals – to survive (measured by cash flow), to succeed (measured by quarterly sales growth and operating income by division) and to prosper (measured by increased market share and return on investment). In addition, the innovation and learning perspective of a case study example had four goals – *technology leadership* (measured by time to develop the next generation), *manufacturing learning* (measured by process time to maturity), *product focus* (measured by percentage of products that equal 80% of sales) and *time-to-market* (measured by new product introduction versus competition) (Kaplan and Norton, 1992: 76).

Kaplan and Norton also described how vision, strategy and a BSC could be linked, providing the example of how Rockwater (a part of Brown & Root/Halliburton) and Apple used the BSC to plan long-term performance (Kaplan and Norton, 1993). During the mid-1990s, Kaplan and Norton expanded their BSC methodology beyond objectives and measures to also include targets and initiatives. This led to the BSC becoming both a strategic and operational instrument (Kaplan and Norton, 1996), able to be used for project performance management. More recently, Kaplan and Norton have further developed the BSC into a strategy mapping tool that can be used to specify goals in terms of plans and initiatives. Kaplan and Norton's Strategy Maps have helped to further clarify cause-and-effect links (Kaplan and Norton, 2000; 2004c), which is particularly useful when seeking support for initiatives that are linked to leading performance indicators. The combination of cause-and-effect links and leading performance indicators allowed Kaplan and Norton to include the definition of intangible assets within Strategy Maps, including the use of traffic light reporting to graphically represent performance (Kaplan and Norton, 2004a).

While Kaplan and Norton were working on the whole-of-business perspective, others were applying these ideas to project environments, for example in IT and ICT projects (Stewart and Mohamed, 2001; Stewart *et al.*, 2002; 2004). Stewart (2001) describes how the BSC was applied on a case study project to evaluate project health (based on the project phases of initiation/conceptualisation, planning, implementation/ execution and closeout). In addition, Norrie (2006) describes how the BSC was used by a range of not-for-profit projects to filter project proposals and cull non-strategic projects, resulting in the realignment and reassignment of scarce resources to strategic projects.

By improving project monitoring, communication and control, Kaplan and Norton's BSC has clearly been shown to be of value to both in-house and outsourced projects.

Hypothetical example of a BSC

TheSource P/L is an organisation that helps local councils within a densely populated region of Australia to source sub-contractors, suppliers and skilled contract management staff for small-scale projects ranging from construction and maintenance that many builders or facilities management organisations may not be interested in, through to sports events, product launches, and production of marketing and community interaction communication services. The stated vision of TheSource is 'To provide local council organisations with a best-in-class level of procurement decision-making infrastructure support'.

The organisation's mission is 'To maintain knowledge and intelligence of how to deliver best-in-class procurement decision-making infrastructure support and to continue to deliver that level of service to local councils who do not have the background or resources to do so'. The organisation's key objectives are as follows:

1   to remain a sustainable and viable business so that they can deliver their mission;
2   to develop and maintain business processes to support their mission which are at least equal to best-in-class procurement support infrastructure in the best resourced PM organisations;
3   to ensure access to knowledge about best-in-class innovations that ensure that the organisation continuously learns from the experience of itself and best-in-class PM organisations and
4   to maintain a commitment to delighting customers and other stakeholders with the results of service delivery whilst also maintaining a commitment to continuous improvement.

TheSource started its journey with a handful of experienced senior managers who were retrenched during the mid-1990s from a large city council, when cost cutting demanded a drastic head-count reduction. These founding members had experience in building and maintenance, event management, IT and social service delivery. All founding members had gained tertiary-level academic qualifications and remained passionate about learning and innovation. The founding members also had a strong social conscious and strong belief in the value of the TBL. They especially wished to assist smaller councils that did not have broad business and environmental sustainability skills, to offer their communities public services and facilities that met TBL goals. The founding members of TheSource formed a collective that evolved over a two-year period into a consultancy organisation similar to the TGC organisation described in Chapter 8 (Miles and Snow, 1995; Miles *et al.*, 2005). As part of their commitment to maintain and improve their service delivery, core members of TheSource conducted a series of stakeholder workshops to investigate stakeholders' needs and identify gaps in the existing service delivery process and systems. Following the workshop, TheSource core staff developed a strategic plan which included the definition of a small number of critical objectives for each of the four BSC perspectives, along with a matrix of goals, measures, targets and corresponding initiatives. The matrix was validated by reviewing it with the most articulate stakeholders. A sample BSC is presented in brief in Figure 6.3.

When defining goals, it is important to focus on 'the critical few' (Murray and Richardson, 2000). Based on interviews with key executives from a sample of 20 organisations from large companies (some being subsidiaries of global corporations) operating in Canada, Australia and Chile, about their strategic planning and performance measurement practices, Murray and

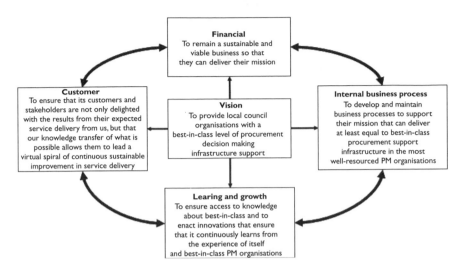

Figure 6.3 BSC showing strategy and high-level objectives.

Richardson concluded that successful strategy implementation was more likely to occur when organisations and staff concentrated on the critical few, high-impact initiatives. Their findings can be considered to be quite logical, since a long 'shopping list' of initiatives would prove to be quite distracting and overwhelming, with the potential for people to focus on what they want to do rather than necessarily tackling high-impact initiatives. In addition, in an environment comprising many concurrent initiatives, it may prove almost impossible to track the true impact of each initiative.

The aim therefore, should be to identify and analyse difficult performance issues into a consolidated list of the top 3, or at most top 5, ranked and prioritised issues. This allows the organisation to maintain focus on the critical few issues that matter most to business sustainability and more easily recognise cause-and-effect loops between the measure and a sustainable outcome. It is worth noting that the definition of 'few' is context sensitive and therefore should be determined according to the number of initiatives required to achieve an organisation's vision, mission and goals. Table 6.2 provides an example of how TheSource defined its learning and growth perspective in terms of the initiatives that would be implemented, monitored and reported upon.

As a result of their strategic planning exercise, TheSource defined 2–3 initiatives per year per BSC perspective. This was done with the expectation that as the whole group of engaged stakeholders (including TheSource staff) became familiar with this approach, the number of improvement initiatives could be increased and the measures, targets and initiatives would become better articulated and designed forming a virtuous circle of improvement.

Table 6.2 Illustration of learning and growth perspective

	Goal/Purpose	Measure	Target	Initiative
To achieve our goal, how will we sustain our ability to change and improve?	*Goal*: To engage our stakeholders in a process of feedback improvement through joint learning workshops *Purpose*: To lead to actual improvement	Number (*n*) and quality (1 = very low, 5 = very high) intensity Impact scale: *immediacy* – (1 = 1–2 years to 5 = within 1–2 months) *Improvement scope* – (1 = cherry picking easy successes to 5 = tackling fundamental effective delivery flaws)	3 cherry picking initiatives with low impact as a trial within first 6 months, with 3 high-intensive fundamental exercises in second 6 months	Cherry picking: 1 – establishing the format of details about a beta version information base on plumbers, electricians, and jobbing builders 2 – incentive system trial that helps people make sense of identified competitive advantage knowledge

Figure 6.4 illustrates the process that was followed. It must be acknowledged that TheSource was established under relatively unusual circumstances. As redundant employees from several councils, the founding members had already developed considerable social capital in the form of trust and respect, intimate customer knowledge and process knowledge gained by working for their previous employers who had now become their clients. Similarly, the founding members had developed considerable social capital with their supply chain including people they had previously contracted via their employer-councils' relatively bureaucratic procurement processes. Therefore, the founding members had a ready pool of stakeholders willing to work with them to develop and implement 'a new way of doing business'. Stakeholder discussions varied from the more formal, to the less formal and spontaneous in quasi-social settings. Because core staff had a community-centric ideology and focus, they frequently met stakeholders in social settings where the topic of conversation naturally flowed towards feedback and validation of their projects. Therefore, their combined forms of stake-holder engagement resembled an action learning research project (Peters, 1996; Peters and Smith, 1996; Coghlan, 2001; Smith, 2001; Zuber-Skerritt, 2002; Coghlan and Brannick, 2005).

After five years, TheSource occupied a niche market; being awarded many small works projects without a need to tender for work. The open book policy and open mind approach adopted by TheSource allowed their clients to satisfy all reasonable probity requirements and while the company income was not excessive (key staff earned about 20–30% more than they would have, had they remained in their council positions) the organisation had built up considerable knowledge capital. Staff members continued to maintain and develop relationships with their clients and supply chain.

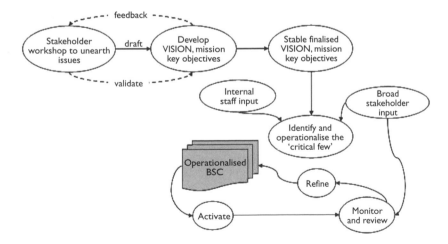

Figure 6.4 Typical BSC operationalisation cycle.

In addition, process improvement initiatives resulted in much previously tacit knowledge, being made explicit. TheSource now has a formidable lessons-learned knowledge base comprising an ICT web portal (see Chapter 7), a well-supported and ICT-enabled COP (see Chapter 8) and sophisticated process knowledge. Since its inception more than ten years ago, TheSource has re-engineered and fine-tuned its business processes and systems to a point which would be nearly impossible to imitate without relatively large investments in time, money and expertise. On this basis, the organisation's vision of business sustainability, expressed in terms of its own position in relation to its competitors has been realised. In addition, TheSource has also accumulated stores of 'goodwill' that have become tangible in terms of their being the first organisation to be contacted by many councils for smaller projects and also an increasing number of larger and more complex projects. This situation has resulted in TheSource now competing with significant PM and Facilities Management operators.

Interestingly, with their move into the sphere of larger and more complex projects, TheSource was approached by some of its 'new competitors' to franchise TheSource business model. Franchise-related negotiations currently under way indicate that TheSource is an organisation that is worth many millions of dollars.

This hypothetical case study illustrates how the BSC, when linked to business strategy and a strong upstream and downstream value chain focus, can provide a vehicle for both a quality culture and innovation culture. BSC measures can be used to make internal procurement and PM activity performance more transparent as well as more focused on sustainability. Used as a reporting tool, the BSC allows clients to gain a better appreciation of all the value elements. When designing reward systems, the BSC also allows probity requirements to be clearly addressed.

Particular aspects of this hypothetical case study lead into the next section of this chapter which focuses on the improved identification and definition of tangible and intangible project outcomes.

Fully realising value – from tangibles and intangibles

Since the early 1990s when Kaplan and Norton were developing and publishing their ideas about a BSC, there has been a veritable explosion of ideas about the true nature of value – perhaps in support of the economic rationalist debate or perhaps providing an alternative view.

In parallel with Kaplan and Norton's ongoing development of the BSC and Strategy Maps, there was a movement under way in Scandinavia and the UK. The Scandinavians had a particular interest in the generation of social capital value through cooperation, knowledge sharing and caring about human workplace issues. Karl Erik Sveiby (1997: 11) compared the visible and invisible, and tangible and intangible parts of a balance sheet. He classified invisible intangible assets into three categories: external structure (brands, customer and supplier relationships); internal structure (an organisation's management, legal and relationship structure, procedures and processes, IT, research and development initiatives, patentable ideas, and corporate knowledge or memory); and individual competence (education, experience, personal networks and so on). Sveiby described how the difference in the total share price (market) value and the book value of a firm can be partially (if not substantially) explained by intangible assets.

Other parallel efforts in Scandinavia included the development of Skandia's intellectual capital (IC) navigator (Edvinson, 1997) and the definition of intellectual capital by Roos and Roos (1997), as comprising organisational, relationship and human resources.

Another strand of related work in the UK, was generated by the accounting professions (Neeley, 2002; Neeley et al., 2002). This work was based on the growing interest in the TBL concept (Elkington, 1997) and acknowledgement that balance sheets were limited to providing information about lagging performance indicators.

Fully realising project value – from tangible and intangible outcomes

The importance of an organisation's tangible and intangible assets combined with projects being procured to implement organisational strategy leads to the need for stakeholders to align the strategic importance of tangibles and intangibles through to the project level (refer Figure 6.5). One means of doing so, is for project stakeholders to use the Outcome Profile™ template[1]

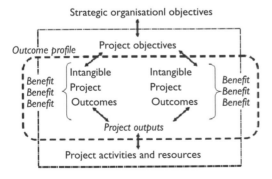

Figure 6.5 Using Outcome Profile™ templates to document the strategic alignment of tangibles and intangibles through to the project level.

to guide them through the process of defining tangible and intangible project outcomes (Nogeste, 2006b).

The Outcome Profile™ template comprises the following subheadings; each for the described purpose:

Outcome name

A commonly agreed descriptive Name for the expected outcome.

Outcome description

A clear and common description of the expected outcome.

Outcome realisation time frame

Specification of the outcome realisation time frame ensures a clear and common understanding of when the outcome can reasonably be expected to be realised – either during or after the project.

Outcome Owner

The Outcome Owner is assigned responsibility for the realisation of the outcome. If the outcome is to be realised some time after completion of the project, then it is impractical to assign the responsibility to the Project Manager.

Benefits

The Benefits of an Outcome are described in terms of the advantages provided by the outcome (Ward et al., 2004: 7) – the underlying reason/s for pursuing the outcome.

Whilst outcomes and benefits are often confused with each other (Ward et al., 2004: 8), they are different. Benefits are only able to be realised as a result of an 'observable outcome' – 'the outcome is needed for the benefit to be realised' (Ward et al., 2004: 54). For example, if an outcome of an Information Technology project is that personnel are able to do their work more quickly, freeing up time, then the ensuing benefit is 'what is actually done with the time that is freed up, since clearly if managers do not find ways to utilise the time released then no benefit will materialise' (Ward et al., 2004: 8). 'Only with the conscious intervention of managers' will an outcome yield business benefits (Ward et al., 2004: 8).

In some cases, project stakeholders may also wish to define potential disbenefits. This will help project stakeholders to agree that the potential disbenefits 'are a price worth paying to obtain the positive benefits' (Ward *et al.*, 2004: 15).

Beneficiaries

The recipients of the benefits (or disbenefits).

Success Criteria

It is important to explicitly define Success Criteria, especially, to avoid multiple and possibly contrary definitions of project success. Project stakeholders may define success in different ways (Shenhar *et al.*, 2001a: 716) by referring to different sets of data, or even when referring to the same set of data, interpret it differently, according to their particular perspective (Rad, 2003). In addition to interpreting data differently, 'the success rating of a project may also differ according to subjective, individual judgement' (Dvir *et al.*, 2002).

Success Criteria may be defined in either quantitative or qualitative terms. It is currently considered acceptable to define the Success Criteria of intangible outcomes in terms of ' "guesstimates" backed up with explanations of assumptions' (Keen and Digrius, 2003) since 'it is better to be approximately right rather than absolutely wrong' (Andriessen and Tissen, 2000). This is an approach which is in keeping with Kaplan and Norton's findings that 'even if the measures (of intangible assets) are imprecise' the simple act of attempting to gauge them 'communicates the importance of these drivers for value creation' (Kaplan and Norton, 2004b).

Outputs

Aligning an outcome with its associated outputs defines the need for the project to generate particular outputs; an approach which is consistent with the UK Treasury Department's Green Book which describes outcomes being able to be expressed in terms of outputs (HM Treasury, 2003: 13).

In addition, it is important to define which outputs are/are not within the scope of the project. For example, if a project is to generate a signed contract, the generation of a contract renewal may be an output to be delivered after completion of the project.

Dependencies

The successful realisation of an outcome, its benefits and outputs will be dependent on a number of factors that need to be clearly defined and documented as dependencies.

Risks

The successful realisation of an outcome, its benefits and outputs will be subject to a number of risks which need to be identified and assessed, along with corresponding mitigation/contingent actions which will need to be incorporated into the project plan. A good starting point for risk identification is to examine the risks associated with previously defined dependencies.

Using the Outcome Profile™ template to define expected project outcomes, benefits and outputs in business outcome vocabulary will increase the likelihood of procuring successful projects because the use of this vocabulary maintains a focus on (business) outcomes rather than (project management) processes (Dallas, 2002). This is an especially important information to provide to project managers and project teams, given the results of recent research studies which have identified that 'Project managers infrequently tie project management outcomes to corporate business outcomes' (Phelan, 2004).

The following three-step method can be used to complete the Outcome Profile™ template.

Step 1 – plan and conduct a stakeholder workshop.
Step 2 – document the workshop report.
Step 3 – use the workshop report as a key input to project planning/ review.

The purpose of each of these key steps is as follows:
Step 1 – plan and conduct a stakeholder workshop.
The purpose of the workshop is for a selected group of project stakeholders to use the Outcome Profile™ template to identify, prioritise and define expected tangible and intangible project outcomes.
Step 2 – document the workshop report.
The workshop report comprises a number of sections including each expected project outcome defined according to the Outcome Profile™ template, an outcomes/outputs cross-reference matrix (refer Figure 6.6) and any additional notes recorded during the workshop.

Output Name	Outcome 1	Outcome 2	Outcome 3	Outcome 4	Outcome n
Outputs Within the Scope of the Project					
Outputs Outside the Scope of the Project					

Figure 6.6 Outcome/output, cross-reference matrix.

The outcomes/outputs cross-reference matrix highlights the relationship between project outcomes and outputs. This cross-reference table illustrates the potential for one output to affect multiple outcomes (Department of Finance and Administration, 2003a) and prevents the situation where the relationship between outputs and outcomes is a 'matter of judgement' (Department of Finance and Administration, 2003b).

Step 3 – Use the workshop report as a key input to project planning/review.

The workshop report comprising the completed Outcome Profile™ templates, outcomes/outputs cross-reference table and any additional workshop notes are used by the project manager to plan/review the project plan.

For example,

1   The completed Outcome Profile™ templates are used to define the project scope and schedule in terms of the project outputs and the activities and resources required to generate them.
2   The outcomes/outputs cross-reference table provides a project reporting framework whereby the progress of individual outputs can be related to the delivery of related outcome/s.
3   The individual Outcome Profile™ detailed risk assessments are combined to become the basis of the project risk register.
4   Issues/Action Items identified during the workshop and documented in the workshop report become the basis of the project issues/action items register.

Whilst relatively simple in structure, this three-step process for completing the Outcome Profile™ template requires significant energy and rigour. It relies upon what Peter Senge (1990) refers to as dialogue; the dialogue between the workshop facilitator and stakeholders helps to reveal/identify, define and prioritise expected outcomes. In addition, the workshop facilitator needs to have well-developed social and relational skills. Therefore, the PM and facilitation experience of the workshop facilitator needs to be taken into account.

A key breakthrough contribution of the Outcome Profile™ template is its ability to guide stakeholders through the process of cross-referencing both tangible and intangible expected project outcomes to tangible outputs. In particular, by explicitly cross-referencing intangible outcomes to tangible outputs, all parties involved have a better appreciation and understanding of cause–effect links between outcomes, outputs and the actions and resources required to develop the corresponding outputs and deliver the expected outcomes. Causal ambiguity is reduced because cause-and-effect loops are easier to understand. This aspect is of particular relevance when considering that causal ambiguity has been identified as one of the principal reasons why best practice and other forms of knowledge transfer is 'sticky' and difficult to transfer (Szulanski, 1996; 2003).

We illustrate the use of the Outcome Profile™ template[2] with one of five action research cases used to develop and validate the Outcome Profile™ template and the three-step process (Nogeste, 2006a).

The CYPRASS project

The CYPRASS Project is a youth-oriented crime prevention project based in an Australian regional town (population 10,000) with the key objective of addressing the risk factors that lead to youth crime. The project is sponsored by a law enforcement agency and overseen by a multi-agency Management Committee.

Success of the CYPRASS project is recognised as being dependent on the delivery of both tangible and intangible outcomes. Relatively satisfied with their definition of expected tangible project outcomes, the CYPRASS Management Committee was keen to improve the definition of expected intangible project outcomes. Therefore, the Management Committee agreed to participate in a doctoral level research study focused on the improved definition and alignment of intangible outcomes and tangible outputs.

The three-step process, previously described, was applied to the CYPRASS Project as follows:

Step 1 – Plan and conduct a stakeholder workshop.

During the stakeholder workshop, members of the CYPRASS Management Committee identified, prioritised and defined the five priority intangible outcomes of

1   youth personal development;
2   networks of positive relationships;

3   cultural change within the law enforcement agency (to accommodate more of a crime prevention mindset);
4   the positive image and reputation of the CYPRASS project and
5   an improved perception of youth by the broader community.

The Management Committee was able to clearly link expected intangible outcomes to tangible project outputs, defining intangible project outcomes in tangible terms. For example, the priority intangible outcome of 'networks of positive relationships' was defined as comprising three tiers of networks; formal links, partnerships and personal relationships.

Step 2 – Document the workshop report

For the CYPRASS Project, the workshop report included an Outcome Profile™ per expected intangible project outcome, as illustrated by the following abbreviated version of the Outcome Profile™ developed for the expected outcome of *Partnerships*.

Outcome name

Partnerships

Outcome description

The purpose of the Partnership outcome is to establish and maintain formal one-to-one inter-organisational sharing links between the CYPRASS programme and other organisations, supported by formal agreements (e.g. Memorandums of Understanding) that define a shared and combined commitment to the provision of youth referral and support services in the local shire.

Partnership organisations will comprise local representation of organisations that have established formal links with CYPRASS. For example, Department of Justice.

Outcome realisation time frame

*   Short (months) to medium/long term (depending on the current status of a partnership)

Outcome owners

*   CYPRASS Management Committee
*   CYPRASS Project Officer

Benefits

1   capability to provide strategic, holistic services based on shared and combined contributions;

2   shared and combined resources, skills and experience capable of providing larger range of services and
3   CYPRASS is seen as a 'networking agent' introducing partners to each other for example could be formalised with 'partner events'.

Beneficiaries

Local at-risk youth, their families, peers and the broader community

Success criteria

Quantitative success criteria
1   number of partnerships;
2   number of partnerships that have delivered personal relationships, for example mentors;
3   increase/decrease in number of partnerships and
4   in-kind resources contributed by partners.

Qualitative success criteria
1   positive testimonials or feedback from partnership members

Outputs

1   Proforma Partnership Agreement, including mention of who/how makes public statements about the CYPRASS programme.
2   Contact List to include which Management Committee member is responsible for managing which partnership/s, including media liaison partnership/s with local media.
3   Management Committee meeting agenda includes standing items for reviewing partnership-related activities, including resource estimates for developing/maintaining partnerships.

Dependencies

1   Management Committee members' time;
2   Management Committee member organisations' support;
3   CYPRASS formal links are capable of delivering a sufficient number and diversity of partner organisations and
4   'Sufficient' number and diversity of partnering organisations.

Risks

In summary, six partnership-related risks were identified. Of these, five are medium risks and one is a Low risk. With the majority of risks being assessed as Medium, partnership-related risks need to be actively managed by the Management Committee to prevent them from becoming High risks and placing the realisation of partnership outcomes at risk.

Step 3 – Use the workshop report as a key input to project planning/review.

By cross-referencing intangible project outcomes to hitherto unplanned ('missing') tangible project outputs, the Management Committee and, more particularly the Project Officer, were able to identify why expected intangible project outcomes were not being realised. The Project Officer was then able to define the resources and activities required to develop the 'missing' project outputs. The process also enhanced the project's risk management process.

Capability maturity models (CMM) and project performance

Organisations that prefer to either pre-qualify potential tenderers, or negotiate with a preferred alliance-type partner, may find it convenient to use a tool that evaluates the organisation's maturity in delivering key tangible or intangible project benefits. For example, if innovation is a key element of a BSC that is important to the client organisation, then it may wish to benchmark the partner organisation with itself. Alternatively, it may wish to monitor its organisational partner to assist that partner to improve its organisational innovation maturity in any given BSC dimension. For example, an organisation commissioning in-house projects may wish to assess the maturity of its internal business units that deliver projects. Whatever the reason for the evaluation, the following section should be of interest.

An early initiator of the CMM concept was developed by a research team from Carnegie Mellon University in the USA supported by the software engineering industry sector (Paulk *et al.*, 1993). Paulk *et al.* (1993: 2) started this work in November 1986. The result was a five-stage model that has been widely adopted and adapted – for example, in assessing construction management processes in the Australian construction industry (CIDA, 1994) and for Knowledge Management and Organisation Learning (the K-Adv) (Walker *et al.*, 2004; Walker, 2005). The most familiar of more recent CMM tools is the PMI's Organisational Project Management Maturity Model (OPM3) (PMI, 2003). Maturity levels are generally described in five stages. *Initial* (1) leads through a disciplined approach to applying best practices to being *Repeatable* (2) Standard consistent application of these processes leads to level (3) *Defined* and once the processes become predictably and routinely applied the maturity level becomes *Managed* (4) Through continuous process improvement it reaches level (5) *Optimised* (Paulk *et al.*, 1993: 6; PMI, 2003: 28). OPM3 is being progressively deployed by organisations concerned with improving productivity and effectiveness of their PM teams.

The (K-Adv) provides us with an example of how a CMM tool might allow us to understand how organisations create competitive advantage through effective use of knowledge.

The K-Adv model envisages a knowledge competitive advantage as flowing from the organisation's ability to better manage its knowledge resources. Figure 6.7 indicates that the pivotal feature of the K-Adv is delivered by people, and only people can effectively manage knowledge so the *'people infrastructure'* is the key element of the knowledge advantage. An effective people infrastructure does not easily occur; it needs to be nurtured by an effective *'leadership infrastructure'* that facilitates and frames an effective people infrastructure because the leadership group controls access to much of the needed resources. With effective leadership in place, a supporting information and communication *(ICT) infrastructure* can be put in place.

Figure 6.8 further illustrates the three infrastructure elements of the K-Adv model. Each of the three elements has two sub-elements, and each of these can be further described in terms of critical attributes. The social capital part of the people infrastructure sub-element for example requires four further attributes – trust and commitment, knowledge creation, knowledge sharing and transfer and knowledge use and 'sensemaking'. Each of these boxes illustrated in Figure 6.8 can be used as the basis of a CMM.

The people infrastructure is the key to the K-Adv, delivering competitive advantage and value because it is this facility which, when supported by leadership and ICT infrastructures, actually delivers value through organisational learning (OL) (Walker *et al.*, 2004; Walker, 2005). In our illustration of how a CMM can be used, we focus on the attribute *'knowledge use and sensemaking'*. This is because it closely links with the BSC 'Learning and Growth' theme that we have followed in this chapter and is further complemented by discussion in Chapter 8.

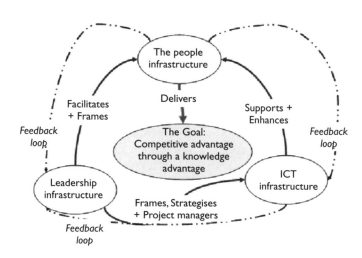

Figure 6.7 K-Adv main elements.
Source: Walker, 2005.

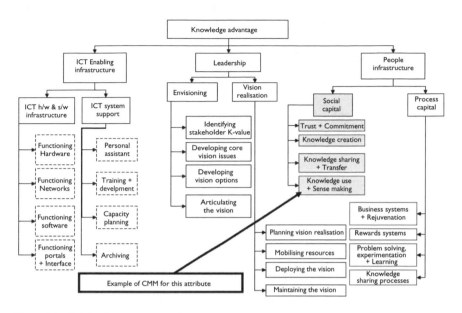

Figure 6.8 K-Adv elements and sub-elements.
Source: Walker, 2005.

The K-Adv tool can be used to facilitate performance measurement in two ways. First, through focusing upon stakeholder goals (prior to, or at the time of, procuring projects) the relevant K-Adv attribute can be evaluated to test the organisation's capability to deliver the identified necessary value. Second, if projects are being undertaken internally, the tool can be used to benchmark and strategically plan how to increase the organisation's CMM level. This process will shortly be explained and illustrated.

The capability maturity level for a specific identified attribute (e.g. for *knowledge use and sensemaking*) can be measured using a scenario or 'word picture' that describes how a particular maturity level may appear to a CMM evaluator. These word pictures can use the CMM matrix description already developed by Walker (2004), or the CMM matrices can be customised to reflect a specific learning and growth BSC initiative identified (as illustrated in Table 6.2). An example follows to illustrate the process.

We will now focus on the 'knowledge use and sensemaking' attribute (box) in Figure 6.8. Walker (2004) identified four performance characteristics for this attribute from the literature: ambiguity and creative chaos; redundancy and thinking; requisite variety; and reflection and curiosity. The key question for this attribute is 'How can we make sense of our knowledge to best use it for competitive advantage?' This can be applied to the 'ambiguity and chaos' attribute measure in terms of the answer 'by providing a demanding stretch challenge in ambiguous terms that provides creative chaos that people respond positively to'. The level of maturity is determined by the organisation through self-assessment and providing evidence that most closely fits the CMM matrix cell. Each cell provides a word-picture that approximately indicates the maturity level. These descriptors were carefully developed from the literature and validated through workshops with stakeholders but we recognise that they will need refinement, customisation and updating to reflect each organisation.

The levels have been determined by the construction of the CMM. Five levels have been broadly identified throughout the model. Level 1, Inactive AWARENESS, is evidenced by 'People seem generally uneasy and unreceptive to unconventional thinking'. Level 2, Pre-active INITIATION, is evidenced by 'Rigid rules and processes make it difficult and demotivating for people to offer creative ideas.' Level 3, Active ADOPTION, is evidenced by 'Small-scale local "skunkworks" initiatives'. Level 4, Pro-active ACCEPTANCE and ADAPTATION, is evidenced by 'The workplace culture appears chaotic with a buzz of new and conflicting ideas being debated and explored.' Level 5, Embedded ROUTINISATION and INFUSION, is evidenced by 'Top management periodically creates crises and facilitates both senior-level management to deliver challenging goals and empowers the coal-face workforce to find delivery strategies.'

This process can be repeated for each of the performance characteristics of the attribute illustrated in Table 6.3 and each attribute indicated in the summary model Figure 6.8 and for each of their identified performance indicators. For simplicity of illustration, we follow with our focus on the social capital sensemaking attribute. The process of benchmarking using the CMM follows by proceeding through each attribute of CMM in the table. The results can yield a current state situational analysis that is useful for auditing and establishing a baseline and understanding the current performance levels for 'lifting' that performance CMM level at a defined future time 'T'.

Figure 6.9 illustrates how the tool can be used for benchmarking and strategic planning. Key relevant leaders in an organisation participating in a K-Adv study would be asked which matrix cell illustrated in Figure 6.9 best describes where they consider themselves to be currently performing and where they would like to be at time 'T' in the future. Gap analysis (identifying the range between the now and preferred future state) reveals the degree of change necessary. Strategic and tactical plans can then be made based on the analysis of how to lift positions in the CMM from the current to desired position. This process can be helpful for planning what is needed and undertaking a feasibility analysis as part of preparing a business plan to instigate the desired change. It is also a useful tool for working with a partner or business unit prior to procuring PM services so that the 'Learning and Growth' part of the BSC can be addressed as part of the

Table 6.3 Sensemaking and its contribution to the K-Adv CMM table

Maturity	Performance & characteristic			
	Ambiguity and creative chaos	Redundancy and thinking	Requisite variety	Reflection & curiosity
QUESTION: How can we make sense of our knowledge to best use it for competitive advantage? Answer →	By providing a demanding stretch challenge in ambiguous terms that provides creative chaos that people respond positively to	By providing sufficient resources to deliver both time and a suitable venue to be able to think and explore mental models and hypotheses	By encouraging people to be open to a variety of views and channels of rich communication	By providing sufficient time and space for people to contemplate and reflect so that they map consequences
Inactive Awareness	People seem generally uneasy and unreceptive to unconventional thinking	The organisation pursues a lean-and-mean approach where all non-core activity has to be justified	A strict code of business determines how things are done within the organisation	Reflection and curiosity is regarded as indulging behaviours
Pre-active Initiation	Rigid rules and processes make it difficult and demotivating for people to offer creative ideas	As a by-product of keeping core ideas-people employed, some level of individual time for regeneration is possible	There is a chaotic ad hoc approach to forms of communicating innovative ideas There are no rules	Reflection and curiosity is supported in theory but in practice is viewed as wasteful
Active Adoption	Small-scale local 'skunkworks' initiatives	A formalised period of 'sabbatical' time-out is resourced through competitive proposal submission	The organisation balances chaos with rigid processes for innovation exploration	People are encouraged to be curious and to reflect but only in their personal time
Pro-active Acceptance + adaptation	The workplace culture appears chaotic with a buzz of new and conflicting ideas being debated and explored	All business units are expected to fund a set resource %age budget to enable new initiatives to emerge	The organisation supports a wide variety of forms of communicating and exploring new ideas	The organisation facilitates presentations by thought leaders to stimulate reflection and curiosity
Embedded Routinisation + Infusion	Top management periodically creates crises and facilitates both senior-level management to deliver challenging goals and empowers the coal-face workforce to find delivery strategies	The organisation sets aside a regenerative investment fund to support initiatives for emerging innovation development across the organisation	The organisation links with outside agencies in strategic alliances and rewards individuals and BUs to collaborate with multi-discipline teams and diverse groups	The organisation hosts and fully supports a corporate university that sponsor action learning internal research as well as participating in cross industry or sector research activities

Current position

People seem generally uneasy and unreceptive to unconventional thinking

Desired position
at time 'T' ▬ : ▬ : ▬ : ▬ :

The workplace culture appears chaotic with a buzz of new and confliction ideas being debated and explored

'Word pictures' describe scenarios to populate the benchmarking matrix

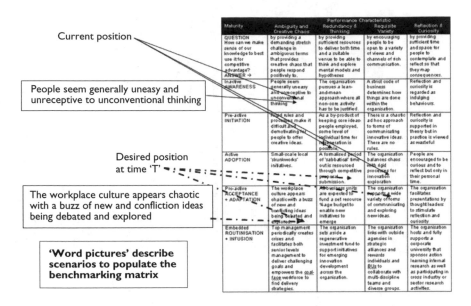

Figure 6.9 Example of benchmarking and gap analysis.
Source: Walker, 2005.

intangible outcomes from the project. This process can also be linked back into an Outcome Profile™ developed using the tools described earlier.[3]

Chapter summary

This section provides some useful examples of tools that can be used to identify and measure expected project outcomes that satisfy stakeholder needs and provide a basis for measuring project performance in terms that have been missing from much of the PM literature to date. The chapter's goal was to provide a transformational view (from the orthodox 'iron triangle' cost, time quality) of project performance. A chapter section identified project success in broader terms than delivering a project on cost, on time and to specified and explicit quality requirements. We purposely went further to include the need to develop a performance management system that meets the need of a broader level of stakeholders, and suggest that even implicit and intangible needs can be made sufficiently explicit and understood that performance measures can be developed to monitor achievement of delivering on these needs.

We have provided insights into how the two clouds illustrated in Figure 6.1 can be defined and addressed. Chapter 3 relating to stakeholder influence on project procurement delivery, clearly indicated the importance of identifying influential high-impact stakeholders whose opinion of how well the project met their needs will shape perceptions of project and PM success. In that chapter, we stressed the need for stakeholder engagement and communication to present useful and meaningful performance status reporting that not only provides a monitoring and control tool but also provides an engagement tool so that relevant joint decision making, where and when appropriate, can effectively take place.

We provided in this chapter examples of three types of tool that can be usefully employed: a BSC; an outcome–output definition and alignment tool; and a CMM tool. We argue that deployment of tools such as these should be factored into the project procurement brief so that project stakeholders and sponsors can better visualise and understand the true level of value they are receiving and can make their judgements of project and PM success on a more reasoned and informed basis.

Vignette

We provided a vignette in the hypothetical example of TheSource in its efforts to develop a BSC for its business. At this point we can add that TheSource has been toying with the idea of more formally linking many of the intangible outcomes that produced 'customer delight' to its BSC measures. It decided to do this so that it could better engage with its customers and supply chain partners and so that it could more fully identify specific (but previously inexplicit) benefits that it has delivered to further place its brand ahead of any of its competitors, especially that it is now considering franchising and thus would need to develop more easily understood procedures to be able to franchise its business idea. Further, it realised that it needed to benchmark its business units as it expanded the scale, scope and reach of the services it could deliver and the CMM appears a reasonable tool to use to do this as well as to develop through gap analysis, plans for future improvement.

Issues to ponder

1   Develop a coarse-grained CMM for the process of procurement performance measurement and try to assess where your current organisation lies in relation to TheSource.
2   Just concentrating upon the BSC for the moment, make a list of the tasks that you think would need to be completed to develop a first draft of a BSC for TheSource and make some preliminary estimates of time, resources and costs likely to be incurred.

3   Again concentrating on the BSC for TheSource, make a list of expected tangible and intangible outcomes for developing BSC benefits.
4   The tools, techniques and processes described in this chapter could present additional management overhead for any organisation, especially a small one like TheSource. To what extent do you think that the cost and management attention required to follow this path is worthwhile? Detail 3 examples.
5   What performance measurement issues would you expect might emerge for a franchisee of TheSource regarding stakeholder engagement and measurement of intangible outcomes that deliver benefits in terms of training and awareness, operational training, development and knowledge sharing?

Notes

1 Refer to www.projectexpertise.com.au/ for details on this tool and how it can be used.
2 For a fuller discussion of the following, readers should refer Nogeste, K. and Walker, D. H. T. (2005). 'Project Outcomes and Outputs – Making the Intangible Tangible.' *Measuring Business Excellence.* 9(4): 55–68. where this material was originally published in more detail than appears in this section of the chapter.
3 See note 1.

References

Andriessen, D. and Tissen, R. (2000). *Weightless Wealth*. Harlow, Great Britain: Pearson Education Limited.
Baccarini, D. (1999). 'The Logical Framework Method for Defining Project Success'. *Project Management Journal*. 30(4): 25–32.
Christenson, D. and Walker, D. H. T. (2003). *Vision as a Critical Success Factor to Project Outcomes*. The 17th World Congress on Project Management, Moscow, Russia, June 3–6: On CD-ROM.
CIDA (1994). Two Steps Forward, One Step Back – Management Practices in the Australian Construction Industry, Sydney: Construction Industry Development Agency.
Cleland, D. I. and Ireland, L. R. (2002). *Project Management: Strategic Design and Implementation*. 4th ed. New York: McGraw-Hill.
Cleland, D. I. and King, W. R. (1988). *Project Management Handbook*. 2nd ed. New York: Van Nostrand Reinhold.
Coghlan, D. (2001). 'Insider Action Research Projects: Implications for Practising Managers'. *Management Learning*. 32(1): 49–60.
Coghlan, D. and Brannick, T. (2005). *Doing Action Research in Your Own Organization*. 2nd ed. London: Sage Footprint.

Cooke-Davies, T. (2002). 'The "Real" Success Factors on Projects'. *International Journal of Project Management.* **20**(3): 185–190.

Cooke-Davies, T. (2004). Maturity and Measurement: What Are the Relevant Questions about Maturity and Metrics for a Project-based Organization to Ask, and What Do They Imply for Project Management Research? PMI Research Conference 2004, London, 12–14 July, Slevin D. P., D. I. Cleland and J. K. Pinto, PMI, **CD-ROM Disk:** 1–15.

Dallas, M. (2002). *Measuring Success.* The 5th European Project Management Conference – PMI Europe 2002, Cannes, France, Davis Langdon.

de Wit, A. (1988). 'Measurement of Project Success'. *International Journal of Project Management.* **6**(3): 164–170.

Department of Finance and Administration (2003a). Specifying Outputs, www.finance.gov.au/budgetgroup/Commonwealth_Budget_-_Overview/specifying_outputs.html, 27 January.

Department of Finance and Administration (2003b). Specifying Outputs: Case Studies, www.finance.gov.au/budgetgroup/Commonwealth_Budget_-_Overview/specifying_outputs__case_studi.html, 17 May.

Drummond, H. (1998). 'Riding a Tiger: Some Lessons of Taurus'. *Management Decision.* **36**(3): 141–146.

Dvir, D., Raz, T. and Shenhar, A. J. (2002). 'An Empirical Analysis of the Relationship between Project Planning and Project Success'. *International Journal of Project Management.* **21**: 89–95.

Eccles, R. G. (1991). 'The Performance Measurement Manifesto'. *Harvard Business Review.* **69**(1): 131–137.

Edvinson, L. (1997). 'Developing Intellectual Capital at Skandia'. *Long Range Planning, Elsevier Science Limited.* **30**(3): 366–373.

Elkington, J. (1997). *Cannibals with Forks.* London: Capstone Publishing.

Flyvbjerg, B., Rothengatter, W. and Bruzelius, N. (2003). *Megaprojects and risk: an anatomy of ambition.* New York: Cambridge University Press.

Harris, F. and McCaffer, R. (1995). *Modern Construction Management.* 4th. ed. Oxford: Blackwell Science.

HM Treasury, U. (2003). The Green Book – Appraisal and Evaluation in Central Government, http://greenbook.treasury.gov.uk/, 6 August.

Ibbs, W. C. and Kwak, Y. H. (2000). 'Assessing Project Management Maturity'. *Project Management Journal.* **31**(1): 32–43.

Johnson, J. (2006). My Life is Failure – 100 Things you Should Know to be a Successful Project Leader. West Yarmouth MA: The Standish Group International Inc.

Kaplan, R. S. and Norton, D. P. (1992). 'The Balanced Scorecard – Measures that Drive Performance.' *Harvard Business Review.* **70**(1): 171–179.

Kaplan, R. S. and Norton, D. P. (1993). 'Putting the Balanced Scorecard to Work'. *Harvard Business Review.* **71**(5): 134–142.

Kaplan, R. S. and Norton, D. P. (1996). Using the Balanced Scorecard as a Strategic Management System, 1.

Kaplan, R. S. and Norton, D. P. (2000). 'Having Trouble with Your Strategy? Then Map It.' *Harvard Business Review.* **78**(5): 167–176.

Kaplan, R. S. and Norton, D. P. (2004a). 'Measuring the Strategic Readiness of Intangible Assets'. *Harvard Business Review.* **82**(2): 52–63.

Kaplan, R. S. and Norton, D. P. (2004b). 'Measuring the Strategic Readiness of Intangible Assets'. *Harvard Business Review*. (February): 52–63.

Kaplan, R. S. and Norton, D. P. (2004c). *Strategy Maps Converting Intangible Assets into Tangible Outcomes*. Boston: Harvard Business School Publishing.

Keen, J. and Digrius, B. (2003). 'The Emotional Enigma of Intangibles'. *CIO Magazine*. (April): 104–107.

Likert, R. (1958). 'Measuring Organizational Performance'. *Harvard Business Review*. 36(2): 74–83.

Miles, R. E. and Snow, C. C. (1995). 'The New Network Firm: A Spherical Structure Built on a Human Investment Philosophy'. *Organizational Dynamics*. 23(4): 5–18.

Miles, R. E., Snow, C. C. and Miles, G. (2005). Collaborative entrepreneurship: how communities of networked firms use continuous innovation to create economic wealth. Stanford, Calif: Stanford Business Books.

Munns, A. K. and Bjeirmi, B. F. (1996). 'The Role of Project Management in Achieving Project Success'. *International Journal of Project Management*. 14(2): 81–87.

Murray, E. J. and Richardson, P. R. (2000). 'Shared Understanding of the Critical Few: a Parameter of Strategic Planning Effectiveness'. *International Journal of Performance Measurement*. 2(1/2/3): 5–14.

Neeley, A. (2002). *Business Performance Measurement – Theory and Practice*. Cambridge, UK: Cambridge University Press.

Neeley, A., Bourne, M., Mills, J., Platts, K. and Richards, H. (2002). *Strategy and Performance: Getting the Measure of Your Business*. Cambridge, UK: Cambridge University Press.

Nogeste, K. (2004). 'Increase the Likelihood of Project Success by Using a Proven Method to Identify and Define Intangible Project Outcomes'. *International Journal of Knowledge, Culture and Change Management*. 4: 915–926.

Nogeste, K. (2006a). Development of a Method to Improve the Definition and Alignment of Intangible Project Outcomes with Tangible Project Outputs. Doctor of Project Management, DPM, *Graduate School of Business*. Melbourne, RMIT.

Nogeste, K. (2006b). Development of a Method to Improve the Definition and Alignment of Intangible Project Outcomes with Tangible Project Outputs. *Graduate School of Business*. Melbourne, RMIT University.

Nogeste, K. and Walker, D. H. T. (2005). 'Project Outcomes and Outputs – Making the Intangible Tangible'. *Measuring Business Excellence*. 9(4): 55–68.

Norrie, J. L. (2006). Improving Results of Project Portfolio Management in the Public Sector Using a Balanced Scorecard Approach. Doctor of Project Management, *School of Property, Construction and Project Management*. Melbourne, RMIT University.

Norrie, J. and Walker, D. H. T. (2004). 'A Balanced Scorecard Approach to Project Management Leadership'. *Project Management Journal, PMI*, 35(4): 47–56.

Paulk, M. C., Curtis, B., Chrisses, M. B. and Weber, C. V. (1993). 'Capability Maturity Model, Version 1.1'. *IEEE Software*. 10(4): 18–27.

Peters, J. (1996). 'A Learning Organization's Syllabus'. *The Learning Organization*. 3(1): 4–10.

Peters, J. and Smith, P. A. C. (1996). 'Developing High-potential Staff – An Action Learning Approach'. *Journal of Workplace Learning.* 8(3): 6–11.

Phelan, T. M. (2004). The impact of effectiveness and efficiency on project success, Hoboken, NJ: Stevens Institute of Technology.

PMI (2003). Organizational Project Management Maturity Model (OPM3) Knowledge Foundation.Newtown Square, Pennsylvania PMI.

Rad, P. F. (2003). 'Project Success Attributes'. *Cost Engineering.* 45(4): 23–29.

Roos, G. and Roos, J. (1997). 'Measuring Your Company's Intellectual Performance'. *Long Range Planning, Elsevier Science Limited.* 30(3): 413–426.

Senge, P. M. (1990). The Fifth Discipline – The Art & Practice of the Learning Organization. Sydney, Australia Random House.

Shenhar, A. J., Dvir, D., Levy, O. and Maltz, A. (2001a). 'Project Success: A Multidimensional Strategic Concept'. *Long Range Planning: International Journal of Strategic Management.* 34(6): 699–725.

Shenhar, A. J., Dvir, D., Levy, O. and Maltz, A. C. (2001b). 'Project Success: A Multidimensional Strategic Concept'. *Long Range Planning.* 34(6): 699–725.

Smith, P. (2001). 'Action Learning and Reflective Practice in Project Environments that are Related to Leadership Development'. *Management Learning.* 32(1): 31–48.

Standish (1994). The Chaos Report (1994), Company Research Report. Dennnis, MA: 14.

Standish (2003). Latest Standish Group CHAOS Report Shows Project Success Rates Have Improved by 50%, http://www.standishgroup.com/press/article.php?id=2, 25 March.

Stewart, R. A. and Mohamed, S. (2001). 'Utilizing the Balanced Scorecard for IT/IS. Performance Evaluation in Construction'. *Journal of Construction Innovation.* 1(3): 147–163.

Stewart, R. A., Mohamed, S. and Daet, R. (2002). 'Strategic Implementation of IT/IS Projects in Construction: A Case Study'. *Automation in Construction.* 11(6): 681–694.

Stewart, R. A., Mohamed, S. and Marosszeky, M. (2004). 'An Empirical Investigation into the Link Between Information Technology Implementation Barriers and Coping Strategies in the Australian Construction Industry'. *Journal of Construction Innovation.* 4(3): 155–172.

Stewart, W. E. (2001). 'Balanced Scorecard for Projects'. *Project Management Journal.* 32(1): 38–53.

Sveiby, K. E. (1997). The New Organizational Wealth: Managing and Measuring Knowledge-based Assets. San Francisco: Berrett-Koehler Publishers, Inc.

Szulanski, G. (1996). 'Exploring Internal Stickiness: Impediments to the Transfer of Best Practice within the Firm'. *Strategic Management Journal.* 17(Winter special Issue): 27–43.

Szulanski, G. (2003). *Sticky Knowledge Barriers to Knowing in the Firm.* Thousand Oaks, CA: Sage Publications.

Turner, J. R. (1999). The Handbook of Project-based Management: Improving the Processes for Achieving Strategic Objectives. 2nd. London, UK: McGraw-Hill.

Walker, D. H. T. (2004). The Knowledge Advantage (K-Adv) Unleashing Creativity and Innovation, Unpublished report draft manuscript. Melbourne: 183.

Walker, D. H. T. (2005). *Having a Knowledge Competitive Advantage (K-Adv) A Social Capital Perspective*. Information and Knowledge Management in a Global Economy CIB W102, Lisbon, 19–20 May, Franciso L. Ribeiro, Peter D. E. Love, Colin H. Davidson, Charles O. Egbu and B. Dimitrijevic, DECivil, **1**: 13–31.

Walker, D. H. T., Wilson, A. J. and Srikanathan, G. (2004). *The Knowledge Advantage (K-Adv) For Unleashing Creativity & Innovation in the Construction Industry*. Brisbane CRC in Construction Innovation http://www.construction-innovation.info/images/pdfs/2001–004-A_Industry_Booklet.pdf.

Ward, J., Murray, P. and David, E. (2004). Benefits Management Best Practice Guidelines. Bedford, United Kingdom, School of Management, Cranfield University.

Zuber-Skerritt, O. (2002). 'A Model for Designing Action Learning and Action Research Programs'. *The Learning Organization*. 9(4): 171–179.

Chapter 7

E-business and project procurement

Derek H. T. Walker, Guillermo Aranda-Mena, Mario Arlt and Justin Stark

Chapter introduction

E-business in its variety of forms that use information communication technologies (ICT) has changed the way that business transactions can take place (Sawhney and Parikh, 2001). E-procurement has drastically altered the value-generation equation by ICT reducing transaction costs (Duyshart, 1997; Duyshart *et al.*, 2003) and facilitating internet commerce (Lawrence *et al.*, 2003). In today's market, many companies present themselves to the world via their internet site; they tender and respond to tenders using web-enabled technologies, manage and control their accounting and information exchange using electronic means, and they also use groupware internet technologies for sharing knowledge, decision making, coordination and project control. Moving from paper-based to object-oriented data models has transformed much of the procurement process and improves supply chain integration.

There has been a clear journey along a value curve over time for many companies that suggests the following trajectory: stand alone data/information processing (for internal applications); communication and information/data exchange between business units within an organisation; processing transaction data – for example, electronic data exchange (EDI) between organisations; relationship marketing processes that bind parties more closely together using customer relationship management (CRM) applications and inter-operability of data exchange; and e-business (Sharma, 2002). Also, the journey takes us from substantially paper-based to fully integrated electronic forms of information exchange, financial transactions, coordination and monitoring, and control of resources and activities. This is illustrated in Figure 7.1.

The literature presents case study examples of e-business for major companies, such as BHP Billiton,[1] one of Australia's largest companies and a major global resource company (Chan and Swatman, 2000), as well as medium-sized construction companies such as Kane Construction[2] in Australia which are typical of many organisations that undertake projects

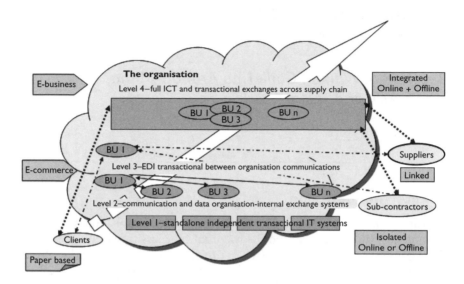

Figure 7.1 E-business trajectory.

using ICT to not only present themselves to potential customers and supply chain partners but also as part of a portal that allows them to interact with their supply chain using e-business applications. This particular organisation uses its web presence to project its corporate image and market its services. Kane Construction's web site[3] provides an example of how this project-based organisation can interact with the world as well as presenting its e-face to prospective clients and supply chain partners who may use e-business as a matter of course.

The relevance of linking e-business to procurement in a book such as this relates to general changes in procurement patterns that includes closer alignment (and hence the need for ICT interoperability) as well as business process alignment. Companies that use ICT as a 'given' may be unlikely to want to deal with organisations that have failed to develop e-business capability or might choose to exploit a superior ICT capability. The rationale is that much of the value that integrating organisations in a supply chain through common platform ICT with interoperable data provides a critical competitive advantage (Sawhney and Parikh, 2001). The nature and extent of this capability varies amongst organisations involved in projects, and so this chapter seeks to: help us understand what is meant by e-business; how it operates so that we can gain a historical perspective and better appreciate this evolving business paradigm; and to be better prepared for e-business demands in a turbulent, demanding and highly competitive global business environment. Finally, government has seen e-business as a strategic interest for procurement and has established or sponsored organisations to help develop standards and to promote e-procurement. The Australian Procurement and Construction Council Inc[4] is one example (Australian Procurement & Construction Council Inc and DOLAC, 1997).

This chapter is presented in three broad sections. A vignette at the end of the chapter presents a scenario and prompts relevant questions flowing from that scenario.

The first section deals with definitions commonly used, a brief description of the evolution of e-business with respect to project procurement, drivers and inhibitors, legal issues and issues about system compatibility (both from a business function and technology viewpoint). The purpose of this section is to familiarise readers with the main e-business issues that relate to project procurement. The second section deals with the role of ICT portals in project procurement. This ranges from technical developments such as wireless, portals and facilitating technological infrastructures to the business systems and regulations that need to harmonise with technology advances to enable e-business to be embraced and to facilitate data-mining. The systems and regulation issues that are discussed revolve around government policy, legal aspects and business process alignment issues that make effective use of advantages that e-business offers. The third section provides readers with a historical perspective of e-business for project procurement. In this section, the trends that developed in the 1990s, when e-business took off as a concept, through to the early part of this decade will be discussed. The discussion continues to the present with a perspective of current examples of how e-business has evolved. Finally, in this section, we speculate about how project procurement may be undertaken in coming years. Of course our experience of history is that change overlaps with few if any schisms. The Stone Age did not end because we ran out of stones (Lovins, 2000). Similarly, while the picture we paint from the 1990s may be familiar to some readers in 2007, others may see the 2010 speculations as short-term plans within their own workplace.

Fundamentals of procurement using e-business

The above section intimated what e-business means and a history of its development was briefly provided. Before proceeding further we will provide definitions and a more detailed history of this trend. First, definitions; there are many derivations of the term e-business that may be confusing, so a few definitions are provided to clarify what we are discussing in this chapter.

Fundamentals

Poon and Swatman (1999) describe e-commerce as sharing business information, maintaining business relationships and conducting business transactions using internet-based technologies. The focus is clearly upon using e-technologies to conduct business transactions where purchases and payments are made electronically. However, using e-commerce does not necessarily gain an organisation a competitive advantage, because it is an enabling or facilitating technology and not a solution, as pointed out by Barnes *et al.* (2004). They undertook a series of case studies of 12 e-businesses and compared how these businesses linked their e-commerce activities with their full business operations such as the front order-entry and procurement of resources area, logistics, operations management, product/service delivery and the strategic alignment of all these aspects. They adapted a 1980s four-stage maturity model of organisational strategic role and con-tribution of the operations function (Hayes and Wheelwright, 1984) to evaluate maturity of e-business. This approach helps us better understand e-business development stages and indicates where various e-business forms fit into the Figure 7.1 e-business trajectory.

Stage 1 relates to inadequate technology holding back operations. Organisations may use e-commerce applications but their technology or its integration with the logistics and delivery side may cause problems, bottlenecks and inefficiencies. For example, an e-business system may be used to generate business (sales, commitments to project or whatever) but this part of the business system does not 'talk' to the production end so delivery suffers. Stage 2 follows industry norms and best practice but as an unthinking follower so the results can be hit and miss – perhaps as good as many but not gaining any clear competitive advantage because the overall strategy has not yet been fully developed or implemented. Stage 3 may have addressed the need to integrate e-commerce with e-operations, perhaps through an enterprise resource planning system (ERP) that is internally focused to address efficiencies and thus derive competitive advantage through a cost advantage. This stage can be described as deploying e-business because business activities are e-enabled in a well-linked manner and linked to the business strategy. Stage 4 represents a more transformational stage where the integration at Stage 3 is present but it allows greater engagement with external stakeholders so that new opportunities are searched for, the linked strategy is more outward looking and leads the organisation to greater differentiation and focus on potential market niches.

Christopher Barnatt (2004) provides a useful historical account of how e-business has emerged. He traces the origins back to a fiction writer William Gibson who in a book written in 1984 (now updated (Gibson, 2004) to reflect upon current developments emerging from the fantasy of the original edition), first coined the term cyberspace to describe the consensual hallucination of an abstract electronic world that human beings could occupy across computer networks. Barnatt (2004: 80) also provides an evolutionary maturity model in four stages. In Stage 1 he discusses how in the early 1990s the electronic frontier first mooted by Gibson could be conceptualised as a business model with the advent of the concept of 'virtual' communities, businesses and teams and the increasing use of *Lotus Notes* as a groupware tool. Businesses still interacted with the client and customer using traditional technologies. During the second half of the 1990s, Stage 2 emerged with the rise and fall of the dotcom. The dotcoms used purely internet-based e-commerce technologies to undertake online business activities while traditional businesses did business with customers offline, and the term 'bricks' or 'clicks' became shorthand ways to differentiate between these two radically different business models. In the aftermath of the dotcom collapse, which was attributed to poor business sustainability issues, the Stage 3 emerged as an integration stage. This stage represents an evolution towards e-business where the online and offline customers were able to conduct business with traditional businesses that had embraced e-commerce and were integrating these technologies with their operational business processes – thus moving towards e-business. Stage 4 represents the current mid-2000s situation where many 'traditional' organisations are now using multiple integrated online and offline interfaces that customers can use to respond to a customer's preferred interaction with these companies. This stage also relates well to the Stage 4 situation described by Barnes *et al.* (2004).

We can see from this discussion, there is a range of possible ways of understanding e-business from its pure focus on e-commerce tools for interacting with a client to procure goods or services to a fully integrated suite of software tools and enabling systems and business processes that allows seamless electronically enabled procurement through delivery processes that have been strategically thought through and aligned to the delivery mode. Thus e-business can be seen to refer to the integration of e-commerce with electronically enabled business processes to deliver goods and services. E-business maturity can be measured in terms of the levels of integration of processes from procurement to delivery and its interaction with development of business strategy from reactive through to proactive.

Forms of e-commerce and e-business

We have introduced some fundamental concepts and this now allows us to understand how e-commerce and e-business may be undertaken. We can now consider some of the many e-procurement derivatives that are explained below. Interaction between customers and suppliers (and all participants within a supply chain) can be enabled through information, files and transactions being transmitted electronically by

- P2P – peer-to-peer networks where individuals deal with other individuals within a network. The company PayPal[5] has recently established a P2P payment system and there are a number of P2P file swap networks in existence.

- B2C – Business to customer. Selling online to individuals or organisations such as Amazon for books CDs and so on.
- B2G – Business to government. This is widespread in tendering and across the Globe this is becoming a common feature. For example, the Commonwealth Government of Australia,[6] whose purpose is for citizens to obtain information as well as be able to transact business and search for tender requests or to respond electronically to tenders and so on.
- C2G – Citizens to government. Many governments have extensive web sites for interacting with citizens to provide information and to receive feedback.
- C2C – Consumer to consumers. In many ways, EBay is a C2C facilitator as its business model is to allow consumers to interact as if in an open (physical) market place swapping, bartering and buying and selling through an online intermediary. Hobbyists and people interested in memorabilia also get together to interact and swap/buy, making transactions through using the e-facilitator.
- C2B – Consumer to business. This may use the reverse auction model offered by a number of organisations including airline discounters where bidders set their price limit and the intermediary facilitates subscribing businesses to decide whether to accept these bids. Some organisations have sophisticated electronic agents which allow consumers to configure their options to order items. Dell[7] is one of many that has this capacity.
- G2G – Where governments can be linked; communication between these entities can be invaluable. In Australia, like the USA, Germany and other countries with a federal and state (or province) constitution, this kind of cross-government electronic interaction is most useful. Governments also interact as mentioned earlier with business in G2B and with consumers in G2C transactions.

E-procurement drivers and inhibitors

E-procurement has created new enablers for *providing information* (about products, services, companies, corporate actions, etc.) and *communication* (audio, video and data in real-time or near-time) in addition to facilitating a *transaction* (purchase, auction or other forms) between businesses and their customers. Through computer networks (extranets – business specific interactions) and the internet in particular, businesses and consumers can obtain information cheaper and faster. In many cases they can communicate at virtually no cost[8] and transact instantaneously.

While e-commerce is typically associated with the emergence of the internet, it dates back to the facilitation of transactions in the 1970s using EDI standards for electronic invoicing. Many banks and their business partners have been extensively using electronic funds transfers since the 1980s as a fast and cost-efficient payment model.

E-procurement can entail various procurement facilities from electronic order systems (whether used in online stores or electronic exchanges), to electronic market places and desktop purchasing systems (DPS).

E-procurement yields a series of benefits:

- reduction in (administrative transaction and back-office) costs;
- improvement of process efficiency;
- shortening of order fulfilment cycle times;
- improved commercial relationships with suppliers;
- lowering of inventory levels;
- improvement in management of the supply chain and
- lower price paid for goods.

However, e-procurement is not a one-size-fits-all approach, due to the following disadvantages:

- significant increase in IT maintenance costs, ongoing management and updates, standardisation of processes between differing systems and general interoperability issues;
- significant upfront cost for enablement;
- IT security risks and
- increase in prices paid for goods is possible, especially for small order volumes.

The main driver for e-procurement is cost reduction achievable through process automation, reduced inventories, identification of and procurement at lowest market prices for commodities and the better use of economies of scale and other effects.

Five e-procurement models and their impact on projects are explained as follows: These present not only how e-commerce transactions are conducted but also the ways in which it enables the procurement transactions to trigger e-business processes for delivery of products or services.

E-sourcing

Strategic sourcing is a systematic process for reducing the total cost of externally purchased goods or services, at a defined quality level. E-sourcing is the automation of this process. It allows identifying new suppliers for a specific category of purchasing requirements using internet technology across spatial boundaries. Benefits of E-sourcing include increased decision-making flexibility and (potentially) lower prices through a broader range of suppliers. However, sales personnel are often more effective at negotiating the best deal possible with a customer than can a customer when confronting an electronic system – especially for non-commoditised offerings with many and varied options available. Also, large companies are unlikely to start undertaking business with unknown suppliers, which to them can constitute a large risk factor, though preferred supplier lists may alleviate this risk.

For the procurement of project services, e-sourcing may not be a relevant approach, due to their characteristic as a 'unique endeavour'. However, project organisations may use e-sourcing for the efficient procurement of commodities required as factor inputs on projects.

E-enabled (reverse) auctions

In a typical auction, a seller puts up an item for sale, multiple buyers bid for the item and the highest bidder will buy the goods at a price determined by the result of the bidding. In a reverse auction, a buyer issues a request for quotations to purchase a particular item. Multiple suppliers quote the price at which they are willing to supply the requested item or service. The transaction is awarded to the supplier that provided a combination of the lowest price or best service delivery (time, quality assurance, etc.).

E-enabled auctions can be found in many industries (automotive, aviation, chemical, construction, health care, food & beverage, agriculture), horizontal marketplaces and can apply to B2B, B2G, B2C and C2C markets. Global e-exchanges like Auto Exchange[9] and AUTOVIA[10] have attracted automotive supplier and customers alike, due to the commoditised nature of the products and services in this industry.

There are many claims for cost savings from e-enabled auctions: One report from the consulting group IDC indicates that 'when asking 68 purchasing managers about the amount of savings generated by using e-procurement applications in the past 12 months, 44% said they were able to reduce their costs by between 1 and 4% of their budget. Another 13% said the range was between 5 and 9%' (IDC and Pang, 2001: 3). However, e-enabled auctions are not the silver bullet of procurement: some of the savings from using reverse auctions are absorbed by higher costs related to errors in supplier data, post-auction negotiation, quality problems of new and unknown suppliers, late deliveries or non-delivery and other factors. A functioning market place for e-enabled auctions should minimise the risks by inviting only pre-qualified suppliers which meet certain criteria regarding quality and the ability to deliver.

The biggest drawback of the approach is that the diminishing buyer–supplier relationships may cause harm in the long run. Since a supplier loses what the buyer gains from a game theoretical perspective where some form of retaliation is likely to happen: add-on services, maintenance and support beyond the contractually agreed service levels and follow-up contracts are likely to be higher priced than in a traditional buyer–supplier relationship. For a project business, this insight is very relevant: if a sustainable relationship is intended, a relationship-minded procurement approach is more likely to succeed. Other project scenarios, such as a one-time transaction, may warrant using e-enabled auctions.

Web-based Enterprise Resource Planning

Web-based Enterprise Resource Planning (ERP) procurement modules use a software system that is based on internet technology to create and approve purchasing requisitions, place purchase orders and record the receipt of goods and services. It describes, essentially, the integration of ERP solutions based on SAP, Oracle, Peoplesoft, JD Edwards (both now absorbed by Oracle) or other platforms, which used to be insular solutions restricted to the company using the ERP system, with its suppliers and customers. Typically, transactions are facilitated through Electronic Data Interchange, using standards such as EDI ANSI X12, EDIFACT and most importantly the current *de facto* standard, XML, which are supported by recent versions of ERP solutions. In order to effectively and efficiently use web-based ERP, functioning ERP systems must be in place, clear buyer–supplier processes (workflows) have to be defined and implemented and a high-volume interaction between the buyer and long-term suppliers is meaningful, to recover implementation cost and realise efficiency gains. Project organisations, which meet these criteria will benefit from web-based ERP.

As a result of web-based ERP integration efforts, buyer–supplier relationships are typically strengthened, which makes this e-procurement approach suitable for relationship businesses.

In a Delphi study of 23 Dutch supply chain executives on the impact of ERP on supply chain management, that is of itself an important procurement choice consideration, reveals some interesting findings that are relevant for future e-business considerations (Akkermans *et al.*, 2003: 284). Their findings report that there are five main issues facing these supply chain executives:

1   further integration of activities between suppliers and customers across the entire supply chain;
2   on-going changes in supply chain needs and required flexibility from IT;
3   more mass customisation of products and services leading to increasing assortments while decreasing cycle times and inventories;
4   the locus of the driver's seat (the group 'in charge') of the entire supply chain and
5   supply chains consisting of several independent enterprises.

Their study revealed an only modest role for ERP in improving supply chain management due to the current (at the 2003 time) focus of ERP as being on integrating and harmonising business processes within an organisation rather than across the supply chain. This, they maintain, presents considerable inhibitors for innovation within firms; however, they flag a move towards ERP re-focusing itself on a supply chain with provision for facilitating either common standards or common interface protocols that allow different firms within a supply chain to gain advantage from information, data and knowledge flows across the chain. This interesting aspect may be more dominant at the time of publication, or shortly thereafter, of this book. The promise that ERP systems offers is that those firms which link into an ERP-enabled supply chain may be able to be both innovative and effective – they may be able to further stretch improvement boundaries while maintaining benefits of data, information and knowledge transfer. For a useful review of the history and analysis of critical factors relating to ERP, readers can refer to Al-Mashari et al. (2003). They present a model of a taxonomy for ERP critical factors (Al-Mashari et al., 2003) that is somewhat similar to other studies into the diffusion of innovation, for example, see Peansupap and Walker (2005).

The second main finding is that the panel experts saw only a modest role for ERP in improving future supply chain effectiveness and a clear risk of ERP actually limiting progress in SCM (Akkermans et al., 2003). ERP was presented as offering a positive contribution to only four of the top 12 future supply chain issues that were identified in their study: (1) more customisation of products and services; (2) more standardised processes and information; (3) the need for worldwide IT systems; and (4) greater transparency of the marketplace. Implications for subsequent research and management practice are discussed. The following key limitations of current ERP systems in providing effective SCM support emerge as the third finding from this exploratory study: (1) their insufficient extended enterprise functionality in crossing organisational boundaries; (2) their inflexibility to ever-changing supply chain needs, (3) their lack of functionality beyond managing transactions and (4) their closed and non-modular system architecture. These limitations stem from the fact that the first generation of ERP products had been designed to integrate the various operations of an individual firm. In modern SCM, however, the unit of analysis has become a network of organisations, rendering these ERP products inadequate in the new economy.

The key lesson that can be gathered from the ERP literature is that it has substantial potential to influence procurement decisions based upon e-business from integration and harmonisation that can gain strategic ICT and IT-related competitive advantage to participating organisations (and that includes the client in an outsourcing procurement arrangement or internally where projects are being managed from inside an organisation).

E-MRO

Similar to web-based ERP, e-MRO (Maintenance, Repair and Operating) enables non-product-related MRO supplies. Channels for e-MRO include direct or DPS, standardised catalogues (e.g. at buyer's site), broker intermediaries (Application Service Providers, which purchase goods on the buyer's behalf within certain price limits) and supply-side shop systems.

Like web-enabled ERP, e-MRO is suitable for processes that show high standardisation, little human agent involvement and a supply-side focus on price.

Advantages of e-MROs include the following:

- straightforward set-up for standardised purchasing processes of e-goods;
- potential realisation of significant cost savings and
- upgrade of purchasing agents to fulfil more complex and demanding tasks.

Downsides are the following:

- no personnel interaction between contracting parties;
- no discretion over quality and prices;
- no complex decision making possible;
- content rationalisation leads to potential loss of valuable product information and
- across-the-board lack of Universal Product Codes (UPC).

For project organisations, the implications are the same as for web-enabled ERP.

E-informing

E-informing describes the gathering and distributing of purchasing information both from and to internal and external parties using web technology. In this context, channels like email, brochure-ware web sites, permission marketing, online newsletters, listserv discussion groups, online chat rooms are used. Typical problems of e-informing surround reliability and relevance of information found. A number of online research, data and news services, which emerged from the market research community, have established information services about companies, their performance, feedback from customers and records of corporate action which allow much faster evaluation of existing and potential suppliers.

E-informing is critical for the project procurement process as it enables inexpensive and immediate access to information about business partners.

E-procurement implementation

This should follow general PM best practices and can be summarised as comprising five distinctive phases which should be considered for any e-procurement implementation:

Defining objectives and a strategy for the effort

As with any business improvement effort, objectives and a strategy should be defined upfront. Is the effort merely geared at cost savings? How will this change impact on the supplier relationships of the company? Which products and services should be bought/sold through e-procurement channels, and which should not? Can we achieve further process improvement and reduce the number of defects in the procurement process? These and other questions should be answered prior to pursuing the transformation effort.

Establishing a business case for gains from e-procurement

As discussed before, business benefits from e-procurement will not materialise for every procurement scenario. A quantitative business case will help identify the merits and measure their achievement during and after the improvement project.

Process re-engineering

As mentioned before, the primary benefit from e-procurement implementation may reside in the improvement of internal or buyer–supplier processes, since these processes typically involve frequent manual interaction and may be barely standardised or even defined. A re-engineering effort for these processes should be conducted, with the emphasis on optimal interaction between buyer and supplier, with information technology as a constraint. A lesson learned from e-procurement projects is that they are often assigned to a chief information officer (CIO) or chief technology officer (CTO) of a company and therefore IT-centric. Implementing existing inefficient processes into a new technology platform will only partially (or not at all) realise the potential gains from e-procurement enablement. E-procurement bares the potential to radically improve process management through workflow definition and (automated) workflow management.

Technical implementation

The implementation of e-procurement solutions, such as web-enabled ERPs can come at substantial cost. Provided the company already possesses a modern ERP system, e-procurement capabilities will be embedded or easy to add through additional modules or (web) services. If, however, an older or proprietary ERP solution is used, substantial custom development of e-procurement functionality and especially interfacing capabilities to electronic markets or the systems of the business partners may be required, which may destroy the business case for the implementation. Lastly, training of the procurement stakeholders, ongoing support and maintenance also have to be considered as part of the implementation cost. As in any IT implementation, user acceptance, component and integration testing will be required to assure a functioning platform from Day One.

Change management

Lastly, process and system changes as described bring significant change to an organisation. In order to achieve acceptance of the e-procurement solution, change management plans have to be developed and executed to achieve adaptation of the e-procurement solution by the procurement staff. Since change management always constitutes a challenge, this component of the effort can be critical to the success of the project.

Legal, ethical and security issues

Undertaking e-business transactions with partners or procuring goods or services using e-business entails considering legal and ethical obligations (though these may be complicated by jurisdiction issues). In theory, trading using e-business should be conducted under the same rules and regulations as with paper-based, word-of-mouth or telephonic systems. The main complication that e-business presents is centred on copyright, ownership of data/information and matters of censorship.

Another aspect is fraud. Fraud was mentioned in Chapter 4 and reference was made to the KPMG survey.[11] Wilson (2004: 2) argues that fraud can be understood by considering the Cressey's Fraud Triangle model that posits fraud to occur where there is opportunity, pressure and rationalisation. Pressure comes from the fraudster's motivational drives, be that greed, need, malice or ego. Rationalisation derives from the individual's ethical stance and as explained in Chapter 4 people may be moral chameleons and thus join in a fraudulent act with others rationalising it to be an act that is congruent with the norms of the group or, as is more commonly the case, it may be an individual who rationalises the act as 'pay back' at someone or an organisation for a perceived injustice or because of some other sense of the act being 'not immoral, unethical or illegal'. These two drivers depend upon the individual. The third part of the triangle relates to opportunity and that has procurement and e-business ramifications. A procurement system may afford opportunity for corrupt or unethical practices such as collusion or bribery. E-business affords the opportunity to hack into systems, manipulate transactions and commit other fraudulent acts. It is this reason, and the widening vulnerability of individuals and organisations through electronically linked devices, systems and data bases that behoves us to ensure that procurement systems design-in and enact appropriate security and protection measures; these go beyond what is legally required to what is sensible and prudent.

A further issue is the degree to which locking in a supply chain to a specific set of tools or approaches can restrict competition or collaboration. In an interesting example, while not related to e-business but sharing common issues, a study of 30 projects associated with the Taipei Mass Rapid Transit (MRT) programme showed that the government's deliberate restrictive procurement practices locked out local contractors for bidding, and when these practices ceased, costs for the same type of work dropped at an average of 26% (Wei *et al.*, 1999). The issue related to pre-qualification criteria which favoured large multi-national contractors with high levels of technical and management expertise that should have provided excellent value for money as well as knowledge transfer. However, due to political factors at the time, many global companies did not find tendering for the work attractive so that the pool of potential bidders was drastically reduced, thus leading to higher bids. A similar danger presents itself when a common set of e-business tools are required to be used that may restrict potential supply chain partners' ability to collaborate through these strict pre-qualification hurdles being in place.

Finally, there are valid concerns about the ownership and right of use of e-collected meta-information such as details about searches undertaken using search engines. Indeed, it has been speculated that the meta-data about each individual using the search engine Google currently comprises a massive (as far we know) untapped information base that can be mined for sale as marketing data for e-businesses (Carlisle, 2005). Organisations such as Boeing have already seen the value in selling meta-data derived from facilities management (FM) and maintenance feedback information to clients thus transforming itself from a manufacturer, to a services and FM organisation and then to consultant and targeted information/knowledge provider. In the final transformation it uses a range of data mining techniques to use historical repair, maintenance and other data about its customers' use of its products and services to enable it to develop new information and knowledge-based business activities (Szymczak and Walker, 2003). To enable customer data, in whatever form that takes, to be used either for project effectiveness improvement or for additional business opportunities requires acknowledgement of data-information ownership together with access and use rights. This project procurement concern should be addressed, to facilitate informed consent of knowledge and information to be shared.

Compatibility and interoperability issues – information modelling

This section discusses and explores how procurement choices can change the relationship between parties in a supply chain so that what was intended (the concept and design of a change action, product or service) is effectively translated into delivery of the expected outcome. Often in construction and manufacturing, for example, the process required by the procurement choice imposes barriers to creativity stemming from the original concept to be transformed into a product that retains the creative value envisaged. We will now explain how compatibility or incompatibility between sub-systems (i.e. interoperability issues) can enhance or inhibit the delivery of value encapsulated in the conceptual design being transformed into the change, product or service project outcome. The core issue from this perspective is: How in e-business can the supply chain use a common platform for developing design information that can be used seamlessly in the production phase? The solution to this problem that can transform the way that projects are designed and physically delivered is through a common information model. To do this, all the supply chain must be working on the basis that information is shared and automatically translated from one user to the next. This is an interoperability issue. Recent research and development work on the development of information modelling is now presented to be explain this, using the construction industry as its focus.

Building Information Modelling (BIM) is a fundamentally different approach to project design and implementation (or architecture/engineering and construction, from a building industry perspective). It can be said that the traditional way of documenting information has not fundamentally changed in the last 500 years, when symbolic language and orthographic drawing methods where used to design and construct the buildings of the Renaissance period. Orthogonal methods provided the communication means for master builders to represent compressed information. Drawing was then introduced as a method of projection in which an object is depicted on a surface using parallel lines to represent its shape onto a bi-dimensional plane. With the advent of Computer Assisted Design (CAD) in the 1980s (arguably computer-assisted drafting or drawing – but not design) a new means to produce and re-produce orthogonal drawings emerged. Although it offers clear improvements on using ink and paper, drawing as the communication medium has changed only from physically using drawing paper on drawing boards to producing digital CAD-drafted drawings. This new CAD medium has provided improvements for access, replication and sharing of building geometrical information; however, CAD drawings are still generated using straight and curved lines. Thus, while information contained in project CAD drawing documentation has become much more complex and the medium to generate and share documents has evolved, the method to represent it has not substantially changed for over five centuries.

BIM is a method to represent building geometrical information in a three-dimensional (3D) space, thus generating objects instead of orthogonal drawings. In a BIM environment, individual 3D objects also become the repositories of embedded information and attributes related to a particular building material or appliance. Completed project information or documentation can be stored in a single-centralised database or federated databases where information is saved by individual teams in specialised domains. If BIM is generated applying consistent rules and protocols BIM files can be shared across various software packages. This would include the ability to display and modify information in a range of formats according to users' roles within a project or within the organisation. Various software applications could import a single file and view/perform cost tabulations, 2D floor plans and 3D flow diagrams, also interact with performance dashboards where a range of scenarios could be explored with the client or team members.

Therefore, generating BIM in consistent ways is important as it enables the exchange of information across software applications, enabling improved design decisions and collaboration amongst project team members. The Standard for the Exchange of Product Model Data which reached the status of Draft International Standard in 1994 (STEP ISO-10303) is amongst the better known standards in industry. The STEP standard has successfully been implemented in mechanical engineering and manufacturing for over two decades now. However, this standard has proven to be less popular in the building and construction industry – perhaps as a consequence of the more casual contractual nature in which the industry operates. As a response to this, an industry-purpose standard emerged about a decade ago; this is called Industry Foundation Classes (IFCs). The IFC standard was developed by the International Alliance for Interoperability (IAI) and is defined as follows:[12] 'IFCs are data elements that represent the parts of buildings, or elements of the process, and contain the relevant information about those parts. IFCs are used by computer applications to assemble a computer readable model of the facility that contains all the information of the parts and their relationships to be shared among project participants. The project model constitutes an object-oriented database of the information shared among project participants and continues to grow as the project goes through design, construction and operation.'

The IFC has not seen its full potential so far; some impediments might include fear of losing intellectual property (IP) through sharing project information. Originally BIM data access was restricted to a common single data repository – with all its attendant trust and security implications. This is not currently the case. Data can now be locally stored and distributed and accessed according to individual project teams/roles. In this way the IFC standard for interoperability has evolved since the time when there was no internet broadband and the storage mechanisms comprised central data repositories. Trust and security issues remain as current impediments for BIM adoption because designers still fear losing intellectual property.

Since the emergence of the International Alliance of Interoperability (IAI) in 1995, various generations of IFCs have appeared including IFCs 1.0 in 1997, IFCs 2× edition 2 released in May 2003 including the more recent IFCs eXtensible Markup Language (ifcXML) that uses extensible markup language (XML) (Nisbet and Liebich, 2005). This means that the standard for data storage and interoperability is now aligned with today's web development, thus this standard is more likely to reach a wider audience. This should accelerate its pool of potential users across design, client and contracting organisations and into a wider supply chain.

The active promotion of BIM, interoperability and the IFC standard by the EU Software Copyright Directive has been instrumental in promoting BIM and interoperability at a European level. Lueders (1991: 3), in demystifying IP and security issues cites the directive[13] explaining interoperability between computing components as being 'the ability to exchange information and mutually to use the information which has been exchanged'. This does not by definition mean that each component must perform in the same way, or contain all of the same functionality, as every other one does.

In a more global perspective, the IAI has lobbied for and promoted the use and adoption of IFCs (and more recently for internet-based exchange methods IFC ifcXML and the semantic web) to improve representation and interpretation issues (Nisbet and Liebich, 2005). These may include the following:

- Automatic compilation of bills of material in digital form as input data to cost estimations and time scheduling. The quantities can then be used for production preparation and on call deliveries from the manufacturers (savings – shorter time and fewer errors);
- Automatically generated foundation for climate and energy simulation of all the spaces (savings – shorter time, better solutions and energy savings during the lifetime) and
- Fewer co-ordination errors (savings – reduced redesign and re-construction).

The European Interoperability Framework (EIF) definition identifies three aspects to Interoperable BIM (Lueders, 2005):

- Technical – linking up computer systems by agreeing on standards for presenting, collecting, exchanging, processing and transporting data;
- Semantic – ensuring that transported data shares the same meaning for link-up systems and

- Organisational – organising business processes and internal organisation structures for better exchange of data.

Examples of BIM adoption aligned with web technologies include the following:

- Singapore's *'ePlan Check'* a government-driven initiative which automates the review process for drawings and checks planning proposals more accurately and in less time approving or rejecting them. This is because parametric information is loaded into a system (i.e. Government e-planning).
- *'HITOS'* is a Norwegian governmental initiative to contextually and geographically assess building proposals in a BIM, especially on areas of urban context and geographical information systems (GIS).
- Finland also leads in the adoption of BIM and has developed a series of IFC compliant software applications including thermal and environmental analysis software such as Riuska and Ecotect.

In Australia the Cooperative Research Centre for Construction Innovation (CRC-CI) in collaboration with the Commonwealth Scientific and Industrial Research Organisation (CSIRO) has developed and tested a set of IFC-compliant BIM tools and applications. Some include *LCAdesign* for the assessment of greenhouse emissions and embedded energy assessment, especially during the selection of materials and building methods. Other CRC-CI/CSIRO application development includes the 'Contract Planning Workbench' (CRC-CI 2002-056-C)[14] which links 3D models with work schedules (also referred to as 4D models). Specifically, the construction planning workbench automates the production of schedules with visualising building site progression, especially for in-situ concrete structures. A current project includes the BIM Estimator which automates quantity take-off in bridge design for the Queensland Department of Main Roads (CRC-CI 2006-037-D)[15] and these are two amongst several IFC compliant software applications recently developed in Australia.

Other leading BIM development and implementation by private initiatives include Gehry Technologies with SOM architects who are currently implementing BIM methodologies for building design and document management. A current project includes a high rise building in Hong Kong which uses Digital Project (DP) as its BIM platform. DP is based on CATIA by Dassault Systems[16] which is a leading design software, utilised in the aerospace and automotive industries. DP is a software application which performs as CATIA's interface for architects, structural engineers, building services engineers, contractors and fabricators. SMO architects reportedly[17] have improved their relationship with the client and contractors. BIM methods are also enabling SOM to improve communication and collaboration amongst architects, engineers and team members, improving thus design decisions and collaboration amongst team members. These improvements would include the use of BIM as a construction planning and communication tool including the use of work space allocation in building sites. This is done with zoning by colour coding – for instance, work space demarcation and materials. This has direct implications for site safety, engineering communication and logistics.

A recurring question posed by most design practitioners exposed to BIM is whether BIM is worth the trouble. Interestingly, anecdotal reports indicate that BIM will not be an option in the near future but suggest that it is expected to evolve to be the method and the industry standard. The question then becomes how to accelerate its adoption by the industry at large? Impediments such as the fear of losing intellectual property, training issues and cost concerns are amongst the main concerns for design consultants to adopt BIM (see Chapters 8, 9 and 13 for more discussion on innovation and culture implications). However, according to Gallaher et al. (2004) interoperable BIM documentation will be required by most private and public client organisations. Various official reports relate to on the cost of ignoring interoperability among software systems used in the construction industry such as architectural and engineering tools, project programming, costing and scheduling tools (Gallaher et al., 2004). Their report suggests that the cost of non-interoperability is about US$15.8 billion in the US alone. Arguably, those organisations not looking to implement interoperability will have to pay a higher cost. Although there is no hard evidence to back up this statement there will be a cost for ignoring it.

BIM/IFCs compliant software includes Autodesk® Revit®, Nemetscheck® Archicad, Vector Works and Bently® TriForma®. Other standards for Interoperability include Autodesk *Design Web Format* (DWF®) which is a shareware that enables users of other CAD applications that do not offer built-in DWF publishing, such as Bentley® MicroStation® and Solidworks®, to create DWF files. The fundamental difference between BIM and standards such as DWF is in the form and nature of the data because CAD-layered drawings are not understood by BIM as data objects.

Some value propositions for the adoption of BIM are offered:

- spending more time on design and less on documentation;
- intensifying collaboration because BIM models contain information supplied and needed by all participants in a project; until recently BIM implementations mainly focused on using 3D models to improve drawing production, but the real promise of BIM lies in its application across the entire project team, especially in the area of improved building performance;
- automating 2D drawing production;
- automating links between specifications and drawings;

- improving the integrity and accuracy of information;
- improving visualisation and communication, especially for contractors, builders and other stakeholders;
- encouraging the design team to think about an integrated design practice and integrated design process, as this forces all design consultants to think about practical construction implications as they design;
- being an enabler for innovation;
- parametric planning becoming a natural interface with GIS systems;
- intensifying involvement of architects and other designers in construction as they will be modelling virtual buildings;
- improving direct control of computer-aided manufacturing (CAM) and fabrication (previously known as CAD/CAM); this has already been used successfully on the National Museum of Australia, for example (Walker *et al.*, 2003: 251–254);
- improved explorations of form resulting in more innovative yet practical design;
- modelling sustainability and
- mass customisation.

A fundamental change in industry has to happen if BIM is to work. Change can come from: Government enforcement; client organisations identifying added value through its use; the private sector identifying business opportunities through its use; or simply through general education across users and designers. Suggested value propositions for education include meeting the following needs:

- for greater understanding, among students and graduates, of the process implications of design upon construction delivery;
- for integrating design and construction courses;
- for promoting 3D thinking among students;
- for facilitating a culture of collaboration through integrated learning by the various disciplines involved in design and delivery of construction projects;
- for teaching BIM tools and applications for students to be ready to use them;
- for understanding of buildability principles;
- for improving decision-making risk/benefit assessment – define/foresee impacts;
- for introducing project management principles (4D modelling, that is integrating 3D design with project management tools) and
- for students to better understand the construction process and site logistics through simulation and visualisation.

CAD has, perhaps, been the major problem for BIM adoption as drawings have been invariably committed to a hard copy version at numerous (if not at every) stage of a construction project. BIM also supports automation of drawings and blueprints production; for example, there is no need to draw floor plans, cross sections or isometrics but only select and print various views from the BIM model. Another fundamental difference between CAD and web-enabled BIM is the ability to implement new approaches for design assessment and obtaining planning permission.

One of the main fears expressed by many design consultants is risk exposure to losing their design and intellectual property. This has also been discussed and demystified. First, access to data and project information can be restricted by password control in a similar fashion to that currently used by project intranets. Second, from a legal perspective, breaking into systems is no different to a burglar breaking into an office to steal information. Third, BIM and interoperability are not synonymous with cloning – interoperability means that building components, which may differ in functionality, can share information and use that information to function accordingly.

Researchers who have studied this area of interest for over a decade[18] suggest that the key to improve construction productivity relies upon its ability to operate more efficiently and effectively with less information waste and more knowledge sharing. Information being transposed as is currently often the case, even with CAD, results in the potential for miscoding, mistaken interpretation and other errors of transposition. In addition, needless multiple handling of information and data is a waste of time and energy. Similarly, as has been discussed in Chapters 8 and 13, knowledge sharing to achieve innovation is best facilitated through collaborative problem solving – BIM facilitates this. In relation to electronic information management, especially in the form of web-enabled building information modelling, this means that managing and coordinating, not only blue prints, but all documentation and project information from design to property maintenance phases. This presents a unique and new opportunity to diverse industries such as property, construction and project management. It is expected that electronic information management can provide commercial efficiencies through more effective project procurement and supply chain management. This is fundamentally different to CAD, even web-enabled CAD (*Design Web Format* DWF®).

The internet-enabled BIM, as a network-enabled environment, remains in its infancy, but we will see the situation swiftly change in the near future. This principle is now giving rise to many new services previously tied to the physical value chain and has resulted in completely new value chains being created. These include online BIM building product catalogues enabling engineers and architects to build BIM models and run simulation and visualisation scenarios. It also facilitates prototyping and codes and specification checking. Internet developments are offering new means for project collaboration, not just in collocated groups but also in virtual global team-working and supply chain integration in a truly e-commerce environment.

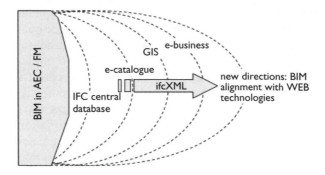

Figure 7.2 BIM diffusion.

Figure 7.2 illustrates how the BIM concept in the Architect/Engineer/
Construction (AEC) and FM professional group has been evolving and
diffusing over time. It indicates BIM evolving from initial IFC central
databases to e-catalogues, then into GIS into e-business as ifcXML
facilitates opportunities for the expansion of BIM capabilities using web
technologies.

To conclude, we argue that the IFCs and the Standard for the Exchange
of Product model data (STEP – ISO) are the most promising standards for
interoperability in the property and construction sectors. The move towards
ifcXML aligns with major web development initiatives and we envisage
that ifcXML will become one of the main enablers for its speedy adoption
and dissemination.

For those interested in learning more about BIM model viewers you can
link to

- IAI Karlsruhe, Germany at www.iai.fzk.de/ifc
- TNO, Delft, Holland, at www.ifcbrowser.com/downloads/BETA/IFC
  Engine Viewer.exe
- DDS, Oslo, Norway at ftp://ftp.dds.no/pub/install/IfcViewer/
  DDSIFCViewer.exe

For further information about interoperability, readers can access the IAI at
www.iai-na.org

The role of IT/ICT portals in procurement

Perhaps one of the most practical developments over the recent years is the
way that web portals have dominated the user interface between system and
user, and user and user. During 1997 and 1998 research into use of ICT by

major construction industry contractors and consultants in Australia and Hong Kong revealed that most parties were only somewhat aware (having heard of or read about) intranets, but very aware use frequently as a matter of course the internet (Walker and Rowlinson, 1999: chapter 8). Now, moving on a decade, they appear ubiquitous for many if not most commercial, government and not-for profit organisations – Arthur Tatnall reports that when undertaking a Google search for the word 'portal', it revealed 35.6 million hits. He explains the term from its archaic roots meaning 'an entrance or gateway' and he discusses nine types of portals (Tatnall, 2004: 3–7) that we expand upon:

- Mega portals that link users to other portals that can be highly extensive. There seems to be a trend for much cross indexing of web links between organisations so that most portals now share this extensive means of linking those who conduct searches or seek information to relevant portal links.
- Vertical industry portals that are based around specific players, such as the construction procurement portal SITE,[19] also these can be part of a value chain gaining access to special areas of business transaction through an extranet of linked internet web sites.
- Horizontal portals when used by a broad base of users across a horizontal market – for example, a site that links a range of businesses in a physical location to promote e-commerce such as Buzewest as described in Pliaskin and Tatnall (2005); a wider scale network portal of this type would be Vicnet[20] the Victorian (Australia) web site that aims to link a state-wide range of businesses as well as other useful community information.
- Community portals often set up by community groups to foster a specific interest – for example, one portal, communitybuilders.nsw,[21] is a site that helps people in communities to set up community activities, explains how to encourage volunteering and provides a range of access to electronic resources and services that do not quite fall into a for-profit activity. These sites might also link to media outlets such as radio or print, or may contain blogs.
- Enterprise information portals (EIP) that are often gateways to a corporation contain the sort of information that interested parties can download such as annual reports, newsletters, an email access point for further information and so on. The global construction group Bovis Lend Lease provides a good example[22] where a whole series of information, particularly under their 'publications' or 'newsroom' areas, is readily available. Apart from immediate responsiveness from requesting a download (annual report, for example) the saving to both reader and host corporation is considerable as it eliminates the need for staff producing and packing off hard copy materials.

- E-marketplace portals, as described earlier in this chapter. Pliaskin and Tatnall (2005: 6) provide an example of the Swiss company ETA SA Fabrique d'Ebauches[23] that manufactures watches as a group of individual companies and presents itself through this portal as a united front to customers as well as its suppliers.
- Personal/Mobile portals, this is a wide and expanding area of development. Pliaskin and Tatnall (2005: 6) note that this technology is being embedded in mobile phone technology. That book, published in 2005, indicates how quickly things have moved as now we see Blackberry™ and many major mobile phone manufacturers offering web through phone, video messaging and a host of services. Some of the more remarkable developments include taking mobile phone technology and personal data assistants (PDAs) and linking them to home appliances such as a robot dog that can wander throughout an area like a watchdog, sense danger (in terms of traces of smoke etc.) and be accessed entirely by a mobile phone portal.[24] Clearly this has major project procurement implications as more integrated and e-connected components restrict the number of providers and force many organisations into highly relationship-based transactions.
- Information portals are moving beyond the printed media, offering e-mags and e-newspapers.[25] The more widespread use of digital ration and TV streaming as well as podcasting has revolutionised many media outlets. The Australian Broadcasting Corporation (ABC) Radio National (RN) has provided transcripts of programmes for several years[26] and they have extended this to listening to programmes that have been set up for streaming as well as automatic links between iPods and the ABC RN web site so that as soon as you link your iPod to the internet it automatically downloads programmes that you have specified as being made available to you. With a trend towards organisations in the supply chain feeling the need to provide current information and communicate 24/7, this technology is most likely to move from being generated by media outlets and also many other organisations.
- The final portal type identified by Pliaskin and Tatnall (2005: 7) is specialised/niche portals that cater for special interest groups. This type of portal merges with the notion of communities of practice (COPs) and ICT tools that support these. An example of one of these is the UK construction industry using the product Sigma Connect that connects the workforce within an organisation or across a range or organisations via the portal to share knowledge and maintain social contact (Jewell and Walker, 2005).

The role and functioning of portals is rapidly changing with new ideas, products and services coming on stream constantly. All of these offer opportunities to rethink how the entire communication aspect of project procurement will function. In Chapter 3 we discussed stakeholders and their influence upon projects. Maintaining contact with stakeholders to provide information and data about projects is already becoming a key competence of PM (PMI, 2003; 2004). Using portals to do this is the next logical step.

E-procurement trends and emerging themes

Organisations have recognised the benefit of e-procurement and spent significant efforts and funds on implementing e-procurement solutions. It is foreseeable that electronic procurement will become 'business as usual' and the term e-procurement will disappear.

Over the past 5–10 years, e-procurement was mostly focused on e-sourcing, achieving cost savings due to the achievement of greater information efficiency and better use of economies of scale. Greater information efficiency, since potential suppliers beyond spatial boundaries could be sought and finding the 'best price' for a certain commodity required less time and search cost.

Similar to B2C price search mechanisms, such as Pricegrabber, Froogle, Epinions and Shopping.com, pricing information for businesses has overall become more transparent. Similar to the B2C search engines, besides net price, additional information on product availability and shipping costs as well as customer satisfaction feedback may be available. Industry portals typically provide contact information, and in some cases, directly allow price queries. There are, however, barriers to the adaptation of the B2C model: often goods are not commodities, but highly customised and therefore not easily comparable. Up-front pricing for custom-engineered goods continues to require a request for proposal (RFP) process in order to define the requirements, which leads to a certain price for a certain scenario. Further complexity emerges from taxes, tariffs and export restrictions in international transactions, which cannot always be anticipated and may significantly impact on cost, scope and timeline for delivery.

A better use of economies of scale through centralised procurement in large corporations or government entities or purchasing alliances for smaller businesses, which allowed generating economies of scale, similar to larger entities.

One example for such use of economies of scale is Marketplace@ Novation, a purchasing alliance of approximately 2,500 health care organizations in the US. According to company sources (www.novationco.com/about/fact_sheet.asp) Novation maintains agreements with over 500 suppliers and distributors of medical, laboratory and safety equipment, capital equipment and services, office supplies and other goods and services. The combined purchasing power of this e-sourcing provider amounts to US $25bn in annual purchases. According to an early study in

2000, the initial 31 hospitals to join this alliance saved approximately $12m annually, a substantial amount of their procurement-related expenses, which spurred rapid growth of the service.

Three major trends could further propel e-procurement:

- Seamless Process Integration;
- Further Reduction of traditional procurement activities and
- Comprehensive Decision Support.

Seamless process integration and process outsourcing

While e-procurement efforts over the past 10 years have focused on purchase transactions, these transactions need to be integrated into the larger picture of the end-to-end supply chain.

For example, a radio frequency identification (RFID)-enabled inventory management system in a warehouse may enable a more responsive and highly automated procurement process. Shelf-mounted readers will detect the removal of a palette containing boxes of a product and report the reduction in inventory to the procurement system. Based on this real-time inventory data and assumptions of future demand, which may result from buying trends, developed from historical and seasonal information, commodities can now be ordered automatically, based on rules defined by the procurement officer. Such rules can look as follows: 'Procure x amount of product y to be delivered no later than on date z at the lowest possible price point, using a preferred vendor list.' Such automation of the actual procurement process allows procurement officers to focus on high-value activities, such as the definition and validation of procurement strategies. Even greater precision will be achieved with a more granular tracking of commodities: as costs for RFID tags and readers continue to decline and large wholesalers and retailers (e.g. Wal-Mart) aggressively push their suppliers towards the adoption of RFID, box- and item-level tagging will allow the most accurate and timely prognosis of product demand.

Other applications, where RFID tags are integrated into the product, will also allow triggering early procurement. For example, tags which monitor wear and tear on mechanical equipment will permit more precise maintenance forecasts; a machine operator would identify early wear on machine tools or motors; a fleet operator could more accurately predict the need for tyre replacements (and orders) based on actual wear rather than mileage-based estimates.

Such a seamless process integration, where human intervention is limited to exception cases and process control will allow further increases in savings from e-procurement, since:

- standardised procurement tasks can be further automated;

- procurement specialists can focus on

  - the definition and implementation of procurement strategies rather than order execution. Such strategies include vendor and product/service selection, standards scenarios for procurement, vendor due diligence and market analysis;
  - exceptional cases, which require special attention, due to large orders, 'unusual requirements', orders which are critical to the success of the business and so on and

- order processes can be accelerated, which in return allows lower inventories and therefore further reduced inventory cost.

From a technical perspective, such process integration is catalysed by web services, which allow the integration among heterogeneous and distributed systems. Due to the nature of web services, which are independent from programming languages, hardware platforms and operating systems, true interoperability can be achieved.

Further reduction of traditional procurement activities

If procurement is an auxiliary rather than a core process, for an organisation, the entire process may be outsourced and managed by a service provider with speciality expertise in e-procurement (see: The Next Level of E-Procurement, Laura Powell, *Public Utilities Fortnightly*; Nov. 2006; 144, 11; 20–21). Besides the core expertise in e-procurement, such a service provider can offer two additional cost-related benefits: (a) through providing additional economies of scale, for example, by establishing eMROs and (b) by executing remaining manual procurement activities in low-cost locations off-shore.

Another change related to traditional procurement activities is the replacement of human negotiating processes through e-negotiations. E-negotiations are executed through electronic agents, who exchange offers (quotes), acquire market information and subsequently engage in negotiations towards an acceptable agreement for detail on e-negotiation strategies.[27] Replacing the human negotiating processes with e-negotiations represents another change to traditional procurement practice.

Comprehensive decision support

Lastly, e-procurement will further accelerate information efficiency of markets, allow better analysis of supplier satisfaction, market prices, terms and conditions and other procurement parameters. In return, this information will allow further fine-tuning of procurement decisions, whether executed by procurement specialists or through electronic agents.

As stated before, procurement specialists will be less involved in transactional processes around the order execution, but rather focus on strategic procurement aspects, such as selection of preferred vendors, default procurement methods and negotiation strategies and so on.

Successful e-procurement initiatives will hinge on the ability to build scalable and adaptable solutions, which will grow and develop with changing business needs and allow for new procurement mechanisms, which cannot be foreseen at this point. In this regard, process maturity and the ability to use ICT effectively and efficiently can become a critical business success factor.

Lastly, decision making will be assisted by continued development of e-tools that measure, monitor and compare procurement options. For example, the construction industry uses a 'bill of quantities' that is a schedule of all items comprising the project scope to assist in cost management at the design stage. This will be extended to provide similar information on project scope in terms of water use impact, embodied energy and CO_2 footprint impact.

Chapter summary

In this chapter we began with Figure 7.1 which illustrated how many organisations are moving along a trajectory of increasing integration of common information, knowledge and data. The current term that most people recognise as describing this process is e-business, which has moved beyond the transfer of financial data and information such as invoices, bills, payments and so on to the transfer of complex models of the project outcome that can be viewed by all participants in a way that suits their needs.

We explained some of the fundamental terms in current use that describe the various forms of e-commerce which is an important part of e-business. We discussed drivers and inhibitors of e-business and also discussed legal, security and ethical issues. The movement of organisations, in supply and value chains, along the e-business maturity continuum led us to discuss an important barrier that is being overcome through great research and development effort into interoperability so that a project can reduce information and wasted management effort caused through re-creating information models at each step in a supply chain. We devoted a substantial part of the chapter to interoperability issues. This is because a common platform for representing an electronic form of project design (that can be presented to supply chain parties in the form, shape and structure that best suits their use of the information) provides an exciting opportunity to substantially eliminate current information duplication and waste practices. It also offers many improvements in visualisation and simulation modelling. A BIM can be used to measure a project's environmental 'footprint'. This could provide a powerful e-business application to facilitate development of the kind of instruments and decision support tools able to determine the basis for a carbon trading currency that (as was mentioned in Chapter 4, carbon emission control is an important emerging triple-bottom-line issue) will need to be addressed in making project procurement choices.

The section on the role of IT/ICT portals linked previous concepts to a tool, portals that allow a convenient interface between users and the various e-business applications. These will be and are, governed by the design of security and access designed by portal developers. Monitoring and tracing user preferences and search patterns can help such systems to customise the appearance of the portals to fit the user's habits as we experience when using standard packages such MS Word which can 'remember' the latest nine documents accessed, for example. However, as was indicated in the section on legal, ethical and security issues, users continually leave a trace of data on preferences and search history that can be mined to target and shape business opportunities. This aspect of business development may not be welcomed by users – customers or supply chain participants. This is because it could represent an infringement of privacy or cooption of intellectual property relating to user habits and preferences, and so procurement systems need to address this important and potentially contentious side effect of e-business applications. This section offered a natural prompt for us to discuss future and emerging e-business trends.

The key theme to this book is project procurement, and a recurring element has been value that can be realised through collaboration and information and knowledge sharing and use. We have endeavoured to discuss the wide and disparate topic of e-business with a focus on how it can be used in a project procurement context. Our research into this interesting and evolving topic area leads us to believe that e-business can enable improved procurement systems; however, the implications for PM and all in the supply chain are that participants will need to be ICT literate and will also need to accommodate fears of sharing data, information and knowledge, so that supply chain management becomes value chain management linked to a procurement choice that demands participants are linked through protocols that allow this to happen.

Vignette

The election of 2010 devastated the government of Howard Shrub, president of 'Emuland', as satirists referred to the country whose government had steadfastly refused to acknowledge rising electoral concerns about climate change, sustainability and offshore outsourcing, that had left many companies hollowed out shells with small numbers of very highly paid senior managers based in regional headquarters and hoards of resentful ex-employees. Many of the recently unemployed had been forced to down-scale their expectations of a career by taking less attractive jobs when changing employment. Their vote was their sole legal remaining means of protest. Shrub retired to his ranch, while many of his erstwhile government colleagues sought refuge in returning to their electoral homes as quietly as possible. The new government, elected with a massive majority based on a platform of radical change, set about trying to implement its political agenda through a series of innovative infrastructure projects. The Public/Private Partnership (PPP) approach was still deemed viable by the new President Greenyard administration but it was considered to need a radical overhaul. A few pilot demonstration projects had been mooted as an innovative way forward, validated by the election promises made in the winning campaign that was still fresh in everyone's minds.

One of the programme of projects that Greenyard announced was the expansion of an integrated public transport system in Sunburne, the capital city of one of the southern states. The vision for that programme was not only to improve transport but also to move the nation towards complete energy self-sufficiency. This was to be a 15-year programme of integrated projects to change the transport culture of the millions of citizens who had overwhelmingly voted for policy transformation. For several years, an increasing number of them had been using gasoline-electric hybrid cars. Also, these commuters had the capacity to use their residential solar panels to power their cars, and home, and feed back surplus energy into the power grid.

A new 200 kilometre mid-city rail beltway was to be constructed to link the spoke-and-hub rail line configuration that currently existed. The beltway was to use an above-ground electric monorail system. Fourteen new stations were to be constructed with park-and-ride facilities for commuters to reach stations by bike or car, the cars having docking stations where they could be recharged. Those participating in residential energy co-generation to feed the electricity power grid could gain benefit from an electronic metering system that calculated debits and credits for parking, offset by power generation from electricity co-generation.

A bill was quickly passed by Greenyard's government affecting construction of the stations. The bill required permits for all new buildings, to be issued only when the owners provided sustainability plans that projected each one's carbon emission 'footprint', specified counter measures, and had an electricity co-generation implementation plan. Generous credits were indicated as an incentive and increasingly severe penalties for non-compliance were proposed over a 10-year period. A further requirement imposed upon those parties in any supply chain that would be involved in the pilot projects (and inferred for most future projects) was that all project design documentation would be based upon a commonly accessible BIM and that a new project portal development was part of the 'project' to link not only the supply chain for the pilot project delivery, but also citizens to be able to gain information from government, and feed back comments and suggestions as it was being developed, delivered and operated.

This programme was linked to an industry renewal strategy that aimed to turn 'Emuland' into a highly e-business and ICT literate nation. A significant barrier to participation in any government project was lack of ability to work to the new procurement policies. The industry renewal policy aimed to refresh the nation's capacity, motivation and ability to develop solar energy farming technologies. It had once been a world leader before the Howard Shrub government had ignored supporting or nurturing this sector, preferring to rely on off-shore solar and alternative energy developments. An education and training renewal strategy was aimed at providing the necessary support to improve the skills of citizens and business.

The pilot projects, particularly the Sunburne Public Rail Transport Beltway project, were recognised as very expensive but valuable learning exercises to trigger ambitious innovation on a scale never attempted before. With an overwhelming political mandate and a realisation that a step change was desperately needed, Greenyard anticipated that his government had at least two, if not three or more, terms of office to clearly demonstrate positive results and radical transformation. Procurement policies held a pivotal place in Greenyard's government strategies.

Issues to ponder

1   Provide examples of three forms of e-business that is inferred in this vignette.
2   Discuss the three most important drivers and barriers to the proposed adoption of e-business as indicated in the programme.
3   The PPP approach would tend to severely narrow the field of potential participants in such an ambitious and bold programme. Provide three examples and discuss their implications of legal, ethical and security issues that you feel may challenge the development of this project.
4   The above project assumes that a BIM can be provided to link project participants. Discuss what you consider are the three most

value-adding advantages that it may produce. Also, highlight three challenges to its use and how these may be mitigated.

5   The project has an extremely long time horizon for its implementation. Discuss the most important (in your opinion) example of emerging trends that may significantly impact upon this vignette programme in terms of FM, collaborative decision making and the balance of on-shore and off-shore supply chain participation.

Notes

1 See www.bhpbilliton.com/bb/home/home.jsp accessed on 27 May 2006.
2 See www.kaneconstructions.com.au/ accessed on 27 May 2006.
3 See note 2.
4 See www.apcc.gov.au/ accessed on 27 May 2006.
5 See www.paypal.com.au/au accessed on 27 May 2006.
6 See www.australia.gov.au/ accessed on 27 May 2006.
7 See www1.ap.dell.com/content/default.aspx?c = au&l = en&s = gen accessed on 27 May 2006.
8 However the cost for establishing the infrastructure for e-business may also be quite expensive, because while the transaction cost might be very low, it is not zero (e.g., procuring market data etc.).
9 See www.autoexchange.com/ accessed on 27 May 2006.
10 See www.autoviauk.com/index.htm accessed on 27 May 2006.
11 KPMG's bi-annual Forensic Fraud survey www.cgs.co.nz/files/Fraud-Survey-2004.pdf accessed on 27 May 2006.
12 Source: www.iai-na.org/technical/faqs.php IAI accessed 20 January 2007.
13 See http://europa.eu.int/smartapi/cgi/sga_doc?smartapi!celexapi!prod!CELEX numdoc&numdoc = 31991L0250&model = guichett&lg = en accessed on 27 May 2006.
14 See CRC-CI 2002-056-C. Cooperative Research Centre for Construction Innovation. Contract Planning Workbench www.construction-innovation.info/index.php?id = 32 accessed on 27 May 2006.
15 See CRC-CI 2006-037-D. Cooperative Research Centre for Construction Innovation Automated Estimator Commercialisation www.construction-innovation.info/ accessed on 27 May 2006.
16 See www.3ds.com/products-solutions/plm-solutions/catia/overview/ accessed on 27 May 2006.
17 See GT 2006 Gehry Technologies – Digital Project. www.gehrytechnologies.com
18 For further details on standards, many papers and reports see www.iai-international.org/Resources/Related_PublicationsSources.html accessed on 27 May 2006.
19 See www.cite.org.uk/ which is described by its home page as CITE is a collaborative electronic information exchange initiative for the UK construction industry where data exchange specifications are developed by the industry for the industry, enabling the industry to move forward together.
20 See www.vicnet.net.au/ its stated reason for being on its home page states 'The Vicnet Division provides community internet services to Victorian not-for-profit organisations. Through Vicnet, the State Library of Victoria delivers information and communication technologies, and support services

which aim to strengthen Victorian communities. Vicnet works with a wide range of not-for-profit community organisations including sporting, recreational, education, multicultural, health and arts groups. It also provides services to the government, education and welfare sectors.'
21 See www.communitybuilders.nsw.gov.au/builder/ accessed on 27 May 2006.
22 See www.bovislendlease.com/ and explore that site.
23 See www.eta.ch accessed on 27 May 2006.
24 See web-japan.org/trends/science/sci030414.html Japanese companies seem to be concentrating heavily on providing innovative services for elderly or disabled people living alone that are linked to mobile phone technology.
25 See www.printdirect.com/ accessed on 27 May 2006.
26 See www.abc.net.au/rn/tranlist.htm and www.abc.net.au/rn/backgroundbriefing/ for downloading sound and podcast files, accessed on 27 May 2006.
27 See the research work of Prof. John Debenham, University of Technolgy, Sydney, for example, www-staff.socs.uts.edu.au/~debenham/papers/ES03.pdf), accessed on 27 May 2006.

References

Akkermans, H. A., Bogerd, P., Yucesan, E. and van Wassenhove, L. N. (2003). 'The Impact of ERP on Supply Chain Management: Exploratory Findings from a European Delphi Study'. *European Journal of Operational Research*. **146**(2): 284–301.

Al-Mashari, M., Al-Mudimigh, A. and Zairi, M. (2003). 'Enterprise Resource Planning: A Taxonomy of Critical Factors'. *European Journal of Operational Research*. **146**(2): 352–364.

Australian Procurement & Construction Council Inc and DOLAC (1997). National Code of Practice for the Construction Industry, Code of Practice. Deakin, ACT, Australia: Department of Labour Advisory Council (DOLAC) Secretariat – Department of Workplace Relations and Small Business.

Barnatt, C. (2004). 'Embracing E-business'. *Journal of General Management*. **30**(1): 79–96.

Barnes, D., Hinton, M. and Mieczkowska, S. (2004). 'The Strategic Management of Operations in E-business'. *Production Planning & Control*. **15**(5): 484–494.

Carlisle, W. (2005). Googlemania. *Background Briefing*. Australia: ABC Radio National, http://www.abc.net.au/rn/talks/bbing/stories/s1556705.htm

Chan, C. and Swatman, P. A. (2000). 'From EDI to Internet Commerce: The BHP Steel Experience'. *Internet Research: Electronic Networking Applications and Policy*. **10**(1): 72–82.

Duyshart, B. H. (1997). *The Digital Document*. Oxford: Butterworth-Heinemann.

Duyshart, B., Mohamed, S., Hampson, K. D. and Walker, D. H. T. (2003). Enabling Improved Business Relationships – How Information Technology Makes a Difference. In *Procurement Strategies: A Relationship Based Approach*. Walker D. H. T. and K. D. Hampson, Eds. Oxford: Blackwell Publishing: Chapter 6, 123–166.

Gallaher, M. P., O'Connor, A. C., Dettbarn Jr., J. L. and Gilday, L. T. (2004). Cost Analysis of Inadequate Interoperability in the U.S. Capital Facilities Industry, Research Report. Gaithersburg, Maryland: U.S. Department of Commerce Technology Administration, National Institute of Standards and Technology (NIST), NIST GCR 04-867 Final Report 2–8: 210.

Gibson, W. (2004). *Neuromancer.* New York: Ace Books.

Hayes, R. H. and Wheelwright, S. C. (1984). *Restoring Our Competitive Edge: Competing Through Manufacturing.* New York: Wiley.

IDC and Pang, A. (2001). eProcurement Ensures Visionary Companies a Place in the New Economy, http://www.oracle.com/applications/procurement/IDC_wp_Compaq.pdf, accessed on 28 May 2006.

Jewell, M. and Walker, D. H. T. (2005). Community of Practice Perspective Software Management Tools: A UK Construction Company Case Study. In *Knowledge Management in the Construction Industry: A Socio-Technical Perspective.* Kazi A. S., Ed. Hershey, PA: Idea Group Publishing: 111–127.

Lawrence, E., Lawrence, J., Newton, S., Dann, S., Corbitt, B. and Thanasankit, T. (2003). *Internet Commerce Digital Models for Business.* Sydney: John Wiley & Sons Australia.

Lovins, A. (2000). 8 October, Natural Capitalism – A Lecture by Amory Lovins. *Background Briefing.* Sydney: ABC, www.abc.net.org/rn/talks/bbing/stories/s196391.htm

Lueders, H. (1991). Interoperability and Open Standards for eGovernment Services. Brussels, The Council of the European Communities. Volume 1 P. 0111: 42–46.

Lueders, H. (2005). Linking up Europe: The Importance of Interoperability for e-Government Services. *EU Commission Staff Working Paper: Initiative for Software Choice Secretariat* Interoperable Delivery of European eGovernment Services to Public Administrations, Businesses and Citizens: 1–23.

Nisbet, N. and Liebich, T. (2005). ifcXML Implementation Guide, International Alliance for Interoperability Modeling Support Group: 1–48.

Peansupap, V. and Walker, D. H. T. (2005). 'Exploratory Factors Influencing ICT Diffusion and Adoption Within Australian Construction Organisations: A Micro Analysis.' *Journal of Construction Innovation.* 5(3): 135–157.

Pliaskin, A. and Tatnall, A. (2005). Developing a Portal to Build a Business Community. *Web Portals: The New Gateways to Internet Information and Services.* Tatnall A. London: IDEA group: 335–357.

PMI (2003). *Organizational Project Management Maturity Model (OPM3) Knowledge Foundation.* Newtown Square, PA: PMI.

PMI (2004). *A Guide to the Project Management Body of Knowledge.* Sylva, NC: Project Management Institute.

Poon, S. and Swatman, P. (1999). 'An Exploratory Study of Small Business Internet Commerce Issues'. *Infomation and Management.* 35: 9–18.

Sawhney, M. and Parikh, D. (2001). 'Where Value Lives in a Networked World'. *Harvard Business Review.* 79(1): 79–86.

Sharma, A. (2002). 'Trends in Internet-based Business-to-business Marketing'. *Industrial Marketing Management.* 31(2): 77–84.

Szymczak, C. C. and Walker, D. H. T. (2003). 'Boeing – A Case Study Example of Enterprise Project Management from a Learning Organisation Perspective'. *The Learning Organization, MCB University Press.* 10(3): 125–139.

Tatnall, A. (2004). *Web Portals: The New Gateways to Internet Information and Services.* Hershey, PA: Idea Group.

Walker, D. H. T. and Rowlinson, S. (1999). Use of World Wide Web Technologies and Procurement Process Implications. In *Procurement Systems: A Guide to Best*

Practice in Construction. Rowlinson S. and P. McDermott, Eds. London: E&FN Spon: 184–205.

Walker, D. H. T., Hampson, K. D. and Ashton, S. (2003). Developing in Innovation Culture. In *Procurement Strategies: A Relationship Based Approach.* Walker D. H. T. and K. D. Hampson, Eds. Oxford: Blackwell Publishing: Chapter 9, 236–257.

Wei, L., Raymond, J. K. and Ahmad, H. (1999). 'Effects of High Prequalification Requirements'. *Construction Management & Economics.* 17(5): 603–612.

Wilson, R. A. (2004). 'Employee Dishonesty: National Survey of Risk Managers on Crime'. *Journal of Economic Crime Management.* 2(1): 1–25.

Chapter 8

Procurement innovation and organisational learning

Derek H. T. Walker and Tayyab Maqsood

Chapter introduction

We have stressed throughout this book that effective project procurement concerns value generation, not simply cost minimisation. The interface between business strategy (see Chapter 5) and procurement often links in well with performance measures (Chapter 6) and both culture (Chapter 9) and how to foster innovation and leverage knowledge and learning – the focus of this chapter. The purpose of this chapter is to help make explicit the link between strategic intent, developing learning strategies through procurement and PM practice and making project procurement choices.

To achieve the objectives set for this chapter, it is separated into three sections, together with a vignette that presents a hypothetical situation along with relevant questions that, if answered, will reinforce learning from this chapter.

The first section deals with business strategy and innovation to identify how innovation and creativity can be important to many organisations and how they may frame strategies to remain sustainable and competitive in delivering stakeholder value. The second section discusses the nature of innovation and learning and provides some recent models that link it to strategic engagement through procurement. The third section focuses upon organisational learning opportunities, its advantages and its link to procurement.

Business strategy and project procurement

In Chapter 5 we discussed the nature of the firm, how organisations develop strategy and how that links to project procurement to gain and maintain competitive advantage. We explained how the resource-based view (RBV) of the firm can shape an organisation's strategy to use procurement strategies as well as its internal development to shape its competitive niche. We also indicated how a sub-set of the RBV argues that learning is a critical resource with which to develop and hone an organisation's dynamic capabilities – so that it can be nimble and lean and quickly respond to competitive forces. In this section we will further explore how strategy develops and results in a procurement decision that defines the extent to which the organisation will integrate itself with its supply chain (including its customers and other stakeholders).

Strategy and innovation level

We briefly explore Porter's 5-forces model to link it to delivering competitive advantage. Porter's competitive strategy (Porter, 1985; 1990) identified five forces that shaped competitive advantage. These are listed as follows:

1   threat of new entrants;
2   bargaining power of suppliers;
3   threat of substitute products;
4   bargaining power of buyers and
5   intensity of rivalry among competitors.

Through analysing the nature and trajectory of these forces, a firm could adopt a competitive advantage by focusing on a narrow or broad target with the aim of positioning itself with a cost advantage (being the lowest cost provider) or a differentiation strategy (having a unique offering or position in terms of brand, reputation and gaining loyalty). Being innovative and a better manager of knowledge than competitors means that if an organisation is seeking a cost advantage then knowledge and innovation can enable it to be efficient and keep abreast (if not ahead) of its competitors. If the organisation is seeking a differentiation advantage, doing something different, providing a (near) unique product or service or having systems and procedures that are efficient, effective and difficult to replicate, then it will likely have a more lasting competitive advantage than having a temporary cost advantage – which may be breached through being readily replicated (or improved upon) or undermined through substituting inputs that are commonly available. Procurement choices are concerned with deciding which organisations and clients a firm will collaborate with to deliver projects so that its business strategy should shape its decision – it may, for example, seek out complementary resources if these are lacking. Also, strategic procurement choices involve looking inwards and outwards and making decisions about whether (1) either an exploration (outward looking) stance to discover new things should be taken or an exploitation of external resources, or (2) to focus upon internal capabilities and resources (March, 1991; Rothaermel and Deeds, 2004).

The Porter model allows us to understand procurement from the perspective of the competitive forces he identified. The *new entrants* challenge revolves around building a competitive advantage that is difficult to breach; however, it may well be the case that a new entrant may have a nascent idea that can be capitalised upon by a dominant player. In this case, it is in the interest of the dominant player to join forces with the 'bright spark' and attempt to develop a sum that is greater than the potential of their parts. Thus alliances, joint ventures and partnering arrangements can make sense so long as the parties have appropriate levels of trust and distrust (see Chapter 9 for more detail on this). The *bargaining power of suppliers* can be accommodated by taking a lean supply chain approach (Womack *et al.*, 1990; Womack and Jones, 2000). Alternatively, a lead supplier position can be taken through looking at the strategy of working with global providers at various levels. Ferdows (1997) provides a model for achieving this strategy based upon global factory production and outsourcing where players advance up the value-adding supply chain. The ethical and business cost/benefit needs to be considered when making a choice. *Threat of substitute products* can and should be seriously considered so that (as in judo) a confrontational force can be not only deflected but turned against the aggressor. In procurement strategy terms, this is concerned with building upon good ideas to match organisational-internal strengths and competencies to flip the substitute idea either out of the field, or to morph it with other innovations and absorb it into a more powerful and sustainable offering. The *bargaining power of buyers* can also be accounted for. One way is to overwhelm them by being a sole supplier or make it very difficult to switch. Alternatively, as has been suggested by Dorothy Leonard-Barton (Leonard-Barton, 1992; 1995; Leonard and Rayport, 1997; Leonard and Sensiper, 1998), a firm can work with its clients and buyers to provide more value than either parties separately providing the end product/service. *Intensity of bargaining power of suppliers* can be strategically dealt with by offering non-obvious advantages to form alliances that can form synergies so as to equalise or reassert a more secure position at 'the table' – where sophisticated negotiation in procurement comes in to play. *Rivalry among competitors* can be handled with a defensive strategy that allows co-opting competitor competence (Prahlad and Ramaswamy, 2000) or domination.

In this section we may have given the impression that being innovative to achieve a differentiation competitive advantage is a path worth taking. Of course, in doing so we are assuming a level of business sophistication that may well be beyond many organisations. However, most organisations state (judging by the rhetoric of the business press and by annual reports) that cost competitiveness is 'king'. Certainly producing value for money requires a keen eye on costs as well as innovative and smart ways of delivering outcomes (both tangible and intangible). So we need to consider the forms of competitive strategy that organisations adopt. These forms differ from the 'schools' of strategy described in Chapter 5 in that they are actions rather than philosophical stances.

Miles and Snow (2003) provided a neat typology of strategic characteristics of a firm's strategy under four organisational types – prospector, analyser, defender and reactors. As Croteau and Bergeron (2001) note, firms choose one type over another depending upon the perceived business environment. *Prospectors* wish to have access to the largest possible market and are highly innovative in bringing change to unsettle their competitors. They introduce innovative offerings and aim to be the supplier of first choice to those organisations that are willing and wish to pay a premium to be 'ahead of the pack'. *Defenders* are less innovative than prospectors, or rather they are less agile, so they need to select niches that they are comfortable in and defend these aggressively. They will concentrate on finely honing their costs and quality of offering so that they fend off competitors as long as is possible to make the cost/benefit worthwhile. *Analyser* firms have a bit of both prospector and defender in them. They tend to resemble the early adopter category of innovator as outlined by Rogers (2003). These organisations ruthlessly select their market segments and position themselves where they can best maximise their outcomes. They are happy to hold second place and learn from the innovators to improve their offerings relative to those challenging them. These organisations believe that they can maintain a strongly competitive position from learning from the mistakes and experiments of innovative prospectors and do so more effectively than their 'also-ran' competitors. The final group are *reactors*. This group expend energy in countering competitors in value-negative or at best neutral ways such as spending money on marketing, public relations, legal battles and

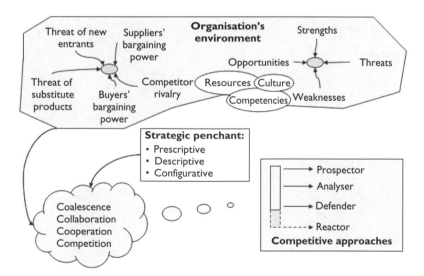

Figure 8.1 Firm's characteristics shaping its procurement decisions.

other superficial ways to try to exclude competitors or reduce the surfacing of innovative ideas that challenge their ability to deploy these ideas. In the first three of this four-typology model, the emphasis is upon innovation and alliancing or finding value-adding ways to deal with change and opportunity. In the fourth example, energy is not aimed at product/service innovation – rather perception management.

In Chapter 5 we discussed the summary of strategic schools of thought offered by Mintzberg *et al.* (1998). This helps us to discern the penchant that influences or dominates an organisation's strategic stance. When that influence is combined with an understanding of an organisation's environment, governed by the competitive forces that Porter (1985; 2001) envisages (informed by an organisation's knowledge of the limits/scope of its resources, competencies and culture to perceive the strengths, weaknesses, opportunities and threats that it faces), organisations can begin making decisions about how its procurement strategy might be shaped. Its natural competitive approach (based upon its perception of its 'world view' and how that affects it) further shapes an organisation's procurement approach. Figure 8.2 illustrates this set of forces that shapes procurement decision making.

An organisation's strategy-making group will see the task of procurement decision making with a penchant for one or more schools of thought outlined in Chapter 5 that will help explain the way that such decisions are made. The result can determine the extent to which it will form relationships with its supply chain. Walker and Hampson (2003a: 45) describe how enterprise networks, partnering and alliances develop. They use the idea of

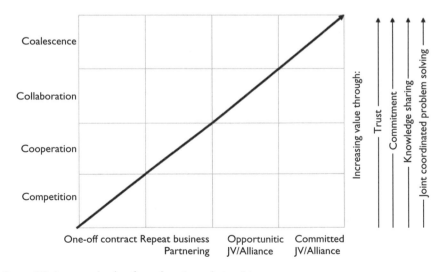

Figure 8.2 Increased value from forming relationships.

a continuum of the degree of objectives alignment between supply chain entities from low to high as providing increasing potential benefits through partnering and cooperation based on Thompson and Sanders (1998: 74).

The lowest level is competition where there is a complete disconnect between the parties' objectives, and often this results in each link in the supply chain trying to take advantage of gaps in scope specification to claim 'extras', and generally devoting a lot of energy into anticipating conflicts or reacting to attacks from others (in the form of claims and counter-claims). This changes to a point where through cooperation; the parties have some limited commonality of objectives. There may be a formal attempt to align the project delivery goals through partnering arrangements as discussed in Chapter 2. This releases some of the reactive energies and reduces the burden of administrative effort in preparing for conflict. Well-defined escalation plans allow problems and conflicts to be better dealt with by those with appropriate authority to make commitments 'stick'. As the alignment increases towards collaboration, there is a substantial overlap of interests, organisation-specific objectives and benefits (including innovation and knowledge generation and transfer) that become more evident. As this relationship becomes more encompassing it can reach a stage of coalescence where genuine mutual objectives are addressed by the parties, and the alignment reaches a point where there is a tight rather than loose fit of goals, methods, processes, infrastructure support and even culture. This is manifested in intense joint problem solving and communication that can improve both innovative capacity and knowledge management (KM) as illustrated in Figure 8.2 here. Tightness of fit between strategy and operational delivery is recognised as being more productive and likely to lead to successfully meeting objectives (Miles and Snow, 1984).

Figure 8.2 illustrates the links between level of cooperation and relationship between organisations in a project supply chain and increased value generated through increasing trust, commitment, knowledge sharing, coordinated joint problem solving and other cultural understanding aspects which permit energy to be directed towards innovation and coordinated effort rather than defensive routines. In this way, organisations are placed closer to strategically acting like prospectors or analysers (Miles and Snow, 2003), not so much by strategising product/service offerings to the market, but rather in their attitude towards their contribution towards the project goals and vision.

Given the holistic nature of this book, it is not surprising that we promote the idea that most organisations that are likely to be sustainable are better advised to follow an innovation and knowledge intensive route to achieve development of products and services that are sustainable business prospects. We have stated our bias as leaning towards innovation and learning and so we will proceed upon that path. To that end it is worth briefly discussing how alliances, joint ventures and close collaborative groups can synergise their efforts towards producing value-adding project outcomes.

Relational-based project delivery – partnering, joint venturing, alliances

In Chapter 2 (procurement choices section) we discussed relational-based project delivery systems broadly under partnering arrangements (often under competitive tendering arrangements), and the formation of joint ventures and alliances. In this chapter we concentrate on these as facilitators of innovation and learning.

The literature on partnering and its advantages is substantial. Numerous studies and reviews have been undertaken in many countries: in the USA (Weston and Gibson, 1993; CII, 1996), in the UK (Bennett and Jayes, 1995; Mathews, 1999; Akintoye *et al.*, 2000; Bresnen and Marshall, 2000; Dainty *et al.*, 2001; Bresnen, 2003) and in Australia (Lenard *et al.*, 1996; Lenard, 1999; Uher, 1999; Walker and Hampson, 2003b). The general consensus is that partnering assists and facilitates an environment in which trust can develop and in which an orderly dispute resolution process can be agreed upon so that cross-organisational teams can work together for joint benefit. However, as was pointed out by Walker *et al.* (2002) in partnering, individual organisations in a project supply chain may sink while others may swim. A principal point of departure that is evident from their case study of the National Museum of Australia (NMA), which was delivered under a project alliance arrangement, was that the alliance locked in (through its rewards and penalty arrangements) that team members only felt consequences of gain or pain from acting *jointly* rather than separately.

Project joint ventures are legal arrangements whereby separate organisational entities form a project team together to deliver projects. Large-scale construction infrastructure projects, common for example in Hong Kong (HK), have tended to be delivered using joint ventures (JVs) of global and local construction organisations. The main reason why individual, often very large, construction companies do so is to offer a customer-focused service package which meets that customer's needs (Walker and Johannes, 2003a). While one motivation for establishing JVs has been to spread financial and other risks, there is a strong motivation to better capture learning so that this hidden and intangible asset may be later applied to improve JV partners' competitive advantage (Walker and Johannes, 2003b). Figure 8.3 illustrates this motivation based upon interviews with nine CEOs/senior executives of JV partners and 40 site management staff on three case studies undertaken in HK (Johannes, 2004). The benefits highlighted that, including intangible benefits such as enhanced reputation/brand image, improved organisational learning and enhanced business opportunities are consistent with the literature dealing with other industries.

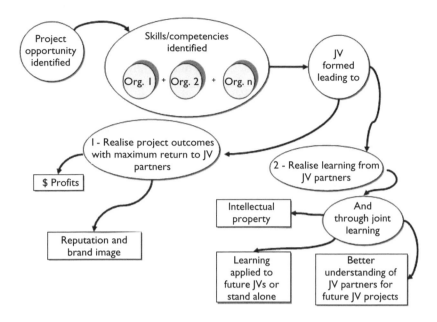

Figure 8.3 JV and learning.
Source: Walker and Johannes, 2003b: 111.

For example, in e-commerce JVs, *originators* develop content, *syndicators* package content and *distributors* deliver the product (Werbach, 2000: 87). Doz and Hamel (1998) provide many examples of advantages of forming alliances which bring together separate skill sets that may lead to JVs in a separate entity that develops new businesses, though they caution us about how one firm may take advantage of another firm in a JV in a hidden competitive ploy to dominate a market. Cross-border alliances and JVs can provide valuable knowledge insights through close interaction between partners. This was particularly the case with US and Japanese alliances in the 1980s and 1990s (Bleeke and Ernst, 1993). An important point made in most of the literature is that JVs and alliances should be managed with a careful power symmetry in mind (based on financial or intangible contributions) so that strategic decision making is not dominated by one party; that way, trust and commitment is more likely to survive.

It is evident from the more current literature (Miles and Snow, 1995; Miles *et al.*, 2005) that the most successful business models that promote agility, innovation and rapid adaptation to turbulence that is so prevalent today, are for organisations to form formal or informal networks. These go beyond the kind of regional alliances found in the Milan region for fashion or Silicon Valley for electronics (Porter, 1998) or *keiretsu* links in Japan or *chaebol* in Korea (Dyer *et al.*, 1998). The current crop of alliance organisations form virtual and global linkages based upon common culture and trust that ensures that contributions made by partner organisations receive due benefit (Miles *et al.*, 2005). These types of organisations have responded to the difficulty in attracting talent and fully utilising that talent in one central organisation, by forming alliances (both formal and informal) as a cluster of atomic-like small firms that focus around a constellation of talented entrepreneurs. This is what Miles and Snow (1995) refer to as the *spherical form* and they cite an Australian Sydney-based firm, technical and Computer Graphics (TCG), as an example of this kind of CAD organisational form that procures its required resources through a loose alliance of like-minded players. This concept moves them well beyond their earlier prospector and analyser type firms to a form that deploys a business model in which it

1 identifies a market niche (much like an analyser would do);
2 finds a development partner (probably as a JV partner);
3 locates a major customer (which may involve the customer being a significant interactive problem-solving contributing partner);
4 involves other TCG firms (part of their alliance network which typically has 5–10 professionals) and
5 extends the triangle in new directions (often new and innovative applications will emerge as part of the creative and innovative process).

This new organisational form has a governance regime dominated by a strict set of ethical codes of practice along with cultural norms (Miles and Snow, 1995: 9). This form of procurement and its governance structure validates the importance of transparent ethical frameworks being established (see Chapter 4) and a strong culture (see Chapter 9) because the alliance works effectively through affective commitment simply because energy is not wasted in dispute resolution; rather, conflict is centred on creative abrasion (Leonard and Straus, 1997) which stimulates innovative ideas. The organisational form typified by TCG overcomes many of the more structured and formal internal and legally complex arrangements identified as causing failures in network organisations during the 1990s – the paradox of coping with a turbulent environment using tightly linked organisational structures (Miles and Snow, 1992).

Figure 8.4 illustrates how the core organisation will need to respond in sourcing its required talent in this networked organisation age. The core organisation will require well-aligned infrastructure – this is not new and has been advocated for a long time (Miles and Snow, 1984; Pettigrew and Fenton, 2000). The key message is that by managing the central elements of the organisational infrastructure, particularly information communication technology (ICT) supporting people to be innovative, the sum is far greater than its constituent parts. Pettigrew and Whittington (2003) argue that

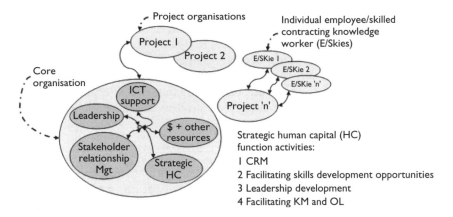

Figure 8.4 Alliance implications for procurement.
Source: Walker, 2005.

managing structures, processes and boundaries (relating to deciding the scope, and scale of what will be internally managed and what will be outsourced) is crucial and that mismanaging one can have a leveraged adverse affect as much as managing all of these three elements well can have a leveraged improvement in performance. Figure 8.4 illustrates its core activities as leadership (stressing both strategy and deployment), stakeholder relationship management, ICT support, financial and resource support and strategic human capital (see Chapter 10 for more discussion on this issue). Four human capital activities are highlighted as needing to be excellent: a customer relationship management (CRM) system that can track individual employees with rare and valuable skills as knowledge workers (E/Skies); facilitating skills development opportunities – these could include a community of practice (COP) approach, mentoring/coaching, knowledge events, training and seminar type events; leadership development to ensure that those in leadership roles in the organisation have sufficient acumen and ability to persuade and inspire others; and an active facilitation of KM and organisational learning (OL), recognising knowledge and innovative capacity-delivery as the most significant asset that this type of organisation possesses. People in small organisations with the type of person required for complex project work of this nature will work on the projects and there will be many projects managed by this type of organisation. More detail of this concept has been presented elsewhere (Walker, 2005).

This section argues that the competitive environment exerts pressures upon organisations that may hold a particular competitive approach (that they tend to respond with) and that coalescence into a committed relationship best serves most of today's knowledge-intensive organisations.

These organisations need to be highly adept at effectively managing their infrastructure, and particularly need a superbly functioning strategic HR department that nurtures not only its employees but also its 'extended family' of regular value contributors. This view sets procurement choice upon a new plane and requires new levels of sophistication of how to relate and collaborate with people and supply chain members. This view moves well beyond the limits of relying on an organisation's legal or contracts administration department to ensure that projects are successfully delivered.

Learning and innovation – a procurement perspective

This section focuses upon how learning and innovation can be encouraged, nurtured and facilitated through an appropriate procurement strategy.

The literature from the USA, UK, Australia and elsewhere confirms that demanding clients and users are good for triggering innovation. In a large-scale study of innovation in the Australian construction industry it is clear that demanding clients expect and, through their explicit or implicit expectations and procurement practices, trigger their project delivery teams to develop innovative products, processes or services (Manley, 2006; Manley and Mcfallan, 2006) a case in point is the NMA project mentioned earlier in this chapter where one of the selection criteria was a 'demonstrated ability to add value and bring innovation to the project' (Walker and Hampson, 2003c: 91). Examples from the UK include a demand-side set of drivers, though the pressure emerges from both the client and the deliverer trying to gain a competitive edge by offering valuable alternatives to those initially expected (Gann and Salter, 2000). Other UK studies also indicate a role for discerning clients and project providers as well as the role of the professions (Winch, 1998; 2005); US studies indicate a similar pressure that arises from construction clients who demand cleverer solutions to their project problems, and professionals who wish to deliver more value (Tatum, 1984; Slaughter, 1998). Added to this is the influence that some of the US literature suggests is exerted by lead users of innovations who undertake beta testing or forms of alliance can influence development of products and processes (Von Hippel et al., 1999). Organisations work with such users through a process termed 'empathic design' where solutions are developed in the field through trial-and-error experimentation and observation of lead users who can help perfect innovative solutions when developing products, services and processes (Leonard-Barton, 1992; 1995; Leonard and Rayport, 1997; Leonard and Straus, 1997). All these examples, from across many continents and disciplines, share a central idea that innovation is unleashed from the creative and curious interface between clients and users who wish to explore new and more useful ways to do their job (Leonard-Barton, 1992; 1995; Leonard and Rayport, 1997; Leonard and

Straus, 1997). This supports the inference that if clients choose to include a requirement to demonstrate innovative capacity in delivering projects, that serendipitous value may be generated as an outcome, provided that the client side of the team has the capacity, motivation and open mind to engage in experimentation.

Recent research provides a useful model that helps to explain the process. At first Figure 8.5 appears somewhat crowded and complex but it is composed of several simple ideas that are highly relevant to PM procurement choices. It acknowledges that the external environment with its inherent competition and opportunities and threats exerts a powerful dynamic that drives motivation to innovate. Opportunities for access to knowledge repositories exist for firms through their professional bodies, universities and shared sources such as COPs. These sources push/offer access to research outcomes and other knowledge forms such as theoretical frameworks and case study data, often with an expectation or hope that the organisation will reciprocate by feeding back some of the firm's internal knowledge bank into that source for validation or enhancement through collaborative investigation. Thus, there are both knowledge push-and-pull forces exerted between the external source and the organisation. The organisation possesses existing knowledge through its people and processes, and this is

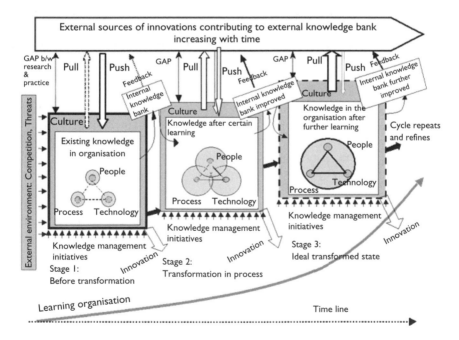

Figure 8.5 An innovation trajectory.

Source: Maqsood (2006b: 174).

enabled by its technology infrastructure. Knowledge is mainly forced into the organisation with much effort as a push force with very weak pull forces allowing knowledge to be fed back to the outside world for validation and enhancement. This represents a state in which there is a low level of organisational learning influence from outside. While this may be adequate for short-term survival if the organisation's internal knowledge capital base is strong, it will inevitably miss signals of external change and turbulence, say, from substitute offerings by competitors or players outside the 'industry' that could enter the competitive field. In this sense an organisation may be either a prospector, defender or reactor in the Miles and Snow (2003) typology. An organisation at this stage is also missing valuable opportunities to innovate through empathic design as indicated by Leonard and Rayport (1997) and Von Hippel *et al.* (1999).

Stage 1 illustrated in Figure 8.5 indicates the stage that most organisations are currently experiencing. People, processes and technology may be poorly aligned so that there are barriers to the generation and flow of knowledge, and while it is used, it may be sub-optimally applied – this is indicated by the separation of these three elements and the weak link by dotted lines. Processes may be traditional for the industry with little experimentation and/or they may be somewhat *ad hoc*. Technology support could include widespread use of groupware and portals as discussed in Chapter 7 but the aim of its deployment would centre on data processing rather than information processing and coordination, storage and retrieval of explicitly recorded information rather than developing and extending communities of practice. People would be linked to technology and processes through more formal means in standard ways with little experimentation on how to 'tweak the system', and there may be a general satisfaction with the way things work when compared to past times when IT was solely about data processing and processes were recorded in 'the manuals'. The organisational culture is inward looking and hence the barriers to outside influence are impeded by an impervious barrier – indicated by the thick boundary lines. Organisations more acutely aware that they should continuously improve would be undertaking limited experiments in using groupware as a KM tool such as Lotus Notes, Yellow Pages for finding and sourcing within-firm experts etc. In a recent study of three of Australia's leading construction contractors, observed KM initiatives included development of portals for information retrieval, limited within-firm communities of practice for ICT use enhancement and information and knowledge storage through the portal for aspects such as safety reporting, lessons learned and other explicit knowledge repositories (Peansupap, 2004; Peansupap and Walker, 2005a).

The greater the number and more intense the scope of these KM initiatives the greater is the chance that the organisation is moving towards or entering the Stage 2 development. Experimentation and reflection on experiments undertaken are key elements in an organisation increasing its capacity (to assimilate external knowledge) to absorb further change and embed learning moving towards being a learning organisation (Cohen and Levinthal, 1990; Gann, 2001; Caloghirou *et al.*, 2004). This is because experimentation and actively seeking external knowledge assists in: (1) developing language and meaning (i.e. jargon, 'in-terms' and technical language) that can be shared easily with outside sources; (2) being better able to categorise and conceptualise external knowledge and thus be in a position to internalise this knowledge and apply it to the workplace; (3) better able to discern patterns and the potential value of external knowledge (Cohen and Levinthal, 1990).

The degree to which an organisation is oriented towards seeking external knowledge varies according to the type of organisation. According to Meacheam (2006), an organisation's KM stage of development can be categorised by analysing it in a 9-cell matrix. On the vertical axis he creates three rows representing the 'knowing cycle': (1) the way it makes sense of knowledge received, transferred and generated; (2) the way it creates knowledge and (3) the way that it uses knowledge for decision making. On the horizontal axis he constructs three columns representing support structures identified by Croteau and Bergeron (2001): the organisations *infrastructure* (its technology, physical assets that allow people to interact and tools); its *infostructure*, the rules and processes that allow people to communicate; and *infoculture*, the background knowledge that actors have embedded about their world and workplace and the way that the organisation encourages or restricts knowledge transfer. Meacheam (2006) argues that using the Miles and Snow (2003) typology of prospector, analyser, defender and reactor can help explain how and why different organisational types seek knowledge from external sources in different ways. This typology is useful in assessing where an organisation's maturity level may lie in terms of Figure 8.5 as well as explaining what level best fits that organisation. For example, a prospector will be seeking knowledge about what external markets fail to currently deliver and what internal innovations it can hone, reshape or further develop to aggressively lead its competitors. An analyser would need to carefully seek external knowledge from the market and identify potential for finding a specialised niche market for its products/services and learning as an early adopter (Rogers, 2003) how it can improve upon lead innovators. A defender organisation would be concentrating to a large extent on knowledge that it can muster to stay in the market and lower costs and raise barriers to entry by others, perhaps through continuous improvement and incremental improvements in its offerings. A reactor organisation is poor at knowledge absorption and finds itself continually under threat and perhaps uses external and internal knowledge to mount campaigns to fend off competitors, use spin-doctoring to advertise its offerings in the most favourable light possible and restrict what it may view as dangerous or subversive leakage of proprietary knowledge from the organisation. Meacheam (2006) offers insights that helps to explain an organisation's likely trajectory along the Maqsood (2006a) model presented in Figure 8.5.

Stage 2 of the Figure 8.5 model indicates an organisation that has reduced the imperviousness of the boundaries between the organisation and the outside world and has also better integrated its people, processes and technology. There is stronger evidence of knowledge creation, transfer and use for problem solving as well as a greater degree of feedback of knowledge to be tested and validated by external sources such as academia, professional associations, clients and competitors (perhaps through undertaking benchmarking exercises). This stage indicates a more or less equal push and pull of knowledge. KM initiatives are more prevalent and intense in scope and these propel the organisation's learning and develop its absorptive capacity.

At Stage 3 an idealised condition is contemplated. The people, processes and technology is completely seamless and embedded. The boundaries are highly permeable and there is a greater export of knowledge for validation than is being imported. This model appears similar to the spherical organisational form advanced by Miles and Snow (1995; Miles *et al.*, 2005) as discussed earlier. While this appears to many organisations a dangerous and insecure way of managing its critical knowledge assets, it may be that such a state provides greater advantages for rapid innovation and commercialisation than its downside of being vulnerable to exploitation by its competitors. As Miles and Snow (1995: 9) indicate, organisations that operate in this way need to have a strong culture, clear governance and cultural protocols, filtering mechanisms to exclude predatory organisations, and as indicated by Walker (2005) in Figure 8.4, a superb strategic human capital management system.

The main implications for procurement choices discussed in this section relate to how organisations may present themselves as a learning partner of first choice in alliances or in developing their internal dynamic capabilities and knowledge resources. Further, the models presented thus far indicate that organisations that work closely together (either by collaborating on projects or developing innovations through synergising their particular skills base) need to align their development and use of ICT. Before we leave this part of the discussion we should look at how some of the underpinning technologies such as ICT may best be diffused. Results from a recent study undertaken by Peansupap (2004) for his PhD, extensively published in the literature with Walker, provides revealing insights into the drivers of, and barriers to, ICT diffusion that operate at the organisation (macro), project (meso) and individual (micro) level (Peansupap and Walker, 2005a,b; 2006a,b). The study surveyed 117 management staff from three large Australian construction organisations and was supplemented by a follow-up qualitative study that assists us understand how three large Australian construction contractors, that routinely use ICT in a project environment, diffuse ICT innovation.

Innovation diffusion is defined as the process in which a new idea, concept or technology has been introduced throughout a social system over a time period (Rogers, 2003). The term 'ICT diffusion' is defined in this chapter as the process by which an ICT application is adopted and implemented by an organisation until its expected users accept and transfer knowledge of how to use these ICT applications throughout the organisation.

Wolfe (1994: 407) identifies three streams of innovation diffusion research. Diffusion of innovation (DOI) research addresses patterns of how innovation spread throughout a studied group of adopters. Organisational innovativeness (OI) addresses the determinants of how innovation occurs – focused upon the organisation. Process theory (PT) addresses the process of innovation and how and why adopters carry out innovation. Stages of innovation have been classified in various ways. He notes 10 stages: idea conception, awareness, matching, appraisal, persuasion, adoption decision, implementation, confirmation, routinisation and infusion (Wolfe, 1994: 410) while Rogers (2003: 199) offers five stages: knowledge, persuasion, decision, implementation and confirmation. The outcome of the study reported upon in this chapter was specifically focused upon PT for the actual implementation stage of the ICT application's deployment.

Using factor analysis, the Peansupap (2004) study identified 11 factors influencing ICT diffusion in these organisations. These fell into four categories – management (M), technology (T), individual (I) and environment (E). Also the model indicates (thick-lined elements in Figure 8.6) how ICT diffusion takes place from initial adoption to actual implementation through six key ICT diffusion processes and how each factor impacts upon those diffusion processes:

1   developing new business practices/processes;
2   organisational adoption of the ICT decision;
3   preparing for the initial use of ICT applications;
4   reinforcing the actual use of the ICT applications;
5   clarifying benefits of ICT application use and
6   developing a positive perception towards ICT and ICT diffusion.

The extent of the influence of these factors is presented in Table 8.1.

Results from two phases of the study are presented in Table 8.1. The 11 factors are grouped by factor categories as illustrated in Figure 8.6. The rounded mean values are based on 1 = very low to 5 = very high from the survey results of the three Australian organisations, a government department, a global consultant and a major contractor. The three case study results for the factors are based on qualitative research based upon interviews with a representative sample of staff from three major construction contracting firms, the major contractor in the survey was included in the qualitative study. All organisations are based in Australia

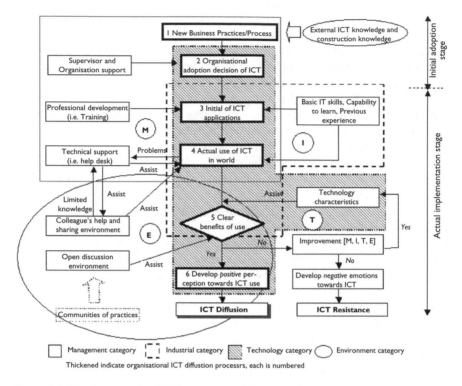

Figure 8.6 The framework of ICT innovation diffusion with an organisation.

and the contractors typically turn over in excess of AU$1 bn per year and all three have experience of using IT for well over a decade. An implication that can be drawn from this study is that there is consistent support for the nature of ICT diffusion amongst these ICT-sophisticated users to be seen as requiring:

1   users that have already experimented with IT applications and thus have built up an absorptive capacity for grasping how to use new ICT tools and have developed some confidence in their ability to tackle and overcome problems when learning how to use new ICT tools;

2   management needs to financially support and resource an ICT help or assistance infrastructure that includes training, help desk facilities, one-on-one 'hand holding' and providing a reward system even if this is focused on less tangible rewards such as promotion prospects;

3   an ICT infrastructure that provides real and tangible benefits from its use and is sufficiently free from technical problems to not hinder or frustrate users. Users commented negatively about ICT infrastructure when it tended to relate to low speed/low bandwidth access on site or when technical issues interrupted their access and use of ICT tools and

4 A workplace culture and environment that supports collegial help and knowledge exchange, deters or at least not rewards knowledge/information hoarding and allows the safe 'space' for criticism and feedback when poor performance is evident for management support or technical support.

It becomes clear from the above research study that if a procurement approach is intended to include the need for a supply chain to engage in the shared use of common or well-interfaced ICT tools outlined in Chapter 7 e-business, then participating organisations will need to

- source staff with requisite entry-level ICT skills;
- have a minimum ICT infrastructure that includes management support as well as technical tools, equipment and support;
- develop and maintain a workplace culture that is rich in sharing knowledge and information and allows dissenting voices to be assertive in demands for support to use available tools and foster enthusiastic commitment to achieve benefits through collaborating with users of this essential ICT infrastructure.

Table 8.1 Comparison of ICT user experience of resulting innovation diffusion factors

Factor description	Survey rounded means	Case study org A	Case study org B	Case study org C
Individual (I) factors	3.8			
F3: Supporting individual/personal characteristics	4.2	• • •	• • •	• • •
F2: Clear benefits of ICT use	4.1	• • •	• •	• • •
F9: Positive feelings towards ICT use	3.6	• • •	• •	• • •
F10: Negative emotions towards ICT use[1]	1.8	•	• •	•
Environment (E) factors	3.9			
F6: Supporting an open discussion environment	3.9	• •	• •	• •
F8: Collegial help	3.9	• • •	• • •	• • •
Management (M) factors	3.2			
F5: Supervisor and organisational support	3.8	• • •	• •	• • •
F1: Professional development and technical support	3.3	• • •	• • •	• • •
F7: Supporting tangible and intangible reward	2.4	•	•	•
Technology (T) factors	3.5			
F4: Supporting technology characteristics	3.7	• • •	• •	• • •
F11: Frustration with ICT use[1]	1.7	•	• •	•

Notes
1 Negative factors, therefore a low value implies high emotions towards ICT applications.
• Low level of perceived present factors.
• • Medium level of perceived present factors.
• • • High level of perceived present factors – comparison between results of case study of three large global ICT sophisticated contractor organisations.

The main gap in the discussion in this chapter so far relates to helping us understand how to better manage knowledge and its effective transfer that acts as an effective trigger to spur innovation.

Organisation learning, KM and innovation

Many of the underpinning assumptions expressed in earlier sections in this chapter, relate to how organisations undertake KM through the creation, sharing, transfer and then use of knowledge in a value chain. However, we have not yet explained the KM process in this book. Fortunately, there is a large and growing body of literature and several world class academic journals that a key word search at a university library will readily locate.

While many 'facts' about objects may be said to be independently out there (e.g. a sun that rises each day, etc.), it is only people who construct and create knowledge in their minds. We could say that much knowledge is recorded explicitly, the knowledge we gain from manuals, protocols and procedures, for example. However, even that type of knowledge is too dense and rich to be fully recorded and there are always acceptable exceptions to rules and workarounds. We could say that such codified knowledge is partial knowledge or that manuals merely contain information and data not knowledge. Polanyi (1997) made the famous often quoted statement that we know more than we can say and referred to knowledge as being *tacit*, that is unspoken and embedded. His illustration of this was our ability to instantly identify a familiar face out from a crowd and being unable to explicitly and fully explain how we do that. A business example would be how we solve a particular problem, use a protocol or procedures manual or even to operate many software packages.

A further knowledge-type concept is *self-transcending knowledge* that lies beneath tacit knowledge and is knowledge that is not embodied – an example of this is when an artist 'sees' a painting or sculpture and knows its form but the actual artefact remains within the stone or paints and materials until liberated by the artist (Scharmer, 2001) – a procurement solution design could fit this metaphor where knowledge of the end product exists in part in a project's drawings or specifications but the artefact is not tangible until liberated by those involved in its production to 'liberate it' through their self-transcending knowledge process.

One of the early KM theorists, Nonaka (1991), offered some early KM concepts. He explained KM in terms of a spiral process referred to as the SECI cycle where tacit knowledge is shared between people in a social setting (S); this knowledge is then made explicit (E) through discussion of examples and metaphor that hone in on 'details'; it is then combined (C) with existing explicit codified knowledge in notes, memos and procedures and manuals; and finally it is institutionalised (I) where it is formally reframed by an organisation and embedded in the culture as 'the way that things are done around here'. This all occurs in a setting where such knowledge sharing is encouraged and fostered in a knowledge sharing space or environment – a '*ba*' (using the Japanese term) (Nonaka and Konno, 1998). This view of how knowledge is created, captured, transferred, used and then reframed in a KM cycle is more directly applicable to small groups and individuals in its philosophy than for organisations.

When applying this thinking more broadly to larger organisations, we can use the model developed by Crossan *et al.* (1999) who envisaged the KM process in a larger organisational context as comprising the '4Is'. *Intuiting* happens when experiences, images and metaphors help to socialise an individual's knowledge through expressing feelings and other tacit knowledge. This is then processed by others through *Interpreting* this knowledge and making sense of it; it is further refined through *Integrating* this new knowledge with the organisation's existing knowledge base and then finally *Institutionalising* it by embedding it into its culture and routines. All this happens at the individual, group and organisational level with the potential for insights and re-framing of knowledge to grow meaning rather than have it locked up as tacit individual knowledge. The process dynamic is driven by a feed forward 'push' of knowledge and feed back 'pull' set of forces. This looks at knowledge organisationally as a series of stocks and flows (Bontis *et al.*, 2002) in which knowledge reserves are accessed and used, restocked and replenished within an organisation and sustained by the organisation through people, also with the organisational impact being more obvious than with the SECI model (Nonaka, 1991; Nonaka and Takeuchi, 1995). This Crossan *et al.* (1999) model, however, while being further refined than the SECI model, failed to recognise the powerful institutional and cultural forces at play. For example, peer pressure exerts a supportive or restrictive force at the individual level. Group norms and leadership styles similarly affect the feedback and feed forward flows at the group level on individuals and also can force change upwards. Similar forces along with corporate governance (see Chapter 4) have a significant impact at the organisational level through providing discipline and domination drivers (Lawrence *et al.*, 2005).

The net effect of considering these forces is that a 4Is model, with the impact of power and influence considered, provides us with a clearer picture of the forces and pressures at work in managing knowledge. This is illustrated in Figure 8.7.

Influence during the intuition process can be episodic – affected by behaviours such as moral suasion, reward systems derived through negotiation. It has implications regarding the way that champions of ideas emerge or are made available for those with intuitive ideas than can lead to successful productive innovation. Ideas need champions to argue and negotiate for often scarce resources and they need to be backed by expertise and a culture that readily brings these together. Integration is also episodic

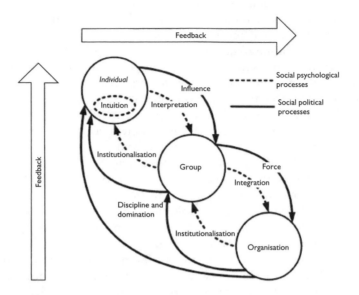

Figure 8.7 The 4Is plus power.

Source: Lawrence *et al.*, 2005: 183.

and requires force to regulate available behaviours to 'push through' innovative ideas which may require organisational routines to facilitate the flow of ideas from a group, or filter them in some way that helps contrast weak ideas with powerful ones. The organisational systemic response to institutionalisation of innovative ideas and knowledge is through discipline and domination – procedures and protocols that should ideally enhance knowledge feedback and feed forward.

The addition of the social political elements help make the Figure 8.7 model more useful for us in attempting to design an optimal system to maximise the potential for knowledge to flow more freely. In the previous section we discussed an organisation's *infrastructure* (its technology, physical assets that allow people to interact and tools); its *infostructure* (its KM practices); and *infoculture*, the background knowledge that actors have embedded about their world and workplace and the way that the organisation encourages or restricts knowledge transfer (Croteau and Bergeron, 2001; Meacheam, 2006).

A weakness in the above model that restricts our full understanding of the KM and innovation diffusion process is that knowledge is notoriously 'sticky' – difficult to transfer and tends to stick to the person or group generating it. KM is about not only creating knowledge, but its effective transfer for further refinement and use. The above model makes the assumption that knowledge naturally flows but it does not indicate the flow speed/rate. Szulanski (1995; 1996; 2003) undertook a PhD on the stickiness of knowledge and identified seven sources of knowledge stickiness: Source lacks motivation (unwillingness to share knowledge); Source lacks credibility (the source lacks authority, expertise or is perceived as unreliable or untrustworthy); Recipient lacks motivation (doesn't care); Recipient lacks absorptive capacity (has not the background to perceive cause-and-effect links, lacks underpinning knowledge or experience in experimentation to know how to use the knowledge); Recipient lacks retentive capacity (forgets vital details); Barren organisational context (the culture or governance structure inhibits knowledge sharing) and Arduous relationship between source and recipient (lack of empathy, trust or commitment to collaborate in the task of sharing knowledge).

He concluded from testing his model using canonical correlation analysis of a data set consisting of 271 observations of 122 best-practice transfers in eight companies that contrary to conventional wisdom blames primarily motivational factors, his findings show major barriers to internal knowledge transfer are knowledge-related factors such as the recipient's lack of absorptive capacity, casual ambiguity, and an arduous relationship between the source and the recipient (Szulanski, 1996). These factors relate to individual's and organisation's ability to build and sustain social capital that could be designed into a procurement system.

Nahapiet and Ghoshal (1998) argue that social capital relates to the structural links and ties that facilitate a network of people who have the required quality of knowledge to make it worth their interacting to manage knowledge and that they have a cultural match that allows sharing and transfer to take place. This is determined by the way they interact – how they access and share knowledge, their motivational drivers, the value that is expected from their interaction, and their ability to share knowledge. This links to knowledge stickiness which determines the rate, scope and scale of knowledge transfer. Figure 8.8 was developed from the above theory and adapted by Manu and Walker (2006) to enable benchmarking of knowledge transfer among teams of foreign aid experts and a local construction supply chain on a Pacific Island aid project. The model serves to illustrate how elements of social capital form part of the general three infrastructures identified by Croteau and Bergeron (2001) are relevant to designing a procurement system with these three structures in mind.

The thrust of this section has been focused upon exploring how KM occurs. People jointly solve problems and in doing so use and re-frame knowledge, and they tend to transfer this to colleagues within and outside their organisations. One facilitating structure for this has been a COP, described as 'a group of people, informally bound together by shared expertise and passion for a joint enterprise' (Wenger and Snyder, 2000: 139).

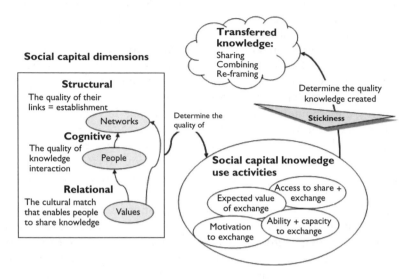

Figure 8.8 Social capital and knowledge transfer.

Source: Manu and Walker (2006: 480).

This notion of cooperation and collaboration to create and share knowledge, particularly in a non-formal structure that was not initiated by organisations following a KM strategy, is an interesting construct that has been internalised by firms. Also a COP has been shown to be a powerful situational learning tool with small groups working on projects and reflecting upon their experiences (Orr, 1990; Sense, 2003; 2004; 2005). COPs have evolved into forms that include their original informal arrangements but now also include more formally structured and organisation-initiated forms (Storck and Hill, 2000; Wenger *et al.*, 2002). COPs not only can be effective KM initiators but they also can significantly contribute to the delivery of social capital that is an important intangible asset for organisations as it facilitates KM activities and can help build strong organisational cultures to assist effective governance. Examples of a construction organisation establishing a COP software infrastructure tool in the UK are provided by Jewell and Walker (2005) and more informal COPs that involved ICT in Australia (Peansupap and Walker, 2005a). It is evident from these case studies that considerable social capital was being formed through that process.

This chapter has focused upon how KM can assist in innovation in project environments. Several models were presented to illustrate the important underpinnings that are required to build social capital that assists KM and can help reduce stickiness of knowledge so that it can be readily shared and used. The main point made is that KM can facilitate innovation that in turn can provide value that would not otherwise be derived from projects. We stress that collaborative procurement arrangements best suit this model and we also suggest and urge those making procurement choices that they should design the procurement system to maximise collaboration and easy knowledge transfer across the supply chain. The authors develop this argument more specifically elsewhere (Maqsood *et al.*, 2007). While it may be argued that relationship-based procurement approaches are more likely to achieve this end, it may be possible that more cost competitive arrangements could achieve the same end. However, when considering the organisation type referred to at the start of this chapter (prospector, defender, analyser and reactor) it is unlikely that reactors have the capacity to understand the subtle nature of how the infrastructure, infostructure and infoculture can be used to provide innovation and value beyond cost reduction strategies.

Vignette

Helen, Stakeholder First Developments (*SFD plc*) Pacific Region's newly appointed Director of Procurement commissioned a management consultant's report into how to best encourage innovation within the company and to strategically develop its business development planning and implementation. Six months ago, the organisation had lost a long-term alliance deal with a dynamic and the highly prized client *PharmaXL* who had been establishing pharmaceutical research and production facilities across the region. The stated reason for SFD narrowly missing out to a rival *SmartBuilt* (that they had been unaware even existed) was because that their rival was more proactively innovative and this better matched *PharmaXL*'s corporate supply chain management philosophy. SFD had always felt confident that its numerous awards and prizes for innovative construction and design by the Master Builders' organisations and several professional bodies was proof of its leadership in this area. Losing out to this 'unknown' company had been a shock.

The consultants report indicated that SFD was patchy in its uptake of ICT innovations and other operationally technical innovations in construction technology. While SFD had some great supply chain partners who relished challenges, most preferred to have contract responsibilities well spelled out, particularly with regard to trial and error on innovations. The SFD board had several solid engineers who had been energetic leaders in the 1970s and were justifiably proud of the awards and excellence prizes that they had won in recent years and they felt that they had much to offer the 'youngsters' in the organisation – if only everyone could find some extra time....

SmartBuilt had been in existence for only five years and was a small organisation with a CEO and three principals with 60 permanent employees, based mainly in their Vancouver head office. The three principals comprised a Chief Operations Officer (COO), Chief Knowledge Officer (CKO), and Chief Relationships Officer (CRO). The Vancouver head office provided total ICT support and they seemed to have a string of alliances with construction contractors and engineering consultants all over the globe. They had very small dormitory offices that maintained a presence in a number of cities: San Francisco, Toronto, Princeton NJ, Singapore, Shanghai, Melbourne, Buenos Aires, Santiago, Helsinki, St Petersburg, Stockholm, Panama, The Hague, Cape Town, Bangalore, Dubai and São Paolo. They also seemed to be linked with several leading universities, and several of the senior staff, described as project managers, had been co-authors on about 15 keynote conference presentations, journal papers and book chapters with others from their university collaborators, as well as some authors who appeared to be associated with some very heavy corporates that included *PharmaXL* and two of its competitors.

Helen read the report with deep interest so that she could make some sensible suggestions to the board tomorrow. She had tapped into her university friends and network of colleagues from Arup engineering, Skanska and Bechtel. Most had heard of *SmartBuilt* and had some interesting observations. Trying to pin this organisation down was like nailing jelly to a wall. It had an alliance with many more specialist PM groups, individuals with wide global experience, but this all seemed to have crept up over the horizon since 2000. Its web site and annual report seemed a bit unconventional as it stressed that it only worked with clients and supply chain partners who were prepared to join in a knowledge-impact evaluation system which *SmartBuilt* administered. There appeared to be a pre-alignment (pre-project start) agreement on identifying improved innovative practices and the development of some kind of points system that worked like currency or assets but had future potential returns, add-on value and transformational value; it was unclear from the web sites how this unusual system worked. One thing that was clear, *SmartBuilt* seemed to work with a common group of supply chain partners and clients, and the project teams floated in and out from somewhere almost at call. The more that Helen tried to fathom how this organisation operated, the more confused she became. This seemed to present a totally different business model to any she had previously encountered.

Issues to ponder

1   What level of alignment of process, people and technology does *SmartBuilt* appear to have achieved relative to SFD?
2   Discuss three vital clues that are missing from the consultant's report that would help Helen better understand how an innovation culture was developed by *SmartBuilt*.
3   Discuss three clues that suggest how *SmartBuilt's* supply chain and customer partner choices might have evolved.
4   Describe the top three innovations that *SmartBuilt* appears to have been developing.
5   Draw up a plan and justify how Helen may counter the competitive threat that SFD faces when trying to align with other key customers or supply chain partners not currently aligned with *SmartBuilt*. What are the options facing Helen and SFD's board?

References

Akintoye, A., McIntosh, G. and Fitzgerald, E. (2000). 'A Survey of Supply Chain Collaboration and Management in the UK Construction Industry'. *European Journal of Purchasing & Supply Management.* 6(3–4): 159–168.

Bennett, J. and Jayes, S. (1995). *Trusting the Team.* Reading, UK: Centre for Strategic Studies in Construction. The University of Reading.

Bleeke, J. and Ernst, D. (1993). *Collaborating to Compete – Using Strategic Alliances and Acquisitions in the Global Marketplace.* New York: John Wiley.

Bontis, N., Crossan, M. M. and Hulland, J. (2002). 'Managing An Organizational Learning System by Aligning Stocks and Flows'. *Journal of Management Studies.* 39(4): 437.

Bresnen, M. (2003). *The Seven Deadly Paradoxes of Partnering (Seven Deadly Sins?).* TG23 – Culture in Construction: International Conference on Professionalism in Construction – Culture of High Quality, Hong Kong, 27–28 October, Fellows R., pp. 1–15.

Bresnen, M. and Marshall, N. (2000). 'Partnering in Construction: A Critical Review of the Issues, Problems and Dilemmas'. *Construction Management and Economics.* 18(2): 229–238.

Caloghirou, Y., Kastelli, I. and Tsakanikas, A. (2004). 'Internal Capabilities and External Knowledge Sources: Complements or Substitutes for Innovative Performance?' *Technovation.* 24(1): 29–39.

CII (1996). The Partnering Process – Its Benefits, Implementation, and Measurement. CII, *Bureau of Engineering Resources.* University of Texas at Austin, CII Source Document 102–11.

Cohen, W. M. and Levinthal, D. (1990). 'Absorptive Capacity: A New Perspective on Learning and Innovation'. *Administrative Science Quarterly.* 35(1): 128–152.

Crossan, M. M., Lane, H. W. and White, R. E. (1999). 'An Organizational Learning Framework: From Intuition to Institution'. *Academy of Management Review.* **24**(3): 522–537.

Croteau, A.-M. and Bergeron, F. (2001). 'An Information Technology Trilogy: Business Strategy, Technological Deployment and Organizational Performance'. *The Journal of Strategic Information Systems.* **10**(2): 77–99.

Dainty, A. R. J., Briscoe, G. H. and Millett, S. J. (2001). 'Subcontractor Perspectives on Supply Chain Alliances'. *Construction Management & Economics.* **19**(8): 841–848.

Doz, Y. L. and Hamel, G. (1998). *Alliance Advantage – The Art of Creating Value Through Partnering.* Boston: Harvard Business School Press.

Dyer, J. H., Cho, D. S. and Chu, W. (1998). 'Strategic Supplier Segmentation: The Next "Best Practice" in Supply Chain Management'. *California Management Review.* **40**(2): 57.

Ferdows, K. (1997). 'Making the Most of Foreign Factories'. *Harvard Business Review.* **75**(2): 73–88.

Gann, D. (2001). 'Putting Academic Ideas into Practice: Technological Progress and the Absorptive Capacity of Construction Organizations'. *Construction Management and Economics.* **19**(3): 321–330.

Gann, D. and Salter, A. J. (2000). 'Innovation in Project-Based, Service-Enhanced Firms: The Construction of Complex Products and Systems'. *Research Policy.* **29**(7–8): 955–972.

Jewell, M. and Walker, D. H. T. (2005). Community of Practice Perspective Software Management Tools: A UK Construction Company Case Study. In *Knowledge Management in the Construction Industry: A Socio-Technical Perspective*, Kazi, A. S. (Ed). Hershey, PA: Idea Group Publishing, pp. 111–127.

Johannes, D. S. (2004). *Joint Venture Contracting Relationships Between Foreign and Local Contractors in the Construction and Engineering Industry of Hong Kong: Implications of Understanding Collaborative Practice.* PhD Thesis. Melbourne: RMIT University, School of Management.

Lawrence, T. B., Mauws, M. K., Dyck, B. and Kleysen, R. F. (2005). 'The Politics of Organizational Learning: Integrating Power into the 4I Framework'. *Academy of Management Review.* **30**(1): 180–191.

Lenard, D. (1999). 'Future Challenges in Construction Management: Creating a Symbiotic Learning Environment'. *Journal of Construction Procurement.* **5**(2): 197–210.

Lenard, D. J., Bowen-James, A., Thompson, M. and Anderson, L. (1996). *Partnering – Models for Success.* Adelaide: Australia Construction Industry Institute Australia.

Leonard-Barton, D. (1992). 'The Factory as a Learning Laboratory'. *Sloan Management Review.* **34**(1): 23–38.

Leonard-Barton, D. (1995). *Wellsprings of Knowledge – Building and Sustaining the Sources of Innovation.* Boston, MA: Harvard Business School Press.

Leonard, D. and Rayport, J. F. (1997). 'Spark Innovation Through Empathic Design'. *Harvard Business Review.* **75**(6): 102–113.

Leonard, D. and Sensiper, S. (1998). 'The Role of Tacit Knowledge in Group Innovation'. *California Management Review.* **40**(3): 112–132.

Leonard, D. and Straus, S. (1997). 'Putting Your Company's Whole Brain to Work'. *Harvard Business Review.* **75**(4): 110–121.

Manley, K. (2006). 'The Innovation Competence of Repeat Public Sector Clients in the Australian Construction Industry'. *Construction Manangement and Economics.* **24**(12): 1295–1304.

Manley, K. and Mcfallan, S. (2006). 'Exploring the Drivers of Firm-level Innovation in the Construction Industry'. *Construction Management and Economics.* **24**(9): 911–920.

Manu, C. and Walker, D. H. T. (2006). 'Making Sense of Knowledge Transfer and Social Capital Generation for a Pacific Island Aid Infrastructure Project'. *The Learning Organization.* **13**(5): 475–494.

Maqsood, T. (2006a). *An Investigation into the Role of Knowledge Management in Supporting Innovation for Construction Projects.* PhD Thesis. Melbourne: RMIT University, School of Business Information Technology.

Maqsood, T. (2006b). *The Role of Knowledge Management in Supporting Innovation and Learning in Construction.* PhD Thesis. Melbourne: RMIT University, School of Business Information Technology.

Maqsood, T., Walker, D. H. T. and Finegan, A. D. (2007). 'Facilitating Knowledge Pull to Deliver Innovation through Knowledge Management: A Case Study'. *Engineering Construction and Architectural Management.* **14**(1): 94–109.

March, J. G. (1991). 'Exploration and Exploitation in Organizational Learning'. *Organization Science: A Journal of the Institute of Management Sciences.* **2**(1): 71.

Mathews, J. (1999). Applying Partnering in the Supply Chain. In *Procurement Systems: A Guide to Best Practice in Construction*, Rowlinson, S. and P. McDermott (Eds). London: E&FN Spon, pp. 252–275.

Meacheam, D. W. (2006). *The Effects of the Miles and Snow Organisational Type on the Knowledge Management Methods of Organisations.* PhD Thesis. Sydney: Macquarie University, Graduate School of Business.

Miles, R. E. and Snow, C. C. (1984). 'Fit, Failure And The Hall of Fame'. *California Management Review.* **26**(3): 10–28.

Miles, R. E. and Snow, C. C. (1992). 'Causes of Failure in Network Organizations'. *California Management Review.* **34**(4): 53–73.

Miles, R. E. and Snow, C. C. (1995). 'The New Network Firm: A Spherical Structure Built on a Human Investment Philosophy'. *Organizational Dynamics.* **23**(4): 5–18.

Miles, R. E. and Snow, C. C. (2003). *Organizational Strategy, Structure and Process.* Stanford, CA: Stanford University Press.

Miles, R. E., Snow, C. C. and Miles, G. (2005). *Collaborative Entrepreneurship: How Communities of Networked Firms Use Continuous Innovation to Create Economic Wealth.* Stanford, CA: Stanford Business Books.

Mintzberg, H., Ahlstrand, B. W. and Lampel, J. (1998). *Strategy Safari: The Complete Guide Through the Wilds of Strategic Management.* London: Financial Times/Prentice Hall.

Nahapiet, J. and Ghoshal, S. (1998). 'Social Capital, Intellectual Capital, and the Organizational Advantage'. *Academy of Management Review.* **23**(2): 242–266.

Nonaka, I. (1991). 'The Knowledge Creating Company'. *Harvard Business Review.* **69**(6): 96–104.

Nonaka, I. and Konno, N. (1998). 'The Concept of "Ba": Building a Foundation for Knowledge Creation'. *California Management Review.* **40**(3): 40.

Nonaka, I. and Takeuchi, H. (1995). *The Knowledge-Creating Company*. Oxford: Oxford University Press.

Orr, J. (1990). *Talking About Machines: An Ethnography of a Modern Job*. PhD Thesis. Ithaca, NY: Cornell University.

Peansupap, V. (2004). *An Exploratory Approach to the Diffusion of ICT Innovation a Project Environment*. PhD Thesis. Melbourne: RMIT University, School of Property, Construction and Project Management.

Peansupap, V. and Walker, D. H. T. (2005a). Diffusion of Information and Communication Technology: A Community of Practice Perspective. In *Knowledge Management in the Construction Industry: A Socio-Technical Perspective*. Kazi, A. S. (Ed). Hershey, PA: Idea Group Publishing, pp. 89–110.

Peansupap, V. and Walker, D. H. T. (2005b). 'Exploratory Factors Influencing ICT Diffusion and Adoption Within Australian Construction Organisations: A Micro-Analysis'. *Journal of Construction Innovation*. 5(3): 135–157.

Peansupap, V. and Walker, D. H. T. (2006a). 'Information Communication Technology (ICT) Implementation Constraints: A Construction Industry Perspective'. *Engineering Construction and Architectural Management*. 13(4): 364–379.

Peansupap, V. and Walker, D. H. T. (2006b). 'Innovation Diffusion at the Implementation Stage of a Construction Project: A Case Study of Information Communication Technology'. *Construction Management and Economics*. 24(3): 321–332.

Pettigrew, A. and Fenton, E. M., Eds. (2000). *The Innovating Organization* (Series The Innovating Organization). Thousand Oaks, CA: Sage.

Pettigrew, A. and Whittington, D. (2003). Complementarities in Action: Organizational Change and Performance in BP and Unilever 1985–2002. In *Innovative Forms of Organizing*, Pettigrew, A., R. Whittington, L. Melin *et al.* (Eds). Thousand Oaks, CA: Sage, pp. 173–207.

Polanyi, M. (1997). Tacit Knowledge. In *Knowledge in Organizations – Resources for the Knowledge-Based Economy*, Prusak, L. (Ed). Oxford: Butterworth-Heinemann, pp. 135–146.

Porter, M. E. (1985). *Competitive Advantage: Creating and Sustaining Superior Performance*. New York: The Free Press.

Porter, M. E. (1990). *The Competitive Advantage of Nations*. New York: Free Press.

Porter, M. E. (1998). 'Clusters and the New Economics of Competition'. *Harvard Business Review*. 76(6): 77–90.

Porter, M. E. (2001). 'Strategy and the Internet'. *Harvard Business Review*. 79(3): 63–78.

Prahlad, C. K. and Ramaswamy, V. (2000). 'Co-Opting Customer Competence'. *Harvard Business Review*. 78(1): 79–87.

Rogers, E. M. (2003). *Diffusion of Innovation*. 5th Edition. New York: The Free Press.

Rothaermel, F. T. and Deeds, D. L. (2004). 'Exploration and Exploitation Alliances in Biotechnology: A System of New Product Development'. *Strategic Management Journal*. 25(3): 201–221.

Scharmer, C. O. (2001). Self-transcending Knowledge: Organizing Around Emerging Realities. In *Managing Industrial Knowledge – Creation, Transfer and Utilization*, Nonaka, I. and D. Teece (Eds). London: Sage, pp. 69–90.

Sense, A. J. (2003). 'Learning Generators: Project Teams Re-Conceptualized'. *Project Management Journal.* **34**(3): 4–12.

Sense, A. J. (2004). 'An Architecture for Learning in Projects?' *Journal of Workplace Learning.* **16**(3): 123–145.

Sense, A. J. (2005). *Cultivating Situational Learning Within Project Management Practice.* PhD Thesis. Sydney: Macquarie University, Macquarie Graduate School of Management.

Slaughter, E. S. (1998). 'Models of Construction Innovation'. *Journal of Construction Engineering and Management.* **124**(2): 226–231.

Storck, J. and Hill, P. A. (2000). 'Knowledge Diffusion Through "Strategic Communities"'. *Sloan Management Review.* **41**(2): 63–74.

Szulanski, G. (1995). Appropriating Rents from Existing Knowledge: Intra-Firm Transfer of Best Practice. *Institut Europeen d'Administration des Affaires (France)*, INSEAD – The European Institute of Business Administration.

Szulanski, G. (1996). 'Exploring Internal Stickiness: Impediments to the Transfer of Best Practice Within the Firm.' *Strategic Management Journal.* **17**(Winter): 27–43. [Special Issue]

Szulanski, G. (2003). *Sticky Knowledge Barriers to Knowing in the Firm.* Thousand Oaks, CA: Sage Publications.

Tatum, C. B. (1984). 'What Prompts Construction Innovation?' *Journal of Construction Engineering and Management.* **110**(3): 311–323.

Thompson, P. J. and Sanders, S. R. (1998). 'Partnering Continuum'. *Journal of Management in Engineering – American Society of Civil Engineers/Engineering Management Division.* **14**(5): 73–78.

Uher, T. (1999). 'Partnering Performance in Australia'. *Journal of Construction Procurement.* **5**(2): 163–176.

Von Hippel, E., Thomke, S. and Sonnack, M. (1999). 'Creating Breakthrough at 3M'. *Harvard Business Review.* **77**(5): 47–57.

Walker, D. H. T. (2005). *Having a Knowledge Competitive Advantage (K-Adv). A Social Capital Perspective.* Information and Knowledge Management in a Global Economy CIB W102, Lisbon, 19–20 May, Franciso L Ribeiro, Peter D E Love, Colin H. Davidson, Charles O. Egbu and B. Dimitrijevic, DECivil, **1**: 13–31.

Walker, D. H. T. and Hampson, K. D. (2003a). Enterprise Networks, Partnering and Alliancing. In *Procurement Strategies: A Relationship Based Approach*, Walker, D. H. T. and K. D. Hampson (Eds). Oxford: Blackwell Publishing, chapter 3: 30–73.

Walker, D. H. T. and Hampson, K. D. (2003b). *Procurement Strategies: A Relationship Based Approach.* Oxford: Blackwell Publishing.

Walker, D. H. T. and Hampson, K. D. (2003c). Project Alliance Member Organisation Selection. *Procurement Strategies: A Relationship Based Approach*, Walker, D. H. T. and K. D. Hampson (Eds). Oxford: Blackwell Publishing, chapter 4: 74–102.

Walker, D. H. T. and Johannes, D. S. (2003a). 'Construction Industry Joint Venture Behaviour in Hong Kong – Designed for Collaborative Results?' *International Journal of Project Management, Elsevier Science. UK.* **21**(1): 39–50.

Walker, D. H. T. and Johannes, D. S. (2003b). 'Preparing for Organisational Learning by HK Infrastructure Project Joint Ventures Organisations'. *The Learning Organization, MCB University Press. UK.* **10**(2): 106–117.

Walker, D. H. T., Hampson, K. D. and Peters, R. J. (2002). 'Project Alliancing vs Project Partnering: A Case Study of the Australian National Museum Project'. *Supply Chain Management: An International Journal.* 7(2): 83–91.

Wenger, E. C. and Snyder, W. M. (2000). 'Communities of Practice: The Organizational Frontier'. *Harvard Business Review.* 78(1): 139–145.

Wenger, E. C., McDermott, R. and Snyder, W. M. (2002). *Cultivating Communities of Practice.* Boston: Harvard Business School Press.

Werbach, K. (2000). 'Syndication – The Emerging Model for Business in the Internet Era'. *Harvard Business Review.* 78(3): 85–93.

Weston, D. C. and Gibson, G. E. (1993). 'Partnering – Project Performance in US Army Corps of Engineers'. *Journal of Management in Engineering, American Society of Construction Engineers.* 9(4): 410–425.

Winch, G. M. (1998). 'Zephyrs of Creative Destruction: Understanding the Management of Innovation in Construction'. *Building Research & Information.* 26(5): 268–279.

Winch, G. M., Ed. (2005). *Managing Complex Connective Processes Innovation Broking* (Series Managing Complex Connective Processes Innovation Broking). London: Ashgate Press.

Wolfe, R. A. (1994). 'Organizational Innovation: Review, Critique and Suggested Research.' *Journal of Management Studies.* 31(3): 405.

Womack, J. P. and Jones, D. T. (2000). From Lean Production to Lean Enterprise. In *Harvard Business Review on Managing the Value Chain*, Review, H. B. (Ed). Boston: Harvard Business School Press, pp. 221–250.

Womack, J. P., Jones, D. T. and Roos, D. (1990). *The Machine that Changed the World – The Story of Lean Production.* New York: HarperCollins.

Chapter 9

Culture and its impact upon project procurement

Steve Rowlinson, Derek H. T. Walker and Fiona Y. K. Cheung

Chapter introduction

How is organisational and national culture relevant to making project procurement choices? This is a question that is not, in the view of many of us concerned with the study of procurement, asked frequently enough and answered sufficiently clearly to impart the required sense of relevance.

Projects are delivered by people. Effective PM and procurement of all the people-related elements of a project, and programmes of projects, is a competence that has a critical impact upon project success that cannot be dismissed. Aspects of project leadership are embedded in many of the chapters of this book.

Underpinning leadership is an ability to understand people, understand their values and traits and understand how this may affect the way that the process of PM is conducted. Understanding what makes people 'tick' and indeed how whole systems of people, processes and technology interact is about understanding peoples' and organisations' values and the deepest formative assumptions about what is fair and reasonable. This is vital to undertaking stakeholder analysis (Chapter 3); making sense of how ethics and governance influence procurement choices (Chapter 4); comprehending the basis of what strategic approach, or school of thought, is most useful to explore procurement decisions (Chapter 5); developing and enforcing a performance measure system that is relevant and fair (Chapter 6); how to encourage value generation through innovation (Chapter 8); and how to attract, enlist and maintain the most talented team to deliver projects (Chapter 10). It also strongly links with case studies in Chapter 12 and Chapter 13. Understanding what makes people tick is about understanding their cultural perspective, what they see as attractive incentives and how they see 'fair play' so that they can trust and be confident that they can make a contribution that is worthwhile, valued and engenders their commitment. Culture, when looked at from this perspective, is a vital component and the lifeblood of an organisation.

This chapter is structured in three parts. This introduction section posed the question why culture is central to project procurement and indicated how this may be so. The second section provides some insights into the meaning and definition of cultural value terms that will frame the rest of the chapter. We argue that trust and commitment is vital to any commercial or indeed internal team working relationship in which effort is expended for rewards and benefits so we need to be clear what we mean by these terms and their relevance to project procurement. We also discuss culture from its national and organisational perspective because of cultural orientation and underpinning assumptions and because PM is undertaken using teams of cross-cultural composition. In the third section, we will then explore how understanding culture can be effective and useful when negotiating, either as a part of a procurement process or as part of a mutual adjustment process within teams, and as part of a dispute resolution system devised to deal with misunderstandings, disputes and clarification of ambiguous situations. PM is a process that is undertaken within a fog of ambiguity (at least in terms of the detail) and turbulence because it is about change. We summarise the chapter and then end the chapter with a vignette.

The significance of culture

Culture refers to the way of life of a group of people: patterns of behaviour that are seen to be useful and valuable to the people concerned and worthy of being passed on from one generation to another. Schein (1985) explains culture in terms of operating at three levels of depth. At the superficial level it is expressed by artefacts such as badges, logos, a printed mission statement or a project charter. This level hints at espoused values, what that community claims to be best represented by. At a deeper level lies the espoused value of that community. This may be explained as 'respect for employees' or 'customer focus' or 'can-do' mentality. However, while that provides a more explicit example of what to expect from (and is expected of) that community, there is a deeper layer of embedded behavioural traits that epitomises culture that is underlying these assumptions. These can be illustrated by observed and ingrained behaviour such as 'not letting the team down' so that people appear from nowhere to help when a crisis or difficulty arises.

   Figure 9.1 illustrates culture in its generic systemic form. There is a world of cultures. Nations share some common humanitarian values and, as indicated in Chapter 4, central ethical tenets can be traced across nations. History and specialisation has fostered the emergence of professions and a range of professions generally can be seen to work within organisations and, as discussed in Chapter 4, professions are defined by their culture of their ethical standards. Professions cross organisational and national borders. Project team groups also exist within and across organisational and national boundaries. It can be seen from this representation that cultural diversity extends across many boundaries. People work within the

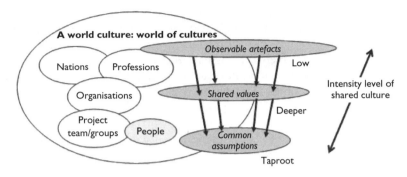

Figure 9.1 A systems view of culture.

complex network of competing and common interests and belief systems. For each 'culture' there is, as Schein (1985) explains, varying intensity of shared culture from the superficial to the profound. At the superficial level, artefacts such as badges or uniforms or 'norms' can be observed that are underpinned by values that are represented by those artefacts. Examples of these are 'an innovative approach', 'respect for the customer-professional-clan member', or 'delivering the lowest cost/superb quality', etc. These are founded upon assumptions of what is 'good' or 'bad'.

Culture takes time to evolve. Assumptions feed values that trigger symbolic representation in artefacts whereas artefacts can be more readily switched and changed. It takes time to change or to evolve into a set of 'new' culture values, and changing assumptions is something that can only be internally realised – often through a difficult and complex process of persuasion. It is very difficult to define each type of culture on its own. No matter which type of culture we look at, they all overlap with each other and cannot be separated from one another. Culture needs to be looked at as a whole. American anthropologists, Kroeber and Kluckhohn (1952) presented 160 different definitions of culture. Bodley (1994) defines culture as involving what people think, what they do and what they produce. Jenks (1993) argues the term 'culture' became widely accepted due to the changes to the structure of social life during the industrial revolution.

Culture is thus broadly defined as the dominant set of learned, shared and interrelated behaviour within a society (Vecchio *et al.*, 1992). Terpstra and David (1985) also see culture as a 'compelling set of symbols, which provides a set of orientations for members of a society'. This is in tune with many other recent texts on culture, for example, in Brown (1998), so the basic view of culture has remained consistent over recent decades. When taking these orientations together, solutions are provided to problems that all societies must solve if they are to remain viable. As Punnett (1998) suggests, culture is interrelated and provides orientation to people.

Owens (1987) and Schein (1985) see organisation culture as patterns of shared beliefs and values that evolve into norms of behaviours, which are adopted in solving problems. Rousseau and Parks (1993) view culture as layers of elements, which all lie along a continuum of subjectivity and accessibility. These layers range from physical manifestations of culture to fundamental assumptions, which are hidden deeply in the centre of culture and are most difficult to uncover (Thorsdottir, 2001). Hofstede (1991: 7–10) argues that the process of engraining cultural values, the manifestation of culture, is embedded through cultural practices being immersed within a culture. Symbols such as words and gestures carry specific meaning recognised by those within that culture. There is usually a set of heroes who possess highly prized characteristics that represent the idealised values held by that culture. Culture is shaped by the collective activities that are essential rituals that bind the members of the culture. Those outside the culture may see all this as peculiar practices that provide a set of markers that identify a particular culture. We see this at all the levels illustrated in the right-hand side of Figure 9.1. Trompenaars (1993) uses the analogy of culture as an onion, which has many layers. The products of the basic values are on the outer layer, and assumptions are deep inside the onion, which reflects Rousseau's view on culture. Schein (1985) proposes three levels in organisation culture – surface (by means of structure and rules of conduct), organisation values and underpinning assumptions, as also suggested by Wood *et al.* (2001).

Culture is knowledge, behaviour and value, and evolves over time. Culture is built over time (Handy, 1985). National culture has an effect on organisational behaviour. It creates divergent mindsets regarding characteristics of interaction and assumptions defining the boundaries of behaviour (Vecchio *et al.*, 1992). National culture assumptions are also made during business and workplace interaction. For example:

- Australians would expect open and honest communication, and are not afraid of confrontation;
- Chinese custom expects that there should be no direct confrontation with superiors; relationship and job security are highly valued and
- The English exhibit a strong sense of pride and show politeness to others, even though a strong sense of distrust might exist between parties.

In essence, national culture influences the attributional basis for evaluating behaviour and the consequences of behaviour (Vecchio *et al.*, 1992).

As shown in Figure 9.1, the surface aspects of organisational culture relate to observable culture. These are the methods that the group has developed and which have to be learnt by new members (Wood *et al.*, 2001), where the behaviours are clearly visible. Observable aspects of the culture emerge from the collective experience of its members. Some of these aspects may be observed from day-to-day practice; others may have to be discovered by asking subordinates about past events and the history of the culture. Going further down, the next level in culture is shared values between members in an organisation. Shared values play a critical part in linking people together. They also provide a set of orientations for members of a society. These orientations, when taken together, provide solutions to problems that all societies must solve if they are to remain viable. Punnett (1998), suggests that culture is interrelated and provides orientation to people. However, the shared values may not be agreed by every member in the organisation, but they have all been exposed to the values and have often been reminded that they are important (Wood *et al.*, 2001). While at the deepest level of culture, there are common assumptions or truths developed and shared by members through their joint experience in the organisation. Culture is learnt, shared and interrelated.

In searching for a practical way of understanding how systems work or what makes people 'tick', we can use these models of culture to analyse and make sense of what we see so that we can build effective environments in which to shape action in the desired manner. The value intended to be delivered through this section is that it should provide a pathway to help us understand the forces at play in organisations and groups. In doing so, we are better prepared to design procurement systems or make procurement choices and should better predict likely outcomes to actions triggered by the procurement process. As we have stressed throughout this book, recent PM procurement trends and ambitions relate to proactively, rather than reactively, influencing outcomes through gaining a holistic view of procurement as being a change agent. Culture lies at the core of managing change. Further, all members of a project team constantly engage in negotiation of what to do, when to do it and what resources (including time, energy and favours as well as more obvious resources such as money, equipment etc.) can be traded. In any negotiation, understanding the value proposition of 'the other' is both vital and culturally bound (Fisher and Ury, 1983; Lewicki *et al.*, 1994; Thompson, 1998; Ting-Toomey and Kurogi, 1998; Munns *et al.*, 2000; Williams and Miller, 2002).

Organisational culture is an interconnected web of relationships (Tierney, 1988). Handy (1985) identifies four primary forms of organisational culture based upon the work of Roger Harrison (see, for example, Harrison and Stokes, 1992):

1 Power culture – which is configured as a web with the primary power at the centre and is frequently found in entrepreneurs;
2 Role culture – in which functions and professions provide the structural pillars to support the overarching top management and is typical of public service organisations;

3   Task culture – in which the structure can be represented as a net and is job or project orientated, such culture can usually be found in a matrix organisation and
4   Person culture – in which people interact and cluster relatively freely.

Although national culture and organisational culture cannot be studied in isolation from each other, there is a clear distinction between the two types of culture. It is important to understand the connections between organisational culture and national culture because organisational culture frequently derives from national culture. Many of the shared beliefs and values that develop in organisations can be traced to commonly held assumptions in society (Wood *et al.*, 2001). Organisations seek a set of common beliefs and values amongst personnel which accord with the beliefs of the organisation, while people retain levels of individuality which impact on organisational performance (Liu and Fellows, 2001). National culture is where people in the same nation share the same beliefs, values and practices. It creates divergent mindsets regarding characteristics of interaction and assumptions defining the boundaries of behaviour (Vecchio *et al.*, 1992). National culture is learnt early in life without awareness. It is reflected in behaviour and reinforced by rules and procedures. For example, in Germany, business plans would be for 20 or even 50 years in the future, whereas in America, the business planning intervals are often much shorter.

Organisational culture is beliefs, values and practices shared by most members of an organisation. Like national culture, each organisation has its own, unique culture. Members of an organisation need to learn their own culture. When one moves to a new organisation, one needs to learn the rules of the new culture. Compared with national culture, organisational culture is acquired at a later stage in life at a conscious level. Also, values change when the top management introduces new beliefs and attitudes to the subordinates. National culture originates from the root and is more difficult to change than organisational culture as illustrated in Table 9.1 originally presented in Cheung (2006: 35).

National culture is reflected in organisational culture. Members of an organisation would resist plans to impose a culture which does not reflect, or which goes against their national values (Laurent, 1986). Although organisational culture values are learnt later in life at the workplace, they also have an influence on behaviour, just as national culture does. The extent to which the individual is influenced by the organisational culture depends very much on how long he/she has stayed with the organisation (Mead, 1998). Organisational culture has a much stronger influence on a long-term employee than a 'job-hopper', who moves rapidly between companies. Similarly, supply chains linked into long-term relationships may develop a programme or project culture. After all, culture takes time to 'take root'.

Table 9.1 Comparison between national culture and organisational culture

National culture	Organisational culture
• Share same beliefs, values and practices • Nation • Learnt early in life • Learnt subconsciously • Difficult to change	• Share same beliefs, values and practices • Organisation • Learnt at a later stage in life • Learnt at conscious level • Changes when moving to a new organisation or when new beliefs or attitudes are introduced by top management

Culture and value

Sharing common values lies at the very heart of culture. Hofstede (1991) argues that culture is a system of values. Indeed, if culture is viewed as an emergent social process, values are very important because the values of the collective form the foundations of an emerging culture (Root, 2000). However, as with culture, values are very complicated and cannot be easily defined. In human society, the removal of value means no value judgements and so no quality (Pirsig, 1974; Root, 2000). All that remains in an imaginary world of no art, variety or trade, is pure science, mathematics, philosophy and logic. Of course, there are certain concepts within the mind that are not affected by the removal of the 'value' concept because they are not part of the physical perceived reality; yet other concepts such as love, which has a strong relationship to objects in the world, are greatly affected (Root, 2000). Value is a set of relationships between the mind (self) and the world (Robinson, 1964; Pirsig, 1974; Morrill, 1980). If values are purely subjective, they can then be equated with the satisfaction of a perceived need or an object being of interest (Perry, 1954). It was argued by Pirsig (1974) and Robinson (1964) that value then just becomes a term for anything the individual likes, and is meaningless within a social group. On the other hand, if values are objective, they can be described as self-existent essences whose reality is independent of the feelings of the observer (Morrill, 1980). Again, Pirsig (1974) argues that the world can be split into subjects and objects; quality and values cannot be measured objectively due to their subjective nature and cannot be subjective because they are linked to the object. Thus, value is neither subjective nor objective. Root (2000) suggests existentialism provides a way out of this paradox by quoting Magee (1988) 'the objects of our consciousness do exist as objects of consciousness for us'. Thus, culture may be viewed as an object as long as it is treated as our consciousness and is real in its effects (Root, 2000).

In sociological terms, an individual is 'cultured' when the valuing process is conscious. Taking a child as an example, the valuing process of a child is flexible and depends highly on context; food is valued highly when the child is hungry but once fed is negatively valued (Rogers and Stevens, 1967). At the early stage, values are driven by the child as an organism, representing a clear approach to values. As the child develops, the valuing process becomes dependent on the external environment. The child learns the good and the bad from others and absorbs them as his/her own values. The valuing process changes from organism-driven to being dependent on the norms of behaviours/values expected by others (Rogers and Stevens, 1967; Root, 2000). Such ideas align with Terpstra and David's (1985) belief – culture is learnt.

However, the mechanism for acquiring values is dependent on the collective level of mental programming (Hofstede, 1980), which distinguishes people from one group to another, by classifying groups. It does not only focus on an individual, but also individuals with similar cultures would cluster together and form groups of different cultures. Some successful organisations share some common cultural characteristics. Organisations with strong cultures possess a broadly and deeply shared value system (Mead, 1998; Wood *et al.*, 2001).

Dimensions of culture

Cultures vary in their underlying values and attitudes (Wood *et al.*, 2001). The way people think about such matters as achievement and work, wealth and material gain, risk and change may influence how they view work and their experiences in organisations. Using an example of Western and Asian culture, a Western manager may value autonomy and therefore not provide detailed operational directions to his Asian subordinates. Based on the findings of Hofstede's study in early 1980s, Koreans can be described as collectivists, whereas Americans are individualists. In this case, subordinates may look up to the manager for direction and treat him (usually male) with total respect. Because of cultural differences, when the subordinates do not perform appropriately, the manager may attribute this to their incompetence, and the subordinates may come to believe the manager cannot be trusted. Such consequences then act to reinforce cultural stereotypes and in turn lead to more divergence between cultures and increase misunderstanding (Vecchio *et al.*, 1992).

One of the most ambitious studies of how cultural differences relate to organisational issues was undertaken by Hofstede in 40 different countries. He developed a framework that offers an approach for understanding value difference across national cultures. Based on his findings and evidence in the field of cultural differences, Hofstede (1980) identifies four (later five after visiting Hong Kong) cultural dimensions. The five dimensions used to differentiate between cultures are

1   *Large or small power distance*: Power distance reflects the degree to which a society accepts a hierarchical system and unequal distribution

of power. Large power distance indicates larger inequalities between the members in these societies with power and those without. People in these societies also find it normal that usually a small number of people have much power while most of the people have less power. In countries with small power distance, the basic idea is that, in principle, everybody is born equal and the principles of, say, social democracy dominate.

2   *Masculinity versus femininity*: Masculinity reflects the degree to which a society defines achievement in terms of success and the acquisition of money or material possessions. Some societies with sharp and strict divisions turn masculine; others when the divisions are loose and blurred turn feminine. In masculine societies, people admire the ones who have success. In feminine societies, people pursue a different set of values such as relationship orientation, concern for quality of life, modesty and caring.

3   *Individualism versus collectivism*: Individualism reflects the degree to which a society values independence from group membership. It is concerned with the form and manner of the relationship between an individual and others in the society. In individualist cultures, people are supposed to look after themselves and their direct families only, people are expected to conform; while in collectivist cultures, people belong to a larger group (e.g. in-group, extended family, etc.) which takes care of their interests in exchange for loyalty and relationships are more tightly structured.

4   *Strong or weak uncertainty avoidance*: Uncertainty avoidance reflects the degree to which a society tolerates ambiguous situations and the extent to which it has created institutions and beliefs to minimise or avoid these situations. In some societies, people are socialised to accept ambiguity and uncertainty and do not feel threatened. In others, uncertainty is seen as disruptive, and makes people psychologically uncomfortable. Strong uncertainty avoidance societies reduce uncertainty and limit risk by ordering and structuring things, imposing rules and systems.

5   *High and low Confucian dynamism*: Confucian dynamism is associated with the teaching of Confucius. This dimension is the degree to which people in a country emphasise values associated with the future over values that focus on the past or present. In societies with high scores on Confucian dynamism, people tend to be pragmatic, future-oriented, and focusing on obligations and tradition; whereas in those with low scores, people tend to be normative and short-term oriented, quick results are expected and people are more concerned with stability.

Further studies on cross-cultural impact on leadership and managing people were undertaken across 61 nations under the global leadership organisational behaviour and effectiveness (GLOBE) study (Ashkanasy *et al.*, 2002; Gupta *et al.*, 2002; Gupta *et al.*, 2002; House *et al.*, 2002; Szabo *et al.*, 2002; Javidan *et al.*, 2005). These extended the type of questions asked by Hofstede. The GLOBE study investigates values that participants perceive along the dimensions for both the 'as is' and 'should be' situation. This distinction is important because a person may be in a situation where they find themselves in a high power distance work environment (the as is situation) and they may feel that the work environment should exhibit lower power distance characteristics. The investigation of the gaps between 'as is' and 'should be' values revealed that national cultures could be examined in terms of nine rather than five dimensions: performance orientation, future orientation, assertiveness, power distance, humane orientation, institutional collectivism, in-group collectivism, uncertainty avoidance and gender egalitarianism. The survey involved thousands of middle managers in food processing, finance and telecommunications industries in these countries, GLOBE compared their cultures and attributes of effective leadership.

While it is beyond the scope of this chapter to discuss the details of this important research, it is worth our considering that the additional dimensions described as future orientation and performance have been dealt with in this book under Chapter 5 (Strategy) and Chapter 6 (Performance measurement) supports our inclusion of these topics as being relevant to designing a procurement choice system to deliver projects. Clearly national cultural values have a bearing on the details of a procurement system design and how choices may be made.

We should also consider how national culture may influence organisational culture from the point of view of what it means to be a team player. This would apply within organisations and in supply chains across organisations serving a project or programme. Gibson and Zellman-Bruhn (2001: 247) conducted an investigation using five different metaphors for teamwork (military, sports, community, family and associates). They derived these from the language team members used during interviews in four different geographic locations of six multinational corporations. Results indicated that after controlling for gender, team function and total words in an interview, use of the teamwork metaphors varied across countries and organisations. They analysed specific relationships between national cultural values and categories of metaphor use between dimensions of organisational culture and categories of metaphor use. Their results revealed patterns of expectations about team roles, scope, membership and objectives that arise in different cultural contexts. Their study provided an insight relevant to this book.

A strong implication of these findings is that multinational managers cannot assume that their own conceptualization of teamwork will be shared. Team members in different nations or organisational contexts are likely to have different expectations for how the team will be managed... our research challenges the assumption that the meaning of teamwork is commonly held across contexts and thus represents a first step toward a cultural contingency framework for the meaning of teamwork.

(Gibson and Zellman-Bruhn, 2001: 300)

It becomes necessary when negotiating within teams, or engaging organisations in project teams as supply chain members, to consider how language, symbols, metaphors and other artefacts represent the workplace culture. These may be either thoughtfully used to take into account culture or they may be poorly used resulting in the project goals being misconstrued, goals misaligned, expectations of actions to be taken being misunderstood. The above aspect could be viewed negatively and fearfully or it could be used positively. Tom Grisham in his doctoral thesis (Grisham, 2006) developed a useful model of using cross-cultural leadership to enhance cross-team and within-team interactions that is applicable to the above discussion. This can be applied in both dispute resolution as well as negotiation – both applicable to this project procurement discussion.

Conflict is placed at the centre of his model (Grisham, 2006: 179) illustrated in Figure 9.2 here. Conflict is addressed through a discourse between parties in which details under discussion are viewed through three sets of lenses. The size of each lens indicates the amount of time appropriate to be spent on each activity illustrated.

In the knowledge lens, in the context of culture, 'the use of metaphors is a critical technique for developing a richer knowledge of cultures (personal,

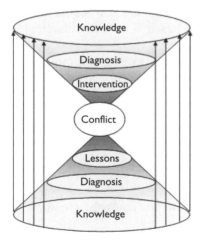

Figure 9.2 The conflict hourglass model.

societal, commercial, etc.). A cultural knowledge of the cultural individuality of the contestants including religion, customs, folklore, music, art, literature, philosophy, language, history, geography, ethics, power, gender, and economic status are critical' (Grisham, 2006: 178). The diagnosis lens involves active listening to pick up salient clues and make sense of these clues, thus cultural knowledge is vital here. The intervention lens can be difficult as it requires intervening in a manner that matches the cultural 'norms' of the parties – there may be serious mismatches that present barriers. The process and experience of addressing conflict results in lessons learned to be generated. These lessons need to be diagnosed and the synthesis of this knowledge fed back into building the knowledge base of parties.

As people participate and interact, conflict inevitably occurs. This can range from discussions about meaning and reframing knowledge about perceptions (making sense of actions and other day-to-day problem-solving negotiation) through to serious disagreements requiring resolution. Communication and understanding is a major part of this and is closely linked with cross-cultural competence.

Cultural implications of commitment on value

We discussed in Chapter 3 aspects of trust, commitment and power in terms of stakeholder management. This is highly relevant to issues revolving around the negotiation for resources and the design of procurement processes whether that is undertaken in-house or outsourced. We discussed in that chapter models that explain how trust is generated and applied, different types of trust and how these relate to project phases and how trust and distrust can co-exist. We also illustrated an extension to the typology of commitment offered by Meyer and Allen (1991) and finally we linked that with a discussion of a power typology presented by Greene and Elfrers (1999: 178). We will, therefore, limit our discussion in this chapter section to how culture may influence these concepts.

Many successful organisations share some common cultural characteristics. Organisations with strong cultures often also have a deeply shared value system (Deal and Kennedy, 1982; Wood *et al.*, 2001). Wood *et al.* (2001) argue that unique, shared values have the following benefits, they

1   provide a strong corporate identity;
2   enhance collective commitment;
3   provide a stable social system and
4   reduce the need for formal and bureaucratic controls.

Culture is a system of shared values that distinguish one group of people from another (Fang, 2001), and value can be explicitly or implicitly desirable to individuals or groups, which also influences their selections of various modes, means and ends of actions (Adler, 2002). Value also provides a guide for desired patterns of behaviour and explains why people selectively attend to certain goals while subordinating others. A high level of commitment by members also implies a strong value system within the organisation. Individuals express culture and its normative qualities through the values that they hold about life and the world around them (Adler, 2002).

Affective, continuance and normative commitment should be considered as components, rather than types of commitment because an employee's relationship with an organisation might reflect varying degrees of all three (Meyer and Allen, 1991). The definition of culture given by Liu and Fellows (1996) reflects a similar theme. They argue that culture comprises a shared understanding of what is correct, proper and normal for the conduct of relationships and has both normative and behavioural components. As commonly accepted by scholars, culture defines people's behaviour. A strong commitment to the organisation by an employee results in behaviour such as being more punctual and getting to work on time; more flexible at work hours to the organisation's benefit; willing to take up tasks outside the job specification; more self-motivated; a good team player; and good at developing relationships with subordinates. Although strong commitment to the organisation by employees influences their performance, Meyer and Allen (1991) argue continuance commitment to the organisation is negatively related to job performance. The suggested relation was observed in many studies as being supported (Meyer et al., 1989; Konovsky and Cropanzano, 1991) but also refuted in some studies (Moorman et al., 1993; Angle and Lawson, 1994; Bycio et al., 1995). A strong positive relationship was found between affective commitment and job performance (Ingram et al., 1989; Meyer et al., 1989; Konovsky and Cropanzano, 1991; Moorman et al., 1993; Leong et al., 1994; Bycio et al., 1995). Similar results were found between normative commitment and job performance (Randall et al., 1990; Ashforth and Saks, 1996) (e.g. Randall et al., 1990; Ashforth and Saks, 1996), although the results were weaker than for affective commitment. Rowlinson (2001) points out that organisational culture and structure must be matched if employees are to retain commitment to the organisation.

Clearly, a strong culture and value system can reinforce a remarkably positive view of the organisation and its environment. However, a strong culture can also be a double-edged sword, creating great resistance to dramatic changes in an organisation (Wood et al., 2001). Organisations with strong power dynamics and role prescriptions can impede the promotion of staff empowerment (Goodstein and Boyer, 1972; Serrano-Garcia, 1984; Gruber and Trickett, 1987; Foster-Fishman and Keys, 1997; Cunningham and Hyman, 1999). Implementing changes to an organisation effectively requires strong commitment from top management (Oakland, 1993; Bresnen and Marshall, 2000; Prabhu and Robson, 2000).

Staff empowerment was cited by Walton (1985) as a powerful means to engender commitment. Fawcett et al. (1994) refer to empowerment as the process of gaining influence over events and outcomes of importance to an individual or group. Staff at lower organisational hierarchy levels feel that they are given greater responsibilities in decision making. This may lead to more commitment to the project and to achievement of its missions (McDonough III, 2000), leading to greater job satisfaction and sense of ownership. The decision-making process in projects is shorter since local (hands-on) knowledge is acquired (Griffin and Hauser, 1996). Project staff work in a more cooperative manner as a team and in so doing enhance project process (Moenaert et al., 1994; Jassawall and Sashittal, 1998) and project performance such as timely completion (McDonough III and Barczak, 1991).

In layman's terms, culture is 'how we do things here'. Culture and its interaction with the organisational structure expresses commitment, and has the potential to affect a project's success. In unsuccessful projects, it is often found that people in the project or the organisation do not understand the organisational culture, and the existing culture is often not supportive of their efforts. Also, having supportive workplace culture procedures in place is the first challenge; getting staff in the organisation or the project team to follow it is another issue. We will take contract administration in the construction industry as an example. Traditional contracting has an adversarial effect on the project team; people tend to work in opposition to each other rather than cooperatively as a team. However, as with national culture, the organisational culture also has an effect on the project team through the project culture. Experience shows that project performance is often hindered by the parent organisation. This is commonly seen in a construction project with a high staff turnover, when the parent company removes staff from the project, especially when a new project arises. The new project member(s) need to learn about the project including the problems facing it and its history, as well as building up their relationships within the project team. If the parent organisation has a culture of communicating in black and white terms, with everything being referred back to the contract document; then staff working on the project will tend to act in a similar manner. Rather than concentrating on working efficiently and effectively on the project work, staff may spend a large portion of their time on correspondence, 'fighting the paper war', with letters back and forth between the contractor and the client representative.

Organisational structure also plays a major role in project success. It is the way in which the interrelated groups of an organisation are held together. Yet to develop or to change an organisational structure takes much less time when compared with initiating a different organisational culture. Changes in organisational structure can also be carried out more frequently, for example, by changing the organisation chart and reporting relationships.

Organisational structure can be described as being based on organisation size, number of supervisory levels, number of sections and job titles, number of operating sites, administrative intensity (based on the supervisor–staff ratio, spans of control by top manager and the percentage of supervisors above the bottom-level supervisors), number of divisions and the form of department (e.g. functional, project, product or matrix forms) (Van de Ven and Ferry, 1980; Galbraith, 1995; 2002).

Cultural implications of trust on value

We discussed models of trust and distrust in Chapter 3. The model illustrated in Figure 3.3 showed that trust is defined as the willingness to rely upon the actions of others, to be dependent upon them and thus be vulnerable to their actions (Mayer et al., 1995: 715). Trust is also perceived as a result of an effective collaborative relationship and higher level of partner/customer satisfaction (Zineldin and Jonsson, 2000; Mohamed, 2003). Trust is seen as representing a 'diffuse loyalty to the organisation' which prompts the individual to 'accept, as the occasion demands, responsibilities beyond any specific contracted function' (Parson and Smelser, 1984). Moorman et al. (1993) believe that trust is built up over a series of interpersonal encounters, in which the parties establish reciprocal obligations. However, Gambetta (1998) and his contributors see trust as a precondition of cooperation because partners need some assurance that the other parties will not defect.

The traditional Chinese way of doing business exhibits a high level of trust at an interpersonal level. Trust has a very strong moral content when seen in the Chinese way (see Confucius/Lun Yu).[1] Trust has the meaning of integrity, as well as the meaning of trustworthiness and compliance. In the Records of Historian, it states in a similar way 'by commending the soldier, the empire should trust him (in the sense of trustworthiness and ability) with all his heart once he is recruited'. Trust in this context has the meaning of integrity, as well as the meaning of trustworthiness and compliance (Lau and Rowlinson, 2005: 20).

The implication of high trust is that one would be confident and psychologically secure. One is more relaxed, less suspicious and defensive towards the organisation one is entrusted to (Westwood, 1993). High trust between parties not only reduces the transaction costs, it makes possible the sharing of sensitive information, permits joint projects of various kinds, and also provides a basis for expanded moral relations in business (Brenkert, 1998). Trust may have a direct effect on work group process and performance, and in Dirks' (1999) findings, it is shown that better coordination and greater efficiency are found in a high-trust group and hence better performance. Barney and Hansen (1994) believe that a firm characterised by a culture of trustworthy values and beliefs will often behave with a strong form of 'trust in exchange' relationship.

Wood and McDermott (1999b: 108–109) define trust from a social perspective with the following statement:

> Trust is a multidimensional (Sako, 1992; Ganesan, 1994; McAllister, 1995) multifaceted social phenomenon (Fukuyama, 1995; Misztal, 1996), which is regarded by some as an attitude (Luhmann, 1979; Flores and Solomon, 1998), by others as a personality trait (Wolfe, 1976) and as a vital social lubricant (Luhmann, 1979; Flores and Solomon, 1998; Gambetta, 1998).

Ganesan (1994) and McAllister (1995) identified two dimensions of trust, whereas Sako (1992) and Mittal (1996) have argued for three dimensions of trust, namely competence (behaviour), motives (feelings) and commitment (beliefs) (Wood and McDermott, 1999a). Das and Teng (1998) refer to trust as the expectation of positive motives (behaviour) of the trustee, while Lewicki *et al.* (1999) see trust as positive conduct, where morality comes from the individual. Trust also has a social meaning concerning both individual and organisation. Social trust described by Earle and Cvetkovich (1995) is a bridge from State A (disequilibrium or non-normal) to State B (equilibrium or normal). It constitutes the in-group and out-group theory where people will behave differently in groups, and is culture-specific (Earle and Cvetkovich, 1995; Fukuyama, 1995). Nevis's model of hierarchy of needs (Nevis, 1983) shows that social need is higher in Chinese culture than in Western culture.

Trust is also seen to have different levels, which is an essential precondition for a successful negotiation (Fisher and Ury, 1983). It is interesting to see how and where trust is implied in different cultures. French negotiators may come to the table mistrusting the other party until they can establish an element of trust, while American negotiators may come fully trusting the other unless led to believe that the other person is untrustworthy (Jackson, 1993). Japanese tend to have a tolerance of ambiguity and rely on mutual trust while facing internationalisation of business. Westwood (1993), finds this is a way of avoiding making offensive statements. Many managers in Asian countries negotiate in a subtle and indirect manner to avoid confrontation. The Chinese negotiation process is in an order of preference compromising: (1) avoiding, (2) accommodating, (3) collaborating and (4) competing (Westwood, 1993), with competing being the last resort (Lau, 1999; Lau and Rowlinson, 2005).

Trust is described as calculative, with constituents of self-interest and vested benefits in the economic dimension (Williamson, 1975; 1985).

Williamson (1993) further suggests two other kinds of trust, namely personal and institutional. Personal trust is suggested to be non-calculative and is irrelevant to commercial exchange; institutional trust refers to the social and organisational context on a contractual basis (Sako, 1992; Williamson, 1993). In management, Deming (1994) states that trust is mandatory for optimisation of a system where a network of interdependent components work together to try to accomplish the aim of the system. In organisational management, trust is shown in the form of achieving results, demonstrating concern and possessing integrity (Shaw, 1997), and in an employee's perception of trust in performance appraisal, trust is measured by ability, benevolence and integrity (Mayer and Davis, 1999). These items are measured as antecedent factors (Wong et al., 2000) for trust relationships in order to evaluate trustworthiness between trustor and trustee.

Trust is particularly important when a relationship contains the following elements:

- entering into any form of contract;
- exchange of information;
- uncertainty arising from unforeseeable future contingencies;
- risk sharing;
- a degree of interdependence between agents;
- the threat of missing opportunities;
- acting as a means of enhancing the effectiveness of a relationship that depends upon extensive cooperation at both inter-organisational and intra-organisational levels;
- developing the business relationship to a higher level;
- reaching alternative goals by group members and
- negotiating to avoid confrontation.

Construction project teams are unique entities, created through a complex integration of factors, with inter-disciplinary players, and varying roles, responsibilities, goals and objectives (Goodman and Chinowsky, 1996). Collaboration and teamwork are therefore crucial since sharing up-to-date information between participants leads to minimising errors, reduction of time delays and breaking the widespread rework cycle. Benefits of collaborative, rather than adversarial, working relationships within construction organisations are well documented (Walker and Hampson, 2003a).

Successful collaborative relationships rely on relational forms of exchange characterised by high levels of trust. Higher levels of trust encourage more open communication between individuals (Argyris, 1973; Ruppel and Harrington, 2000) which subsequently increases understanding of other parties' perspectives (Johnson and Johnson, 1989). Hosmer (1994a,b) suggests trust leads to commitment, which in turn leads to enthusiastic cooperation and innovative effort in people. However, it has been shown in the past that the construction industry has a stronger preference for distrust rather than the full benefits of cooperation (Wood and McDermott, 1999a). There is a need for culture change to bring about increased cooperation between parties on a long-term basis. With relational contracting, based on a long-term relationship and trust, a win–win situation can be created for both the client and contractor. The development of trust between organisations is seen as a function of the length of the relationship between them (Bresnen and Marshall, 2000). It is also believed that the construction industry is one that requires more trust between parties due to the high uncertainty in the industry.

Partnering has been criticised as benefiting the clients' side only (Green, 1999). Bresnen and Marshall's (2000) work shows that contractors would absorb extra costs in the interest of maintaining good relationships with the client and increasing chances of gaining future work. Yet, one may ask why contractors are still involved if they gain no benefit. The reasoning behind this may be the global pressure for change. Partnering is seen to be a pre-qualification requirement in recent years.

In essence, should the client attempt to spend as little money as possible to get the work done and the contractor attempts to earn as much as possible from the project the goals and values of each become incompatible. This reflects human behaviour: from an economic point of view all human beings are selfish; that is, contractors compete for the project. Competitive tendering is the most common approach used in the construction industry. The norm behind the tendering process is that the lowest bid wins the contract. The downside of this mechanism is the lower the proposed project cost, the higher the risk for work to be low quality with extra costs incurred during and/or after work. After all, the contractor also runs a business, and would not carry out work which would result in a financial loss. Rather, a minimum amount of profit to be made is set. With a low bidding price on the tender, what is most likely to happen is that cost cutting on items such as labour cost will affect the quality of labour (something which is difficult to state in the contract) with extensive sub-contracting, which makes work difficult to supervise and has a great impact on the quality control of work done. Also, chosen materials suppliers might perform poorly with no guarantee of prompt delivery of materials. Such problems have been addressed by implementing a pre-qualification system – ISO certification.

However, with relational contracting, benefits accrue not just to the client but also to the contractors because of the high chance of future work. Maintaining a good relationship to sustain a long-term relationship can lead to reduced tendering cost, by means of lower transaction costs and this benefits both the client and contractor. A cynical view might be expressed: first-generation partnering suits the public sector as it provides no guarantees of future work but commits the contractor to a non-contractual relationship; alliancing suits the private sector as long-term business relationships and mutual benefits can accrue – in both cases probity issues can be managed as illustrated in the case study in Chapter 12.

Characteristics and benefits of relational contracting include long-term relationship, trust, commitment, interpersonal attachment and cooperative and collaborative attitudes towards problem solving. Partnering, alliancing, Private Public Partnerships and joint ventures are examples of relational contracting approaches and were introduced to the Australian construction industry during the twentieth century.

Studies show that role cultures are barriers to change (see, for example, Foster-Fishman and Keys, 1997; Cunningham and Hyman, 1999). They are particularly unsuitable for a continuously changing and competitive environment. Interactions in the organisation between functional groups and individuals are controlled by rules and procedures, restricting staff empowerment, especially at the lower or team levels. Role culture is often stereotyped as bureaucracy. On the other hand, task culture is job or project orientated. This culture is extremely adaptable; it reacts quickly to changes and is most appropriate where flexibility and sensitivity to the environment are important (Handy, 1985). This culture suits a project team environment very well; after all, the whole emphasis of the culture is on getting the job done. Project teams are formed for specific purposes and can be reformed, abandoned or continued. Individuals in task cultures are found to have a high degree of control over their work, can be judged by results and have easy working relationships within the project team with mutual respect based upon capacity rather than status.

Perceived effectiveness of relational contracting

The following reflections on the effectiveness of relational contracting are drawn from research within Queensland Government's Departments of Main Roads and Public Works and their relationships with their prime contractors.

General observations and comments

Positive correlations exist between the extent of commitment by both contractor and client with high levels of commitment from both parties resulting in a more productive relationship. Indeed, a high level of commitment from all parties is needed in order for the relationship and the project to be successful. Furthermore, unsuccessful relationship management/ partnering projects were found to have one common theme – lack of commitment from all levels, that is, from project director at head office through site management down to the inspectors and foremen at the work site. Such levels of commitment must be built up through effective workshops prior to and during the project process by experienced facilitators.

In this manner, it is important to recognise that personal relationships are very important for successful partnering/relational contracting. Parties become more cooperative, problems are discussed rather than disputed, there is positive problem solving rather than confrontation, and there is sharing of information which leads to reduction of risks and unreasonable claims.

Cooperation with team members is a fundamental requirement for relational contracting. When professionals rate contribution to company success as of great importance for an ideal job there is often a strong linkage with a high score on affective commitment. Engendering a strong emotional attachment to the organisation and contributing to the company's success becomes highly important. Thus, facilitating a common and positive goal alignment to company success between professionals and the organisation is imperative. Relational contracting cannot be implemented effectively if project team members at all levels have no common goal.

A common problem on projects is that cultural barriers to change exist at both management and operation levels. Problems arise when there is a mismatch between the organisational culture as perceived by the professionals and the organisational structure that is implemented. Professionals generally prefer working in a task culture, but often, in bureaucracies, they find themselves working in a role culture. Matrix organisation is particularly suitable to the construction project environment (Bresnen, 1990; Rowlinson, 2001), and such an organisation will only work effectively with a task culture. In the Hong Kong study conducted by Rowlinson (2001), the mismatch between the actual organisation culture and organisation structure is one of the factors that created barriers to implementing changes in the department studied.

Relational contracting can be said to suit, for example, the Australian culture very well. Due to the nature of the Australian culture, professionals are not afraid to express their ideas or disagreement. Direct confrontation between individuals is accepted and is the preferred method of collaboration as well as conflict resolution. Australian professionals have strong individualist attitudes; open discussion of matters is preferred, which has an implication for decision-making styles and problem-solving techniques. Furthermore, compared to, say, their Chinese counterparts in Hong Kong, Australian professionals are not afraid to express disagreement with their supervisors.

However, uncertainty avoidance is an issue that might impact on the efficiency of implementation of relational contracting. Ineffective rules and procedures can be imposed to satisfy peoples' emotional need for formal structure. Studies have demonstrated that professionals prefer a flat organisation structure and have a strong desire for decentralisation, yet also expect a medium level of formality (this may reflect a sentient or professional culture issue with engineers). Although having roles clearly specified assists the relationships between parties, excessive formalisation, rules and procedures do not necessarily contribute to relationship building and productivity and might in fact have a negative effect on the decision-making process. Decision-making processes are often prolonged due to extensive layers of procedures that affect work efficiency. The importance of having both formal and informal mechanisms in place is an important issue to bear in mind in relational contracting projects.

Strong support and commitment from project parties is crucial for project success and implementation of changes. Also, the more the parties are satisfied with their relationships, the more productive their relationships are, and both levels of relationship satisfaction and productivity tend to increase with the degree of personal acquaintance. These organisational structure and management style issues have been highlighted as being necessary for providing an innovative workplace environment (see Chapter 13) and for effective knowledge management (see Chapter 8).

The model illustrated in Figure 9.3 presents an input-process–output model for successful relationship contracting (RC) originally presented in Cheung (2006: 152).

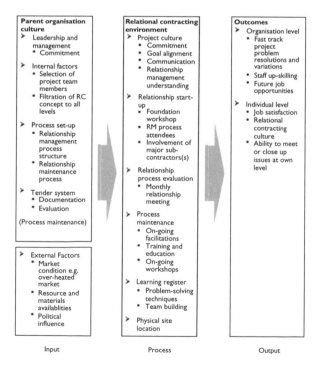

Figure 9.3 Relationship contracting input–output map.

Lessons learned from project alliancing

The following lessons are derived from research (Rowlinson, 2005; Cheung, 2006; Rowlinson et al., 2006) undertaken on the experiences of an Australian infrastructure alliance, the Brisbane Water Environmental Alliance Supply (BWEA) and is supplemented by learning from numerous other alliance projects studied by the authors including the National Museum of Australia project alliance reported in Walker and Hampson (2003b).

Rowlinson et al. (2006), in reporting on the BWEA project, identified a set of skills which could be applied to alliance projects and which were in many ways essential if the alliance was to operate in an innovative and collaborative manner. The first attribute was an ability to work as a team – this ability applied to both individuals and to groups. Second, they identified the importance of stakeholder engagement and communication skills. This engagement needed to take place both within the team and across the teams in order to ensure collaboration, cooperation and understanding. Finally, the ability to think broadly, to think outside of the box, and creatively, to innovate in both project and process, was seen as an attribute which everyone within the team needed to both develop and respect, along with the overriding culture of collaboration and an attitude of 'no-blame'.

A number of process issues were identified as important influences of effectiveness including processes that provide cultural artefacts that symbolically represented the values of the alliance team and the work environment in general. A number of issues were seen as positive influences on the work group culture that was common to the BWEA, the National Museum of Australia (NMA) alliance, as well as being observed by Dr Peter Davis in his PhD research (2006) in which he interviewed some 50 alliance participants (see Chapter 12 in this book). The process issues includes the following:

- specific merchandise and equipment that symbolically focused 'the project' as the prime organisational 'home' of participants delivering the project;
- informal social occasions such as barbecues;
- induction processes and
- team-building workshops.

Of the activities and initiatives that assisted in team development, and contributed to project success, the following appear to be most influential: brainstorming, innovation days, coaching, risk manager role, innovation manager role, facilitation services group, team building sessions and the concept of zero-in targets (a process of focusing on initiatives to be literally zeroed-in on as a level 1 priority). Such initiatives, although generally well accepted can create their own problems such as

- universality (i.e. equity and consistency) of coaching;
- level of involvement in performance measure determination by 'lower' levels of the project hierarchy;
- buy-in (and invitation in) by subcontractors who are often excluded from the 'team' despite being part of this supply chain and significant contributors to innovation and solution building and
- a neglect of documentation and reliance on word of mouth which, in the life of the project and given personnel changes, may lead to ambiguity, misunderstanding and subsequent souring of relations.

It is a commonly held belief that flexibility is an important capability required of individuals for success of projects and that the initiatives above help to develop and foster just such flexibility in both individuals and groups. However, it is worth noting that barriers to flexibility are also present, such as lack of proximity (distance between offices and sites), the budget and how it has been negotiated and 'adversarial' traditional issues such as approval processes for project variations. The move away from the 'old' culture of confrontation and dispute needs constant reinforcement from the (alliance) manager, innovation manager and even the (alliance) psychologist. This all reflects the overriding concern to develop and maintain a positive team culture across organisational boundaries.

It is important for the maintenance of a positive culture that innovation is seen as one of a number of output measures. In order to reinforce this it is important to regularly organise activities such as innovation days in order to continuously develop creativity and innovation which, in most instances, lead to successful outcomes. However, this has to be reinforced by a no-blame culture (Hutchinson and Gallagher, 2003); inevitably, not all innovations are successful and the concept of 'responsible risk taking' has to be recognised and rewarded. Sounding a note of caution, however, many relational contracts can easily revert to being mired in a culture of conservatism where there is not enough on-going innovation, and this is often linked to the time, work-load and work life balance constraints (see the Wivenhoe Alliance case study referred to in Chapters 10 and 13 for an interesting view on this issue).

So, how does one define project and team success? A current and very persuasive view is that project team effectiveness can be demonstrated by

- individuals having developed new skills;
- individuals having obtained knowledge, understanding and a tolerance of other disciplines;
- individuals having been given decision-making opportunities and having developed skills commensurate with this responsibility and
- the team that has worked well together and would like to do it again.

How project success is measured is a more problematic issue as it involves hard criteria, such as dollars and days, and soft criteria, such as stakeholder value attainment and personal development – see Chapter 6 for some further discussion of PM performance measures. Hence, a whole range of criteria are appropriate, such as

- community benefit;
- personal development;
- adopting innovative solutions (perhaps through effective value management);
- high levels of teamwork, *esprit de corps* and empowerment;
- client satisfaction;
- outstanding key performance indicators (KPIs);
- quality attainment;
- on time performance and
- at or under budget.

In addressing the functioning of alliance and relational contracts it is important to bear in mind what members of the project wish to strive to achieve personally in order to cultivate commitment and a culture of collaboration and innovation. It has to be recognised, at the very outset of the project, and proactively managed through facilitation, that many challenges must be faced and dealt with throughout the duration of a project such as

- Barriers between design and construction organisations and processes which often result in frustration and miscommunication. In order to address these key issues, particularly the problem of 'buy-in' by consultants, it is necessary to implement initiatives such as team-building workshops, proximity of the workplace, information- and problem-sharing and dispute-resolution.
- The number of companies and procedures which may mean it is time-consuming and conflict-inducing to standardise systems and procedures for the project. Often, the chosen system will result in a steep learning curve for both project team members and their home office – an inevitable source of both inter- and intra-organisational conflict and
- The decision-making process will be time-consuming, which is, paradoxically, incompatible with the fast pace of the project. This is not a new phenomenon; the introduction of matrix organisations in the early 1960s faced a similar situation enabling people to change their attitudes and accept this multi-organisational, multi-tasking environment as an essential pre-requisite for success.

All collaborative, alliance type organisations face future challenges, such as

- maintaining the momentum of the project and the project team within and between projects;
- commitment of parent organisations. The pulling out of key team members back to the parent organisation before completion is a serious problem in many projects and generates both transaction costs and uncertainty and
- the assimilation of members back into their own organisation once the project or series of projects come to an end. People become used to and 'enjoy' the new culture of collaboration and innovation and return to 'home base' with new expectations and agendas, making re-assimilation difficult.

Chapter summary

This chapter builds upon the discussion of how trust and commitment are core elements that affect the manifestation of national and organisational culture. We discussed the concepts of trust and commitment in depth as these formed the focus to link what may be considered a 'general management' topic to project management theory and PM procurement – the link being the impact of culture upon designing, developing and making procurement choices. We also provided insights from practice, more specifically our research on project alliancing.

A useful way to conclude this chapter is by indicating the benefits of relational contracting. Well-run relational contracting projects are satisfying to work in; they make work enjoyable and lead to enhanced professional development. However, not everyone is suited to relational contracting. Getting the right mix of people in a project team has been identified as one of the most crucial elements for the success or failure of a relational contract. In order to maintain harmony and an effective working atmosphere, senior management and/or the parent organisation must be prepared to remove unsuitable member(s) from the project team. Through a structured context, relational contracting increases both formal and informal communications amongst team members. Higher frequency of communication does not necessarily lead to better team relationships or more collaborative problem-solving behaviour. Instead, good team relationships and collaborative behaviour are found where respect and acknowledgement are expressed amongst team members. The range of participants from across the organisation who take part in meetings in relational contracting projects has a positive impact on the quality of communication and information flow, and is a stimulus to timely and appropriate action.

Informal communication is essential for relational contracting, but it needs to be undertaken in an appropriately structured environment with appropriate procedures.

The degree of match and mismatch between organisation culture and structure has an impact on commitment levels. The fundamental criterion for successful implementation of changes is strong buy-in from top management but buy-in is not only crucial at the senior management level, but at all levels for successful implementation. Indeed, the relational contracting concept needs to filter down to all levels in the project team if team members are to retain commitment and buy-in to relational contracting. There are multiple levels at which relational contracting needs to operate and that each level has its own issues which must be reflected upon. Findings show that a lack of prompt follow-up actions, or not closing up issues, leads to downturn or failure in relational contracting.

In addition, mismatches between organisation and project (temporary intra-organisation) cultures must be recognised and accepted by organisation staff. The parent organisation influences project culture; commitment to the goals and objectives of an organisation is crucial in facilitating successful implementation of relational contracting or organisational changes.

Relational contracting should not be seen as a one-off approach which can be switched on and off as one wishes. It is an overriding philosophy and sea-change in the industry culture. The basic concepts of relational contracting should be promoted as 'business as usual'. Relational contracting needs to be constantly maintained and facilitated to retain effectiveness. A relational contracting maintenance and review process should be set up before a project begins. Facilitation is needed to enable open, blame-free communication and this facilitation must be ongoing throughout the life of a project. The role of facilitator in achieving a relational contracting culture is highlighted in this research.

The involvement of subcontractors in the relational contracting process has been proven to be highly valuable for knowledge exchange and innovations (see the Wivenhoe case study discussion in Chapter 13). Major subcontractors should be introduced to the relational contracting process and be formally engaged in the project organisation structure. Thus, all project parties need to be familiar with relational contracting principles and relational contracting in practice for effective integration. So, education and training are crucial elements for achieving effective relational contracting. A relationship management culture must be championed in organisations through in-house workshops. Relationship management culture and principles should be embedded in people's mindset at an early stage.

Vignette

Vignette scenario

Borat Industries, a recently privatised company formed through the amalgamation of a spin-off oil and gas field development organisation previously owned by a former Soviet Republic Government agency with a US logistics provider, formed an alliance with a major oil company based in Europe together with a major German-owned construction group and a US steel producer operating from its Argentinean subsidiary.

The alliance was established with the aim of negotiating and finalising an agreement to develop a pipeline and a series of pumping stations to transport oil from newly developed oil fields in south-west Siberia and several former Soviet republics to serve the European markets. The project, or rather programme of projects, required state-of-the-art ICT development of remote controls and monitoring systems that would specifically be aimed to detect earth movements caused by seismic, sabotage or terrorist activity. The security systems part of the project required links with government agencies in a number of countries to be crossed by the pipeline.

The 'blue sky' nature of the project required mitigation of political, design, development and marketing risks. For this reason, the principal participants decide that a project alliance would offer the best environment to encourage innovative thinking, appropriately designated risk management and operational effectiveness. The time frame of completion by 2015 at the latest was considered to be extremely tight and so, given its complex and politically sensitive nature, an alliance seemed to offer the greatest chance of success.

The first set of alliance workshops was planned to take place in March 2008 in Dubai, UAE, and all parties have pledged their total commitment to the workshops. The alliance acknowledged that 'Project Glorious Borat' would be the focal point of the programme's identity, and a US public relations firm prepared a presentation to illustrate how the project could present itself, with its icons and symbolic artefacts that might be used to form a single project culture. All partners also agreed that the project would need an educational project to be developed in which systematic training and personal development would help participants and employees to know how to work within this alliance. The first educational project was training for the alliance executive board, which would be convening in Dubai in March 2008. The initiation of the project 'identity' was seen as being a key to setting the cultural environment. This vignette is concerned with the 'project' of mounting the initial Dubai workshops in which the senior executives and organisational sponsors will start to develop a single 'programme/project' culture that will be critical in establishing the cultural environment in which the programme of projects will be delivered.

Issues to ponder

1 Discuss three key national culture issues that may present problems with the 'Project Glorious Borat' programme that would need to be addressed in this first workshop.
2 Discuss three key organisational culture issues that may present problems with the 'Project Glorious Borat' programme that would need to be addressed in this first workshop.
3 What three top priorities for the education programme do you recommend should be addressed? Justify your choice.
4 What three key issues do you think will need to be considered in preparing the alliance to present its programme plan to its major stakeholders?
5 Present three key performance measures that you think would be appropriate for assessing the impact of the first stage of this programme, the initiating workshops in Dubai.

Note

1 For a web site with translation of his writing see www.confucius.org/

References

Adler, N. J. (2002). *International Dimensions of Organizational Behavior*. 4th edn. Cincinnati, OH, Ohio South-Western.

Angle, H. L. and Lawson, M. B. (1994). Organizational Commitment and Employees; Performance Ratings: Both Type of Commitment and Type of Performance Count. *Psychological Reports*. 75: 1539–1551.

Argyris, C. (1973). *On Organizations of the Future*. Beverly Hills, CA, Sage.

Ashforth, B. E. and Saks, A. M. (1996). Socialization Tactics: Longitudinal Effects on Newcomer Adjustment. *Academy of Management Review*. 39: 149–178.

Ashkanasy, N. M., Trevor-Roberts, E. and Earnshaw, L. (2002). The Anglo Cluster: Legacy of the British Empire. *Journal of World Business*. 37(1): 28–39.

Barney, J. B. and Hansen, M. H. (1994). Trustworthiness as a Source of Competitive Advantage. *Strategic Management Journal*. 15(Winter): 175–190. [Special Issue]

Bodley, J. H. (1994). *Culture Anthropology: Tribes, States, and the Global System*. Mountain View, CA: Mayfield.

Brenkert, G. G. (1998). Trust, Morality and International Business. *Business Ethics Quarterly.* 8(2): 293–317.

Bresnen, M. (1990). *Organising Construction.* London, Routledge.

Bresnen, M. and Marshall, N. (2000). Partnering in Construction: A Critical Review of the Issues, Problems and Dilemmas. *Construction Management and Economics.* 18(2): 229–238.

Brown, A. (1998). *Organisational Culture.* 2nd Edition. Harlow, UK Financial Times-Prentice Hall.

Bycio, P., Hyackett, R. D. and Allen, J. S. (1995). Further Assessments of Bass's (1985) Conceptualization of Transactional and Transformational Leadership. *Journal of Applied Psychology.* 80: 468–478.

Cheung, F. Y. K. (2006). Determinants of Effectiveness in Relational Contracting. Master of Applied Science by Research, *Faculty of Built Environment and Engineering.* Brisbane, Queensland University of Technology.

Cunningham, I. and Hyman, J. (1999). The Poverty of Empowerment? A Critical Case Study. *Personnel Review.* 28(3): 192–207.

Das, T. K. and Teng, B. S. (1998). Resource and Risk Management in the Strategic Alliance Making Process. *Journal of Management.* 24(1): 21–42.

Davis, P. R. (2006). *The Application of Relationship Marketing to Construction.* PhD Thesis, School of Economics, Finance and Marketing. Melbourne, RMIT University.

Deal, T. and Kennedy, A. (1982). *Corporate Cultures: The Rites and Rituals of Corporate Life.* Reading, MA, Addison-Wesley.

Deming, W. E. (1994). *The New Economics for Industry, Government, Education.* 2nd Edition. Cambridge, Massachusetts Institute of Technology, Center for Advanced Engineering Study.

Dirks, K. T. (1999). The Effects of Interpersonal Trust on Work Group Performance. *Journal of Applied Psychology.* 84: 445–455.

Earle, T. C. and Cvetkovich, G. (1995). *Social Trust: Towards A Cosmopolitan Society.* Westport, CT, Praeger.

Fang, T. (2001). Culture as a Driving Force for Interfirm Adaptation: A Chinese Case. *Industrial Marketing Management.* 30: 51–63.

Fawcett, S. B., White, G. W., Balcazar, F. E., Suarez-Nalcazar, Y., *et al.* (1994). A Contextual-Behavioral Model of Empowerment: Case Studies Involving People with Physical Disabilities. *American Journal of Community Psychology.* 22(4): 471–496.

Fisher, R. and Ury, W. (1983). *Getting to Yes: Negotiating an Agreement Without Giving In.* London, Penguin Books.

Flores, F. and Solomon, R. C. (1998). Creating Trust. *Business Ethics Quarterly.* 8(2): 205–232.

Foster-Fishman, P. G. and Keys, C. B. (1997). The Person/Environment Dynamics of Employee Empowerment: An Organizational Culture Analysis. *American Journal of Community Psychology.* 25(3): 345–369.

Fukuyama, F. (1995). *Trust: The Social Virtues and the Creation of Prosperity.* Harmondsworth, Penguin Books.

Galbraith, J. (1995). *Choosing an Effective Design – Designing Organisations.* San Francisco, Jossey-Bass.

Galbraith, J. (2002). *Designing Organizations – An Executive Guide to Strategy, Structure, and Process.* San Francisco, Jossey-Bass.

Gambetta, D. (1998). *Trust: Making and Breaking Cooperative Relations.* Basil, Blackwell.

Ganesan, S. (1994). Determinants of Long-Term Orientation in Buyer-Seller Relationships. *Journal of Marketing.* 58(April): 1–19.

Gibson, C. B. and Zellman-Bruhn, M. E. (2001). Metaphors and Meaning: An Intercultural Analysis of the Concept of Teamwork. *Administrative Science Quarterly.* 46(2): 274–303.

Goodman, R. and Chinowsky, P. (1996). Managing Interdisciplinary Project Teams through the Web. *Journal of Universal Science.* 2(9): 597–609.

Goodstein, L. and Boyer, R. (1972). Crisis Intervention in a Municipal Agency: A Conceptual Case Study. *Journal of Applied Behavioral Science.* 8: 318–340.

Green, S. (1999). Partnering: The Propaganda of Corporatism. *Journal of Construction Procurement.* 5(2): 177–186.

Greene, R. and Elfrers, J. (1999). *Power the 48 Laws.* London, Profile Books.

Griffin, A. and Hauser, J. R. (1996). Integrating Mechanisms for Marketing and R&D. *Journal of Product Innovation Management.* 13(3): 191–215.

Grisham, T. (2006). Cross Cultural Leadership. Doctor of Project Management, *School of Property, Construction and Project Management.* Melbourne, RMIT.

Gruber, J. and Trickett, E. J. (1987). Can We Empower Others: The Paradox of Empowerment in the Governing of an Alternative Public School. *American Journal of Community Psychology.* 15: 353–371.

Gupta, V., Hanges, P. J. and Dorfman, P. (2002). Cultural Clusters: Methodology and Findings. *Journal of World Business.* 37(1): 11–15.

Gupta, V., Surie, G., Javidan, M. and Chhokar, J. (2002). Southern Asia Cluster: Where the Old Meets the New? *Journal of World Business.* 37(1): 16–27.

Handy, C. (1985). *Understanding Organisations.* 3rd Edition. London, Penguin.

Harrison, R. and Stokes, H. (1992). *Diagnosing Organizational Culture.* San Francisco, Pfeiffer.

Hofstede, G. (1991). *Culture and Organizations: Software of the Mind.* New York, McGraw-Hill.

Hofstede, G. H. (1980). *Culture's Consequences: International Differences in Work-Related Values.* Beverly, Hills, CA, Sage.

Hosmer, L. T. (1994a). Strategic Planning as if Ethics Mattered. *Strategic Management Journal.* 15: 17–34.

Hosmer, L. T. (1994b). Why be Moral? A Different Rationale for Managers. *Business Ethics Quarterly.* 4(2): 191–204.

House, R., Javidan, M., Hanges, P. and Dorfman, P. (2002). Understanding Cultures and Implicit Leadership Theories Across the Globe: An Introduction to Project GLOBE. *Journal of World Business.* 37(1): 3–10.

Hutchinson, A. and Gallagher, J. (2003). *Project Alliances: An Overview.* Melbourne, Alchimie Pty Ltd, Phillips Fox Lawyers, p. 33.

Ingram, T. N., Lee, K. S. and Skinner, S. (1989). An Empirical Assessment of Salesperson Motivation, Commitment and Job Outcomes. *Journal of Personal Selling and Sales Management.* 9: 25–33.

Jackson, T. (1993). *Organizational Behaviour in International Management.* Oxford, Butterworth-Heinemann.

Jassawall, A. R. and Sashittal, H. C. (1998). An Examination of Collaboration in High-Technology New Project Development Processes. *Journal of Product Innovation Management.* 15: 237–254.

Javidan, M., Stahl, G. K., Brodbeck, F. and Wilderom, C. P. M. (2005). Cross-Border Transfer of Knowledge: Cultural Lessons from Project GLOBE. *Academy of Management Executive.* **19**(2): 59–76.

Jenks, C. (1993). *Culture.* London, Routledge.

Johnson, D. W. and Johnson, R. T. (1989). *Cooperation and Competition: Theory and Research.* Edina Interaction Book Company.

Konovsky, M. A. and Cropanzano, R. (1991). Perceived Fairness of Employee Drug Testing as a Predictor of Employee Attitudes and Job Performance. *Journal of Applied Psychology.* **76**: 698–707.

Kroeber, A. L. and Kluckhohn, C. (1952). *Culture: A Critical Review of Concepts and Definitions.* Boston, Harvard University.

Lau, E. and Rowlinson, S. (2005). The Value Base of Trust for the Construction Industry. *Journal of Construction Procurement.* **11**(1): 19–39.

Lau, H. L. (1999). *Trust as a Human Factor in Management in General and in Construction.* CIB W92/TG23 Symposium, Thailand, Ogunlana S. O., E&FN Spon, pp. 117–126.

Laurent, A. (1986). The Cross-Cultural Puzzle of International Human Resource Management. *Human Resource Management.* **25**(1): 91–102.

Leong, S. M., Randall, D. M. and Cote, J. A. (1994). Exploring the Organizational Commitment-Performance Linkage in Marketing: A Study of Life Insurance Salespeople. *Journal of Business Research.* **29**: 57–63.

Lewicki, R., Saunders, D. and Minton, J. (1999). *Negotiation.* Boston, McGraw-Hill.

Lewicki, R. J., Litterere, J. A., Minton, J. W. and Saunders, D. M. (1994). *Negotiation.* 2nd edn. Sydney, Irwin.

Liu, A. M. M. and Fellows, R. F. (1996). *Towards an Appreciation of Cultural Factors in the Procurement of Construction Projects.* CIB W92 Symposium, Durban, South Africa, Taylor R., pp. 301–310.

Liu, A. M. M. and Fellows, R. (2001). An Eastern Perspective on Partnering. *Engineering, Construction and Architectural Management.* **8**(1): 9–19.

Luhmann, M. (1979). *Trust and Power.* Chichester, Wiley.

McAllister, D. J. (1995). Affect- and Cognition-Based Trust as Foundations for Interpersonal Cooperation in Organizations. *Academy of Management Journal.* **38**(1): 24–59.

McDonough III, E. F. (2000). An Investigation of Factors Contributing to the Success of Cross-Functional Teams. *Journal of Product Innovation Management.* **17**(3): 211–235.

McDonough III, E. F. and Barczak, G. (1991). Speeding up New Product Development: The Effects of Leadership Style and Source of Technology. *Journal of Product Innovation Management.* **8**(3): 203–211.

Magee, B. (1988). *Men of Ideas.* London, BBC Books.

Mayer, R. C. and Davis, J. H. (1999). The Effect of the Performance Appraisal System on Trust in Management: A Field Quasi-Experiment. *Journal of Applied Psychology.* **84**(1): 123–136.

Mayer, R. C., Davis, J. H. and Schoorman, F. D. (1995). An Integrated Model of Organizational Trust. *Academy of Management Review.* **20**(3): 709–735.

Mead, R. (1998). *International Management: Cross-Cultural Dimensions.* Oxford, Blackwell.

Meyer, J. P. and Allen, N. J. (1991). A Three-Component Conceptualization of Organizational Commitment. *Human Resource Management Review.* **1**(1): 61–89.

Meyer, J. P., Paunonen, S. V., Gellatly, I. H., Goffin, R. D., *et al.* (1989). Organizational Commitment and Job Performance: It's the Nature of the Commitment that Counts. *Journal of Applied Psychology.* **74**: 152–156.

Misztal, B. A. (1996). *Trust in Modern Societies.* Cambridge, The Polity Press.

Mittal, B. (1996). Trust and Relationship Quality: A Conceptual Excursion. In *Contemporary Knowledge of Relationship Marketing, Center for Relationship Marketing*, Parvatiyar, A. and J. N. Sheth (Eds). Atlanta, GA: Emory University, pp. 230–240.

Moenaert, R. K., Souder, W., De Meyer, A. and Deschoolmeester, D. (1994). R&D-Marketing Integration Mechanisms, Communication Flows and Innovation Success. *The Journal of Product Innovation Management.* **11**: 31–45.

Mohamed, S. (2003). Determinants of Trust and Relationship Commitment in Supply Chain Partnerships. Unpublished Working Paper, School of Engineering Griffith University, Brisbane, Australia.

Moorman, C., Deshandè, R. and Zaltman, G. (1993). Factors Affecting Trust in Market Research Relationships. *Journal of Marketing.* **57**(January): 81–101.

Morrill, R. L. (1980). *Teaching Values in College.* San Francisco, Jossey Bass.

Munns, A. K., Aloquili, O. and Ransay, B. (2000). Joint Venture Negotiation and Managerial Practices in the New Countries of the Former Soviet Union. *International Journal of Project Management.* **18**(6): 403–413.

Nevis, E. C. (1983). Using an American Perspective in Understanding Another Culture: Toward a Hierarchy of Needs for the People's Republic of China. *The Journal of Applied Behavioral Science.* **19** (3): 249–264.

Oakland, J. S. (1993). *Total Quality Management.* 2nd Edition. Oxford, Butterworth-Heinemann.

Owens, R. (1987). *Organisational Behaviour in Education.* Eaglewood Cliffs, NJ, Prentice Hall.

Parson, T. and Smelser, N. J. (1984). *Economy and Society: A Study in the Integration of Economic and Social Theory.* London, Routledge & Kegan Paul.

Perry, R. B. (1954). *Realms of Value.* Cambridge, Harvard University Press.

Pirsig, R. M. (1974). *Zen and the Art of Motorcycle Maintenance.* London, The Bodley Head.

Prabhu, V. B. and Robson, A. (2000). Achieving Service Excellence – Measuring the Impact of Leadership and Senior Management Commitment. *Managing Service Quality.* **10**(5): 307–317.

Punnett, B. J. (1998). Cross-National Culture. In *The Handbook of Human Resource Management*, Poole, M. and M. Warner (Eds). London, International Business Press, pp. 9–26.

Randall, D. M., Fedor, D. B. and Longenecker, C. O. (1990). The Behavioural Expression of Organizational Commitment. *Journal of Vocational Behaviour.* **36**: 210–224.

Robinson, J. (1964). *Economic Philosophy.* Middlesex Penguin Books.

Rogers, C. R. and Stevens, B. (1967). *Person to Person: The Problem of Being Human, A New Trend in Psychology.*Walnut Creek, CA: Souvenir Press.

Root, D. S. (2000). *The Influence of Professional and Occupational Cultures on Project Relationships Mediated Through Standard Forms and Conditions of Contract.* PhD Thesis, University of Bath.

Rousseau, D. M. and Parks, J. M. (1993). The Contracts of Individuals and Organizations. *Research in Organizations' Behaviour*. 55: 1–43.

Rowlinson, S. (2001). Matrix Organisation Structure, Culture and Commitment – A Hong Kong Public Sector Case Study of Change. *Construction Management and Economics*. 19(7): 669–673.

Rowlinson, S. (2005). *Implementation of Relational Management*, CRC CI Report. Brisbane, 2002–022-A-41: 33.

Rowlinson, S., Cheung, F. Y. K., Simons, R. and Rafferty, A. (2006). Alliancing in Australia – No Litigation Contracts; A Tautology? *ASCE Journal of Professional Issues in Engineering Education and Practice*. 132(1): 77–81. [Special Issue on Legal Aspects of Relational Contracting]

Ruppel, C. P. and Harrington, S. J. (2000). The Relationship of Communication, Ethical Work Climate, and Trust to Commitment and Innovation. *Journal of Business Ethics*. 25: 313–328.

Sako, M. (1992). *Prices, Quality and Trust: Interfirm Relations in Britain and Japan*. Cambridge, Cambridge University Press.

Schein, E. H. (1985). *Organisational Culture and Leadership*. San Francisco, Jossey Bass.

Serrano-Garcia, I. (1984). The Illusion of Empowerment: Community Development Within a Colonial Context. In *Studies in Empowerment: Steps Toward Understanding and Action*, Rappaport, J., C. Swift and R. Hess (Eds). New York, Haworth, pp. 9–35.

Shaw, R. B. (1997). *Trust in the Balance*. San Francisco, Jossey-Bass Publishers.

Szabo, E., Brodbeck, F. C., Den Hartog, D. N., Reber, G., et al. (2002). The Germanic Europe Cluster: Where Employees Have a Voice. *Journal of World Business*. 37(1): 55–68.

Terpstra, V. and David, K. (1985). *The Cultural Environment of International Business*. 2nd Edition, Southwestern Publishing.

Thompson, L. (1998). *The Mind and Heart of the Negotiator*. London, Prentice Hall International.

Thorsdottir, T. (2001). Merging Organizational Culture: Lessons for International Joint Ventures. In *International Business Partnership: Issues and Concerns*, Tayeb, M. H. (Ed.). New York, Palgrave.

Tierney, W. (1988). Organisation Culture in Higher Education. *Journal of Higher Education*. 59(1): 2–21.

Ting-Toomey, S. and Kurogi, A. (1998). Facework Competence in Intercultural Conflict: An Updated Face-Negotiation Theory. *International Journal of Intercultural Relations*. 22(2): 187–225.

Trompenaars, F. (1993). *Riding the Waves of Culture: Understanding Cultural Diversity in Business*. London, Economics Books.

Van de Ven, A. H. and Ferry, D. H. (1980). *Measuring and Assessing Organizations*. New York, Wiley.

Vecchio, R. P., Hearn, G. and Southey, G. (1992). *Organisational Behaviour: Life at Work in Australia*. Sydney, Harcourt Brace Jovanovich.

Walker, D. H. T. and Hampson, K. D. (2003a). Developing Cross-Team Relationships. In *Procurement Strategies: A Relationship Based Approach*, Walker, D. H. T. and K. D. Hampson (Eds). Oxford, Blackwell Publishing, chapter 7: 169–203.

Walker, D. H. T. and Hampson, K. D. (2003b). *Procurement Strategies: A Relationship Based Approach.* Oxford, Blackwell Publishing.

Walton, R. (1985). From Control to Commitment in the Workplace. In *Manage People, Not Personnel: Motivation and Performance Appraisal.* Boston, Harvard Business Review, **63**(2): 77–84.

Westwood, R. I. (1993). *Organisational Behaviour: South East Asian Perspectives.* Hong Kong, Longmans Asia.

Williams, G. A. and Miller, R. B. (2002). Change the Way You Persuade. *Harvard Business Review.* **80**(5): 65–73.

Williamson, O. E. (1975). *Markets and Hierarchies, Analysis and Antitrust Implications: A Study in the Economics of Internal Organization.* New York, The Free Press.

Williamson, O. E. (1985). *The Economic Institutions of Capitalism: Firms, Markets, Relational Contracting.* New York, The Free Press.

Williamson, O. E. (1993). Calculativeness, Trust, and Economic Organization. *Journal of Law and Economics.* **36**(April): 453–486.

Wolfe, R. N. (1976). Trust, Anomie and the Locus of Control: Alienation of U.S. College Students in 1964, 1969, 1974. *Journal of Social Psychology.* **100**: 151–172.

Wong, E. S., Then, D. and Skitmore, M. (2000). Antecedents of Trust in Intra-Organizational Relationships Within Three Singapore Public Sector Construction Project Management Agencies. *Construction Management and Economics.* **18**(7): 797–806.

Wood, G. and McDermott, P. (1999a). Building on Trust: A Co-Operative Approach to Construction Procurement. *Journal of Construction Procurement.* **7**(2): 4–14.

Wood, G. and McDermott, P. (1999b). *Searching for Trust in the UK Construction Industry: An Interim View.* CIB W92/TG23 Symposium – Profitable Partnering in Construction Procurement, Thailand, Ogunlana S. O., E&FN Spon, pp. 107–116.

Wood, J., Wallace, J. and Zeffane, R. M. (2001). *Organisational Behaviour: A Global Perspective.* Australia, John Wiley & Sons Australia Ltd.

Zineldin, M. and Jonsson, P. (2000). An Examination of the Main Factors Affecting Trust/Commitment in Supplier-Dealer Relationships: An Empirical Study of the Swedish Wood Industry. *The TQM Magazine.* **12**(4): 245–265.

Chapter 10

Project procurement and the quest for talent

Beverley-Lloyd-Walker, Helen Lingard and Derek H. T. Walker

Chapter introduction

This book has focused much of its content in one form or another on issues surrounding project outsourcing. In this chapter we concentrate on looking inside project organisations to discuss how they may better present themselves as an employer of first choice. This is entirely applicable to both outsourced and in-house project procurement. We focus our attention on outsourced project procurement strategies from the perspective of the procurement aim as being the supply chain partner and team member of first choice. In our discussion of in-house project recruitment we will also tackle this from the perspective of addressing the value proposition of potential team members. We stress throughout this book that projects are delivered by people. A key procurement issue addressed in this chapter is discussing how organisations can attract and retain talented people who are committed, motivated and highly competent to work on projects. The key question raised is how do we attract and retain such people to be project team members? Also, how do we present our project as being attractive to our supply chain partner teams?

Figure 10.1 illustrates the main thrust of this chapter. It starts with the procurement decision in Chapter 1 that discussed the make or buy insource–outsource decision. It then follows with the influence that negotiation makes as discussed in Chapter 9. Organisational internal labour market policy also has an impact upon how staff are recruited and made available as well as terms and conditions of employment.

This chapter also links with engaging with stakeholder (Chapter 3), particularly project team members; making sense of how ethics and governance may influence procurement choices (Chapter 4); comprehending the importance of adopting a strategic approach to explore procurement decisions (Chapter 5); developing and enforcing a performance measure system that is relevant and fair (Chapter 6); how to encourage value generation through innovation (Chapter 8); and understanding the organisational culture and how cross-cultural teams can best complement what

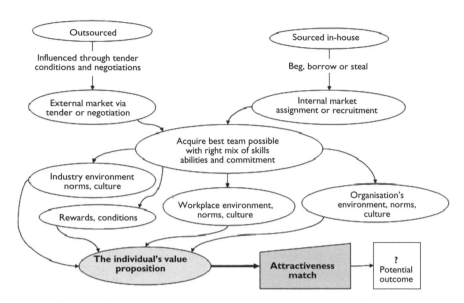

Figure 10.1 Link between value proposition and job attractiveness.

each team member has to offer (Chapter 9). It also specifically links well with the Chapter 13 case study. The main issue, whether projects are delivered in-house or outsourced, in maximising the chances for project success still rests to a major extent with attracting and acquiring the best project team (and supply chain teams) possible with a favourable combination of skills, abilities and commitment to successfully deliver the project. When looked at from this perspective, strategic human resource management (HRM) is a vital component of gathering together the best project resources to contribute to a successful project. Three factors identified in Figure 10.1 are discussed substantially in Chapter 9 from an organisational culture perspective, these are: the industry environment (its culture); the workplace environment (its culture) and the organisation's environment (its culture and also its ethical and governance stance as discussed in Chapter 4). Rewards and workplace conditions will also be more deeply explored. These factors then impact upon the individual's value proposition and the match between that 'want' and what is 'on offer' as influenced by the industry, organisation and workplace environments, determines the attractiveness of the project to individual and teams. While getting the best talent does not ensure project success it does provide a strong potential contribution.

The chapter is structured in three parts. The introduction section poses the question how can project leaders attract the best team to deliver their project. The second section provides some fundamental issues surrounding what the value proposition is for people and the way that people are recruited, integrated into project teams and their talents effectively marshalled. Following that in the third section we then explore how a strategic HRM approach can best deliver the best potential for effective participation in project teams. We will end the chapter with a vignette.

Attracting talent – a people-centric value proposition

In Chapter 8 an alliance model was offered and illustrated (Walker, 2005) in which the need was stressed for high-level strategic human capital development so that a network of willing high achieving PM team members could be readily sourced for projects as and when the need arises. This model assumed outsourcing many PM functions and so many of the PM team members would be contractors – this is the case currently in the IT industry where expert specialists as well as generalists are hired to converge on a project. This situation is typical in the film, TV and entertainment industries (DeFillippi and Arthur, 1998; Hartman et al., 1998). So from a high-intensity outsourced talent acquisition perspective, there is a critical need to be able to source talent and key staff at very short notice. This is because many project authorisations seem to drag on interminably while 'details are sorted out' and then its all-systems-go, and project managers have to magically create a team of top talent. In these cases it is likely that the potential team members who can be recruited are known to the project manager and trust the project senior team and the organisation, as well as wanting to work with them. This requires affective (want to) commitment (see Chapter 3) (Meyer and Allen, 1991). Such people may have motivation through a value proposition that includes experience or an opportunity to gain further trust with a team as part of building their portfolio of experience (Handy, 1997).

A similar situation occurs in many organisations that undertake in-house projects where the project manager may have to beg, borrow and scrounge a team that he/she wants to work on the project (Ensworth, 2001). In this case the challenge facing project managers is to persuade employees with many competing commitments, to agree to be part of a particular project. Sometimes, a project manager may be fortunate and have *carte blanche* to demand resources but even then the issue is gaining affective rather than continuance (need to) or normative (ought to) commitment or even compliance (must) commitment (Meyer and Allen, 1991). However, anecdotal evidence suggests that many project managers assembling teams internally to work on projects that may be seen merely as 'initiatives', are faced with the beg-borrow-steal model of team recruitment. This makes the task of project managers understanding a team member's value proposition even more important than is most likely currently understood.

Further, as illustrated in Figure 8.4 in Chapter 8, current leading edge and future models of sourcing workplace talented staff requires excellence in HRM and human capital development that addresses a more sophisticated value proposition for talented potential staff (Walker, 2005).

Work–life balance – an employee value proposition perspective

Employers naturally seek employees who are highly skilled and specialised or at least highly experienced. More forward thinking organisations support employees' personal and professional growth so this places another level of expectation, the ability and desire to become more resilient, flexible and independent (Waterman et al., 1994). They are making demands about the level of commitment in terms of hours worked, notice, potential travel disruptions to life and they may well be trying to do this through persuasion or, at worst, intimidation to the extent that one's 'career may depend on this'. For example in a recent UK construction industry study of a major national construction company, using an in-depth case study approach, interviewing 15 senior managers and 35 employees at a range of levels within the organisation, by Raidén et al. (2006: 889) they state that formal written contracts of employment required a 42 1/2 hour work week for site-based employees and 37 1/2 for senior office staff, though in practice many of these worked far longer hours as unpaid overtime. This is consistent with an Australian study of similar organisations where 85% of those surveyed worked on average more than 40 hours per week and 40% worked more than 50 hours a week; this has led to a form of discrimination biasing those in the industry (predominantly male) prepared to work long hours (Lingard and Francis, 2005a). It is against this backdrop that some industries such as construction are fighting to remain attractive to the workforce as there are more employees in the generation now in their twenties or early thirties who are more concerned about achieving a work–life balance (Lingard and Francis, 2005a; Raidén et al., 2006).

The changing workforce

Since the middle of the twentieth century, demographic trends and attitudinal changes have had a dramatic impact upon the needs and aspirations of the workforce. One of the biggest shifts has been the entry of women into the paid workforce. For example, in Australia, women's overall participation in the paid work force in Australia has risen from 43.7% in 1978 to 55.5% in 2004 (Australian Bureau of Statistics, 2004b). The most common family form in most industrialised countries comprises two adults in the paid workforce. In 57.5% of Australian two-parent families (with children under 15) both parents were in paid employment in 2003 (Australian Bureau of Statistics, 2004a). Traditional work practices and career paths were predicated upon the gendered notion of a full-time (usually male) breadwinner supported by a full-time (usually female) home-maker. Jacobs and Gerson (2004) note that while individual workers in the USA are actually working fewer hours per week than they have in the past, the fact that, in most relationships, both partners are employed outside the home has created an intense time squeeze on American families. Similarly, a report prepared by the Australian Bureau of Statistics found, that in dual income couples, 70% of all mothers and 56% of all fathers reported that they always/often felt rushed or pressed for time. Moen and Sweet (2004) describe a 'policy lag,' in which there is a mismatch between outdated work and career structures and the contemporary reality of dual earner couples.

This policy lag highlights an important gap in PM procurement practices. Perhaps as a result of these changes in family type, there has also been a substantial shift in the expectation of fathers' involvement in parenting. Russell and Bowman (2000) report that Australian fathers are spending more time with their children than they were 15 years ago and that the relationship between fathers and children is closer. At the same time, 68% of the Australian fathers in the study felt that they did not spend enough time with their children and 57% identified work schedule demands and inflexibility as preventing them from being the kind of fathers they would like to be.

There also appears to be some generational change in workers' expectations with younger workers no longer being motivated by the same rewards as their parents' generation, instead, placing greater value on 'non-standard' work models that enable them to enjoy a more satisfactory work–life balance (Loughlin and Barling, 2001). In the words of one participant in a recent Australian research study of construction industry employees, 'I find it hard to see myself staying in this field if I intend to start a family. Not only do I feel drained each day, it affects my relationships dramatically. It seems as if you don't have the view that you "live to work" you are not committed to your job. Today's generation are looking for self-fulfilment in a job rather than loyalty to a company and security. A balance needs to be kept and at present I don't think that view is easily accepted' (Francis et al., 2006).

In addition, the population is ageing and many couples are delaying childbearing, creating a situation in which many workers now face the situation where they have simultaneous responsibilities for the care of dependent children and elderly family members – the so-called sandwich generation. Indeed, Gorey et al. (1992) predict that, between 1996 and 2041, the number of aged dependants per 100 workers will increase from 18.1 to 34.8.

Conflicting roles

Most employees in the twenty-first century are trying to 'have it all,' combining multiple roles in an attempt to achieve happiness, fulfilment and life satisfaction. However, there is considerable potential for roles to conflict. Kahn *et al.* (1964: 19) described inter-role conflict as 'the simultaneous occurrence of two or more sets of pressures such that compliance with one would make more difficult compliance with the other'. For many adults, paid work and family are central areas of life. Work–family conflict has been defined as 'a form of inter-role conflict in which role pressures from the work and family domains are mutually incompatible in some respect' (Greenhaus and Beutell, 1985: 77). They also identify three dimensions of work–family conflict:

1   *Time-based conflict* – this occurs when time spent on activities in one role, prohibits the fulfilment of responsibilities in another role.
2   *Behaviour-based conflict* – this occurs when behaviour in one role cannot be adjusted to be compatible with behaviour patterns in another role.
3   *Strain-based conflict* – this occurs when pressures from one role interfere with fulfilling the requirements of another role.

Work–family conflict is also known to occur in two directions: from work to family and from family to work, with research suggesting the former is stronger than the latter. Work–family conflict is consistently linked with dysfunctional attitudinal, physical and psychological outcomes which have the potential to cause serious harm to organisations, individuals and families. For example, research has demonstrated work interference with family life to be associated with job dissatisfaction, life dissatisfaction, intention to turnover, general well-being, psychological strain, psychiatric disorders and substance abuse and problem drinking (Netemeyer *et al.*, 1996; Allen *et al.*, 2000; Frone, 2000; Grzywacz and Marks, 2000; Grant-Vallone and Donaldson, 2001; Boyar *et al.*, 2003; O'Driscoll *et al.*, 2003; Hammer *et al.*, 2004). One recent Australian study reports that project-based employees in the construction industry suffer much higher levels of work-to-family conflict than their counterparts who work in the head or regional office, suggesting unique pressures experienced by project personnel (Lingard and Francis, 2004).

Why should organisations care about work–family conflict?

Traditional management strategies were predicated upon a separation of work and family life. Conflict between work and family has been cast as a private matter for individual workers (and their families) to resolve. There is now compelling evidence to indicate that work and family roles should be understood as interlinked: experiences in one domain affecting mood, behaviour and performance in the other. For this reason, the interface between work and home life is increasingly a concern for organisations seeking to recruit, retain and motivate a talented workforce.

Research consistently shows work-to-family conflict to be associated with less positive attitudes towards work. For example, work-to-family conflict is negatively associated with affective organisational commitment (Netemeyer *et al.*, 1996; Thompson *et al.*, 1999) and job satisfaction, reflecting the fact that workers who perceive their work to negatively impact upon their family life are resentful of this interference. Greenhaus *et al.* (2001) suggest that, when work and family life conflict, a logical response is to try to eliminate this conflict by withdrawing from its source. Two forms of withdrawal behaviour that are of interest to organisations are turnover and absenteeism from work. Greenhaus *et al.* (2001) provide evidence that employees who are dissatisfied with the balance between work and family life think more about quitting their jobs. Similarly, Batt and Valcour (2003) report work interference with family to be significantly and positively related to turnover intentions, and employees' perceptions of control over managing work and family to be significantly negatively related to turnover intention. Boyar *et al.* (2003) also found that both directions of work–family conflict predicted turnover intention. Thus, there is substantial evidence to suggest that turnover, which is extremely costly for organisations, may be reduced through providing workplaces that are supportive of workers' family lives. Work–life balance initiatives are also reported to: improve organisations' competitiveness by increasing their ability to attract employees (Casper and Buffardi, 2004), induce employees to exercise discretionary effort in performing their work (Konrad and Mangel, 2000) and help employees to be more productive (Eaton, 2003). Further, Arthur (2003) reports a positive relationship between the announcement of organisational work–life balance initiatives and shareholder returns, indicating that investors view family-friendly firms more favourably. Grover and Crooker (1995) found that employees in companies with organisational work–life balance benefits had higher levels of organisational commitment to the organisation and expressed lower turnover intentions, regardless of whether the employee individually benefited from the policy. Thus it seems that work–life balance benefits have a positive influence on employees' attachment to an organisation because they signify corporate concern for employees and their families.

Burnout

One widely reported outcome of work–family conflict is employee burnout. The most widely accepted definition of burnout conceptualises the phenomenon as a syndrome of emotional exhaustion, cynicism and reduced professional efficacy (Maslach *et al.*, 2001). Burke and Mikkelson (2006: 65) describe burnout as 'a process in which the individual's attitudes and behaviours change in negative ways in response to job stresses…It is a process of progressive disillusionment'. Like work–family conflict, burnout has been associated with the experience of psychological distress, anxiety, depression and reduced self-esteem. To make matters worse, some research suggests that burnout is contagious, spreading to affect colleagues of those who experience it. A recent large-scale survey of the employees of one large private sector construction contracting organisation and one large government construction client organisation was undertaken in Queensland, Australia. The results of this survey (Francis and Lingard, 2004) indicate alarmingly high levels of burnout in the employees of both organisations.

The survey results indicate that employees in both the public and private sector construction industry organisations experience higher levels of burnout than international 'norms'. The mean burnout scores for the survey respondents are provided in Table 10.1, along with a sample of mean burnout scores from a range of different occupations in international research (Maslach *et al.*, 1996). The data reveals that respondents in the construction sample experience higher levels of emotional exhaustion than respondents in the military, technology and management fields. The only occupational group indicating a higher level of emotional exhaustion than the construction employees was nursing. The construction employees also reported higher levels of cynicism and lower levels of professional efficacy than respondents in all other occupational groups.

Lingard and Francis (2005b) explored work–family conflict as a linking mechanism between the job schedule demands and lack of schedule flexibility experienced by many employees and burnout. They report that the relationship between job schedule demands and burnout is actually better cast as an indirect relationship because the perception that work interferes with family or home life acts as a significant pathway link between job schedule demands with burnout. These results have important implications for the prevention of employee burnout because they suggest that job schedule demands create work–family conflict. In the context of this work–family conflict, employees have less opportunity to recover from the physical and psychological impact of work during non-work hours, leading (over time) to burnout.

Table 10.1 Cross-occupational comparison of mean burnout scores

	Construction mean (SD)	Military mean (SD)	Technologist mean (SD)	Management mean (SD)	Nursing mean (SD)
EE	2.76 (1.5)	2.05 (1.23)	2.65 (1.31)	2.55 (1.40)	2.98 (1.38)
CY	2.25 (1.5)	1.63 (1.35)	1.72 (1.14)	1.32 (1.06)	1.80 (1.24)
PE	4.29 (1.0)	4.60 (0.93)	4.54 (1.03)	4.73 (0.88)	4.41 (0.99)

Source: Francis and Lingard (2004).

EE = Emotional exhaustion, CY = cynicism, PE = reduced personal efficacy.

Thus, organisations interested in reducing the risk of employee burnout should consider ways of preventing work–family conflict, for example, through the implementation of family-responsive management interventions.

Work–life balance initiatives

A distinction is drawn between formal work–life balance benefits and informal work–life support. The former are mandated or proscribed by formal organisational policies or programmes while the latter are less tangible and closely related to aspects of the organisational culture (Behson, 2005). There is considerable evidence to suggest that formal work–life benefits, although very important, can be lacking in effect when organisational cultures discourage their use. Work–family cultures have been defined as 'the shared assumptions, beliefs and values regarding the extent to which an organization values the integration of employees' work and family lives' (Thompson *et al.*, 1999: 394) and 'the global perceptions that employees form regarding the extent to which the organization is family-supportive' (Allen, 2001: 416). Work–life cultures are said to be multi-dimensional, comprising the following:

- *Organisational time demands* and the expectation that employees prioritise work over family time;
- *Managerial support and sensitivity* to employees' non-work responsibilities and
- The perception of *negative career consequences* associated with utilising formal work–life benefits.

Within project teams, norms about the working hours that are expected of employees develop. In some cases, long hours spent at the workplace are seen to reflect high levels of commitment and productivity. An example of this is vividly described by Perlow (1998) who investigated the work–life culture of a product development team working in a large US-based software company. Perlow describes how sacrificing family time for work was considered to be an indication of commitment and one 'star' employee routinely started work at 2am and worked through until 5.30pm each day. Within the development project teams, Perlow also reports a culture of 'presenteeism' in which leaving work on time created a bad impression and annual pay increases were always justified in terms of hours spent at work. In performance reviews, top-ranked employees were those who spent between 80 and 100 hours a week at work.

Secret (2000) classifies formal work–life benefits into four categories:

1   Alternate work arrangements;
2   Leave time policies;
3   Child care services and
4   Mental health and wellness services.

Alternate work arrangements include things like teleworking, offering permanent part-time work options, compressing the work week (see case study below), and the modification of start and stop times (sometimes known as flexitime) and job-sharing. Research strongly supports the positive impact of alternate work arrangements on employees' work–life balance. Leave time policies are largely dictated by legislative provisions for maternity, parental and carers' leave and consequently vary from country to country. It is noteworthy that Australia and the USA are the only industrialised OECD countries that do not require organisations to offer paid maternity leave to women. However, it is noteworthy that some Australian organisations are offering paid maternity leave in an attempt to recruit and retain high calibre female workers. Childcare services are more controversial than other categories of work–life initiative because they can be utilised only by people with children. They can include on-site childcare or other childcare services like joint venture childcare centres, employer-sponsored places in local childcare centres, pre-tax credits for childcare assistance, subsidies for childcare costs or childcare information and referral services. Increasingly, as the population ages, care for elderly dependants is likely to become an issue for workers. For this reason, some organisations are introducing elder care services in addition to childcare supports. The final category of formal work–life benefit is mental health and wellness services. These include things like employee assistance programmes offering counselling to employees with work and/or family difficulties.

Lingard and Francis (2005a) report that employees' preferences for work–life initiatives depend on their family structure (e.g. whether they are partnered or single, parents or child-free) and their age. Younger workers (aged between 20 and 29) were more interested in wellness and personal development initiatives while workers aged between 30 and 49 were more interested in childcare services than older or younger workers. Unsurprisingly, childcare services were valued more highly among couples with dependant children than single or child-free couples. Crisis assistance/support was also more highly valued by parents of dependant children than employees in other family structures. These findings are consistent with the view that work–life experiences change over employees' life course with an individual's priorities and preferences changing as they move through different life stages (Moen and Sweet, 2004). The practical implication of the life course perspective is that no 'one size fits all' solution to employees' work–life balance is likely to be found. Instead, organisations need to be flexible enough to adapt and respond to diverse and changing needs of their employees as they move through the life course. Some organisations are, for example, recognising the value of a 'cafeteria' style reward package in which employees can select the salary and benefits combination that best suits their needs.

However, although the introduction of formal work–life benefits is recommended, these benefits will only help organisations to recruit, retain and motivate workers to the extent that they are used (Eaton, 2003). There is evidence to suggest that, in some organisational contexts, workers are reluctant to use formal work–life benefits, fearing negative career consequences or stigmatisation. Veigaa *et al.* (2004) suggest the fear that using work–life benefits will result in being seen as lacking in commitment to one's job and a negative assessment by others is a major factor in the low take-up of benefits in some organisations. Butler *et al.* (2004) report employees who anticipate negative work outcomes are less likely to use work–life policies. In certain masculine work cultures, female employees might be more sensitive to the impact of benefit usage on career consequence. For example, in a study of female construction professionals in Australia, 49.5% of a sample of 109 indicated that they thought taking maternity leave or utilising part-time work options would have a negative impact on their career (Lingard and Lin, 2003). Although, as Bittman *et al.* (2004) note, gendered stereotypes about the characteristics of an 'ideal worker' are equally likely to prevent men from using work–life benefits.

A project alliance case study

Chapter 13 describes innovative project management strategies implemented at the Wivenhoe Alliance Project in Queensland, Australia. Below is an example of how innovative work–life practices were implemented in this case study project. The case study presents evaluation data relating to some of these work practices. This case study reproduces data from a paper by Lingard *et al.* (2007).

The Australian construction industry is characterised by traditional work patterns. There is a strong culture of long hours and weekend work, especially among site-based employees. The industry's rigid adherence to long hours and inflexible work schedules is also believed to hinder its ability to attract and retain talented employees. The Australian construction industry is facing a critical shortage of skilled workers. It is estimated that, if the construction industry is to replace its retiring workers and meet growth demands, between 40,000 and 50,000 new skilled workers will be needed in the next five years (Anon, 2005). The industry's failure to respond to employees' work–life balance expectations threatens to substantially reduce the industry's long-term performance and competitiveness.

There is a growing realisation that an organisation's human resources are a primary source of competitive advantage because, unlike capital, technology and other infrastructure, workforce capability cannot easily be imitated by other organisations. Management theorists argue that employees are capable of performing at high levels when motivated to do so (Pfeffer, 1998). Transformed workplaces in which decision-making power is devolved, information flows are increased and employees strongly identify with the organisation's goals have been called 'high performance' work systems.

These work systems are believed to arise as a result of a number of human resources practices that enable organisations to recruit, develop, motivate and retain talented employees (Way, 2002; Wood and Wall, 2002). High-performance work systems are believed to develop when employees are treated with respect, enhancing commitment to the organisation and increasing trust in management. Higher levels of commitment and trust, in turn, lead to improved individual (and thereby organisational) performance (Wheatley, 1997). High-quality work is a feature of high-performance work systems. Factors influencing the quality of work include the extent to which employees enjoy job decision latitude (Karasek and Theorell, 1990) and autonomy (Parker and Wall, 1998).

This case study explores the question of whether human resource practices implemented to support employees' work–life balance yield benefits for individuals and organisations. In particular, the paper evaluates the effect of the introduction of a compressed work week in a prominent case study construction project: the upgrade of an existing dam in Queensland, Australia. The impact of the compressed work week upon project-based employees' well-being, satisfaction and work–life balance was evaluated. The performance of the construction project using the traditional metrics of time and cost was also subject to post hoc evaluation.

The case study construction project, the upgrade of an existing dam in Queensland, Australia, was delivered by means of a project alliance. The major tasks involved in the project were building an additional spillway for the dam and strengthening the existing concrete spillway by undertaking post-tensioning works. Over 425,000 cubic meters of material were excavated for the new spillway. The project also entailed the construction of a five-span traffic bridge and placement of embankment materials within the additional spillway structure.

The case study construction project was delivered by means of a project alliance (see Chapter 12). The core principle of project alliancing is the achievement of positive outcomes for all alliance members through shared commitment to common project goals (Halman and Braks, 1999; Walker *et al.*, 2002). One defining feature of project alliancing is that participants are selected on the basis of their capability, approaches and systems as well as their commitment, chemistry and the likelihood of their delivering outstanding results (Hutchinson and Gallagher, 2003).

Unlike traditional selection processes, in project alliancing, participants are selected *before* a price is considered. Typically, project alliance objectives extend beyond the traditional emphasis on price, to include the ability to innovate and manage relationships within and between alliance participants. According to Walker *et al.* (2002), there is an important difference between project alliancing and partnering. In partnering, project goals are jointly agreed and dispute resolution procedures are established in an attempt to minimise litigation but participating firms remain independent entities and it is therefore possible for one organisation to gain while another suffers in the same project. In contrast, alliance participants form a cohesive entity which jointly shares the risks and rewards arising during and as a result of the project. As such, poor performance results in penalty to all alliance participants and, conversely, rewards for excellent performance are shared between the participants. Put simply, 'a fundamental design principle of a project alliance commercial framework is that if one participant wins, all win; or if one loses, all lose' (Hutchinson and Gallagher, 2003: 18).

The compressed work week, which involved reducing the length of the working week, but increasing the length of the working day, was introduced part way through the construction phase of the Wivenhoe Dam project. At the commencement of the construction project, the site was operating on (approximately) a 58-hour week, spread over six days. This comprised five 10-hour days (Monday to Friday) plus an 8-hour day on Saturday. Shortly before data collection commenced in March 2005, the site moved to a 5-day week, with working hours extended to 11.5 hour per day on week days. Weekend work was no longer required. By May 2005, the site was operating on winter hours, reducing work hours further to 10.5 hours per day.

Questionnaires were administered to employees at the site in June 2005. The questionnaires were designed with the objective of evaluating employees' reactions to the move from a six to a five-day week.

Employees at the case study project rated their preferences for the compressed work week on a seven point scale, where '1' = 'very strongly prefer five-day week' and '7' = 'very strongly prefer six-day week'. The mean score was 1.79 (sd = 1.55), indicating a strong collective preference for the five-day week. The numbers of workers for each response option are indicated in Table 10.2 and indicates that the majority of both wages and salaried employees expressed a preference for a five-day week. However, a small number of wages staff indicated that they either had no preference or preferred to work a six-day week.

Employees were also asked to indicate the extent to which they believed their work–life balance had changed since the introduction of the compressed week. The change was rated on a seven point scale (from '1' = 'greatly worsened' to '7' = 'greatly improved'). The average rating for salaried employees was 5.94 (sd = 1.39) and for wages employees was 5.87 (sd = 1.32) indicating that, overall, most employees felt their work–life balance had 'moderately improved' since the introduction of the compressed work week.

At the end of the questionnaire, both wages and salaried employees were invited to 'add any further comments' about work–life balance at the case

Table 10.2 Employees' preference for five- or six-day week

	Numbers of staff		
	Combined	Wages only	Salary only
Very strongly prefer 5 day	30	14	16
Prefer 5 day	4	3	1
Slightly prefer 5 day	2	1	1
No preference	3	2	1
Slightly prefer 6 day	0	0	0
Prefer 6 day	2	2	0
Very strongly prefer 6 day	1	1	0

study project. More comments were provided by salaried employees ($n = 15$) than wages employees ($n = 2$). In general, the additional comments provided by respondents strongly supported the introduction of the compressed work week. Most salaried employees who provided additional comments expressed their satisfaction with the compressed work week. For example, one respondent wrote, 'Have enjoyed the five day week – would prefer to put in extra hours during the week if it means having every weekend off.' Another commented, 'The five day week is fantastic. This is the way the whole industry should operate.'

Many salaried respondents also identified 'knock-on' effects for work (e.g. productivity and loyalty to the organisation) and non-work activities (e.g. family activities and domestic duties). For example, one wrote: 'In the last (x) years I've always worked a six day week and was often stressed and tired by Saturday. On this job I've felt very relaxed on the weekend....I've also been able to complete all my jobs around the house. Our crew is happier because their money hasn't changed much and they have a life. I personally think that productivity has been excellent because everybody is fresh and happy.'

Several salaried employees directly linked the five-day week to employee retention. For example, one commented, 'A commitment to pursue a five day week across the business would help people make loyal decisions at project completion...' and another wrote, 'A five day week is what I and my family now demand. I will not work six days again, even if it means changing to another industry.'

The interview data also supported the survey data in indicating that the compressed work week was perceived to improve Wivenhoe employees' work–life balance. Example comments were: 'The five day week has just made it incredible. I've talked to the workers out on site. I mean, they get to spend a whole weekend with their kids and their families now, not just one day' (salaried staff) and 'since Wivenhoe has gone to a five day week it has increased my enjoyable lifestyle substantially' (wages staff).

The link between work hours, work–life balance and productivity also emerged as an important theme in the interview data from Wivenhoe participants. One salaried employee expressed his appreciation of the compressed work week in the following way: 'I was actually contemplating whether the construction industry was not for me. And I was becoming active in seeking other roles. And then the elimination of the Saturday work – really saved that. So if it wasn't for that, [I] probably wouldn't be here at the moment. And, not only had I felt the change, and the huge benefit – my wife has as well. She immediately saw a totally different person on the weekend. So that was really positive. But now I am much happier, much more energetic at work. So I concentrate for longer – well, for the entire time I am here. Whereas before there were times that were non-productive.'

In the Australian construction industry a six-day week is the norm and the move to a five-day week at Wivenhoe was an innovative and unusual measure. Many of the employees interviewed suggested that, under traditional project delivery arrangements, the introduction of the five-day week would be unlikely. Reasons for this include the fact that, under traditional arrangements, the risk associated with the move to a five-day week during the construction stage of the project would be borne entirely by the construction contracting organisation. In this situation, if the five-day week improved performance, the contracting organisation would reap the benefit. However, if the five-day week hindered performance, the contracting organisation would stand to be heavily penalised. For example, a contractor would incur liquidated damages in the case of time over-runs. As the outcome of different alternative work schedules is currently unknown in the construction context there is considerable uncertainty about the impact of a five-day week on project time performance, under traditional arrangements a contracting organisation may be unwilling to take such a risk. However, the commercial arrangement underpinning a project alliance is based upon the sharing of risks and rewards. When interviewed, the Wivenhoe Alliance Project Manager expressed this as follows: '...Then I guess the key for alliancing from that point on – everyone shares in the pain or the gain. So you haven't got a client/contractor mentality where one wins, one loses, that sort of thing. So if we all go well, everyone benefits, all share in savings. If things go badly and off the rails, we will lose. Obviously, when it comes to decision time – you've got a lot of decisions to make, obviously, throughout the course of the project. It helps to align people's thinking to come up with the "best for project" solution.'

In the context of shared risk, the likelihood that innovative solutions to problems will be identified and implemented is greatly increased (Walker et al., 2002; Hutchinson and Gallagher, 2003). Thus, at Wivenhoe, the risk associated with implementing the compressed work week was shared between participants, who also all stood to gain if the measures improved productivity and performance.

The construction work on the case study project was officially completed on 22 September 2005, six months ahead of the scheduled completion date. The project also cost less to construct than originally estimated. Thus, taking the traditional metrics of cost and time, the project was a remarkable success. Owing to the fact that the introduction of the compressed work week on the case study project was subject to a post hoc evaluation it is impossible to determine to what extent the introduction of the compressed work week caused the project's good performance in terms of cost and time. In the absence of a rigorous experimental design, inferences about cause and effect cannot be made. However, the evidence suggests that the site-based employees perceived the compressed work week as being a very important benefit which the employees linked to productivity benefits. Coupled with the project's objective performance data, this provides *prima facie* evidence that alternative work schedules designed to help employees to achieve a better work–life balance are not incompatible with the attainment of time and cost objectives in the context of a construction project.

This section on work–life balance issues provides convincing evidence that current work arrangements have a significant impact upon commitment and productivity levels and also strongly influences the organisational culture. We also highlighted a significant gap in PM procurement practices for sourcing talent.

Strategic HR and the quest for talent

Whether staffing a project internally from the project-originating organisation or externally selecting employees to become members of a new project team, the principles are the same and driven by the knowledge that people are an organisation's greatest asset. Indeed, people are the prime source of competitive advantage today (Kaufman, 1999; Ruona and Gibson, 2004; DiVanna and Rogers, 2005). People and their competencies, knowledge and experience are vital for organisational success in this era when intangible knowledge-based resources, that which only people can provide, are the most likely source of sustainable competitive advantage, (Pringle and Kroll, 1997; Chan *et al.*, 2004). The challenge is to attract, develop, motivate and retain talent (Frank and Taylor, 2004), and the pressure to do so will increase with acute skills shortages already occurring in some professions and trades.

Planning for the right people

Planning for people to ensure the needed skills and knowledge are available across the organisation when they are required is important. Human resource planning has been used to identify staffing needs at the level of numbers and skill types. Today workforce planning and workforce profiling are used to gain a greater level of understanding of the current and future availability of employees or potential employees. In the past succession planning was used to identify potential replacements for the most senior roles in an organisation. Today succession planning is required at all levels of the organisation.

In project environments the source of team members may be internal, external with all or some team members provided by a specialist staffing organisation, met through forming strategic alliances with organisations that employ people with the required skills, or the project or parts of it may be outsourced. With human resources being the key to project success, it is important that current and future supply and demand of this vital resource be constantly tracked. Industry-wide or profession-specific data may be available. For instance, in the USA a report by the University of Texas at Austin used workforce profiles to map likely retirement of engineers, and predicted supply to meet demand for engineers in the coming years. Although only 15% of the current workforce was over 55, their workforce profile showed that the industry had a window of about five years during which the knowledge and social networks of the most senior engineers could be transferred to younger employees before the retirement of the largest portion of the workforce (Davis-Blake *et al.*, 2001). This information provided advance notice of a possible shortage in professional project engineers and the time in which to address the shortfall. Succession planning can now occur, not just at the senior executive level but in a manner that ensures that sufficient numbers of people are available across the range of technical and professional jobs in the organisation to replace those leaving and to meet future project needs. This enables gaps in skills and knowledge to be avoided or at least reduced.

When choosing members for project teams over the next 5–10 years, strategically positioning less experienced staff members with experienced project managers could assist in the transfer of knowledge and provide links to social networks. Researching current enrolment numbers at local universities and colleges would reveal whether there will be sufficient graduates to replace those expected to leave in the coming years. Other plans may be required if, over the coming years, the number of graduates will be insufficient to replace departing staff. We will suggest other possible approaches to ensuring supply for projects throughout this chapter.

Project workforce planning

Workforce planning is required to ensure that current and future knowledge, skills, abilities and other attributes (KSAO) needs are met. When planning reveals, for instance, that the supply of engineers, accountants or computer programmers will not meet future demand, action can be taken to address the situation before it adversely affects organisational performance. This information can quite easily be gained by accessing university and college enrolment data. With a 3–5 year time lag to qualification, and experience required to develop high-level skills, it may be necessary to consider working with preferred contractors or alliance partners to ensure the continuing supply of high-quality personnel. This may involve sponsoring students via scholarships, providing work experience opportunities to accelerate learning and on-the-job experience to ensure supply.

Alternative approaches may involve rethinking future projects – if the supply of structural engineers is currently limited and university and college enrolments indicate that new entrants will not cover those expected to exit the profession, it may be necessary to reconsider the building technologies to be used on a project.

Global competitiveness comes from having world-best staff. Major projects will require consideration of the current and future supply/demand situation. Recent examples demonstrate this. The world-wide move to SAP in the petroleum and chemical industries created a high demand for specialised IT staff. Some of these technicians saw the opportunity to increase their income and moved from one organisation to another, wooed by increased financial rewards and responsibility. They took with them not only their skills but organisational knowledge, much of which could provide competitive advantage. Job changes here were usually between competing organisations, meaning business information could be lost to a competitor. Once the major, required, SAP installations were completed the demand for SAP experts reduced and has now plateaued.

As the new millennium approached we had the Y2K projects across industries. Again, IT workers were in high demand. After New Year's day 2000, the demand for Y2K experts ceased to exist.

Some reduction in enrolments in IT-related courses has occurred across universities and colleges because these changes in demand have been discussed in the media. But there are other major projects which do now, and others that will in the future, require specialised IT skills. London has won the right to stage the 2012 Olympic Games. Many of the stadiums and other facilities required are already in existence or require upgrading only. Others need to be built. One need that will require specialist staffing is that of security. With the current threat of terrorism around the world, highly skilled intelligence staff and IT staff with skills in surveillance technologies and transaction tracking will be required.

A similar situation exists, for example, in the construction industry. When a project is nearing completion, and this is obvious to all team members – how to retain people until the project is delivered then becomes a quandary. We discuss this situation further under retention and rewards.

By careful design of retention and reward strategies and plans, workforce planning is supported.

Organisations that do not plan for their most important resource – people – will not be able to compete into the future. Whether it is internal skills, skills normally supplied via preferred or occasional contractors or those usually provided through alliances, how these needs are going to be met has to be included in long-term planning.

This situation highlights why retaining critical, or key, people is extremely important. Job design, management style and rewards systems can assist in retaining key staff. These people-management activities can also address retention of those working in areas predicted to be in short supply in the future. These people will be in high demand and others will attempt to lure them away. It will be too late when the shortage begins affecting all organisations to begin re-training schemes or to develop reward systems targeting workers in areas of skills shortages.

Attracting and recruiting quality employees

Ensuring that teams contain the right mix of knowledge, skills and abilities requires superior selection skills, whether team members are being chosen from within the organisation or recruited into the firm. A clear statement of the job and of the philosophy of the organisation within which the incumbent will be working will be needed to guide appropriate attraction, recruitment and selection activities.

Recruitment has in the past been viewed as initiating the employment relationship, but with the skills shortages today, attraction is where it begins. Organisations have to consider how attractive they appear to the public, all the time. Many potential employees, and especially those soon to graduate, have in mind the type of organisation they would like to work for, and take the initiative to locate a suitable position with one of their chosen organisations rather than waiting for jobs to be advertised. The organisation's culture, its human resource strategy and people-management philosophy are a large part of its public image. How the organisation does business, whether it is viewed as socially responsible and as having a commitment to sustainability will impact on its image. Potential employees are looking for a personal values/organisation values match.

Employer of choice

The motivation for organisations to become an 'employer of choice' or 'best employer' drives the need to present a desirable image to the public, and especially to potential employees. Shellenberger (1998) claims that best employers or employers of choice provide higher returns for shareholders.

This is not surprising when increased productivity, retention of top talent and overall retention rates are common for workplaces which could be described as 'simply the best' (Hull and Read, 2003).

Becoming a 'great place to work' or 'best company to work for' (van Marrewijk, 2004) requires strong commitment from senior management of the organisation and support for a 'people first' strategy. Two-way communication between employees and management, where the flow of negative information may occur without fear and where the division between management and employees is minimised are also features of best companies to work for. As a result, employees take pride in their work, their team and the company. Van Marrewijk (2004) cites recent research reporting that the top 100 Best Companies to Work for in the USA achieve extraordinarily good financial results. These organisations achieved consistently superior results to other organisations across a range of measures. Credibility, integrity, respect, fairness, pride and camaraderie are all present in Great Places to Work. These organisations are considered socially responsible and have the reputation of being 'good employers'.

Reputation

What attracts a quality potential employee to an organisation? It is necessary to consider the organisation's culture and its employment philosophy in order to become an employer of choice. The culture of the organisation is important. Employees seek a match between their values and the culture and values of the organisation. They want to feel proud of the company they work for (Rushton, 2002). The reputation of an organisation is important when seeking to attract and select high-quality employees; high calibre employees seek to work in companies with high standards (Lipman, 2002). Employees are more likely to remain with an organisation they are proud to work for, and whose values are in line their own values. Corporate reputation has become vital for organisational success and survival; a good reputation will attract creative employees who want to work for an employer whose values are in harmony with their own values (Dolphin, 2004).

The employee value proposition

Research demonstrates that people are attracted to different professions and trades according to, at least to some extent, their personality type. Holland's Typology of Personality (1985) identifies personality characteristics and congruent occupations for each personality type. Using Holland's typology, a better understanding can be gained of the type of person who would best perform a particular job.

It is necessary to consider what the employee value proposition (EVP) is, or what an employee desires (values) in a job and in the organisation in which the job is performed. Value, in all contexts, is defined by the receiver not the giver (Ulrich and Brockbank, 2005). For this reason it is important to listen to what employees want, and in particular to what the type of employee you desire wants. It is necessary to identify the knowledge and skills required to perform the job and the other attributes that will enable the person to 'fit the work group'. For instance, what engineers desire from a job and what nurses desire from a job may differ greatly. Unless a deep understanding is gained of what it is that each of these employee groups desires, and action is taken to develop a workplace environment that addresses the specific EVP of the desired group, efforts to attract, motivate and retain these employees may be misdirected and ineffective.

Understanding the EVP, and using the knowledge from that understanding, assists in attracting, motivating and retaining employees.

Employer branding and becoming an employer of choice

Competition is driving the need for organisations to develop a strategy to locate and retain employees who will enable them to succeed into the future. In an era of skills shortages, the way that organisations manage human capital has become the most critical strategic decision they make. These pressures have created the need for organisations to consider new and different ways of attracting recruits. Increasingly, employer branding is being used by organisations for this purpose.

The staffing process now begins with attraction, followed by the recruitment process, and then selection, where the best applicant is selected. The quality of the attraction effort will affect the quality of the pool of applicants from which the selection choice is made. The best person chosen from a poor pool of applicants will not be good enough today. This is why the quality of attraction activities, strongly supported by employer branding, is extremely important. It influences the quality of the overall staffing of the organisation.

Creating an employer brand is part of the process of becoming recognised as a good employer. It is about developing a way of defining and promoting an organisation's employment attributes to prospective employees, to attract them to apply for vacant positions. It is also a retention tool. It can make current employees aware of the advantages of working for the organisation, and demonstrate that they can be proud to say they work for the organisation (Dessler *et al.*, 2007). Employer branding, therefore, supports both attraction and retention.

Employer branding is used to establish an image that reflects what the organisation stands for by projecting an image of a desirable workplace to attract quality employees. This is done in a way that will ensure that there will be a pool of suitable potential applicants waiting to apply when a position becomes vacant. Suitability is guided by an understanding of personality types and their link to occupations.

Research conducted in Canada found that in the hospitality industry professionals were attracted by messages that stressed issues such as the organisation's growth and diversity. For frontline service employees the message required to attract suitable applicants was one that emphasised their personal role in delivering quality service (Vu, 2006). We need to consider the type of person who is likely to wish to perform the role when attempting to target candidates for different roles. What is required will vary according to group, or in marketing terms, according to market segment.

Firms attempt to differentiate themselves from competitors by highlighting their characteristics as an employer through their employer brand. The organisation's values and the way in which they manage their employment relationship are conveyed through words, signs, symbols or a combination of all of these (Backhaus and Tikoo, 2004). The development of an employer brand begins with the value proposition which is to be embodied in the brand.

The organisation's web site can provide the ideal setting for establishing the employer brand; it is here that statements about values and the organisation's philosophy in relation to its employees, customers and suppliers can be stated. Details of the advantages of working for the organisation may include flexible work arrangements, providing career opportunities and other benefits which make it a great place to work. Living up to these claims may not be as easy as making them. Loss of trust follows if current employees, or new recruits, feel that the statements are hollow or if their experience of the organisation does not match the claims made. New employees are likely to be disappointed and experience disillusionment. This has been found to lead to an increase in employee turnover during the first year of service. What this highlights is the need for organisations to take care in the development of their employer brand. The organisation should truly represent what it stands for and how it treats its employees.

Employer branding can include statements that are extremely effective at attracting applicants; however, at times this may lead to too wide a group of potential applicants. It has been found that after several years of working hard at becoming an 'employer of choice' some organisations in Canada encountered negatives attached to this label. In particular, being viewed as 'the organisation to work for' was found to lead to organisations being inundated with applicants, including unsolicited applications. As a result, considerable time and effort was required to screen and select from an increased number of applicants (Vu, 2006). In addition, after enjoying the benefits of working for an employer of choice employees didn't want to leave their 'best employer'. Low turnover rates are good; no turnover, though, can lead to stagnation and lack of new ideas. Good selection decisions using valid and reliable selection methods will be important when selecting the best from a large pool of applicants. Adequately inducting and training, then managing the performance of new recruits will also be required. Applicants may be attracted to what the organisation has to offer, but not possess the knowledge and skills required for the jobs to be filled. As a result, organisations are realising that they must become 'the employer of choice for employees of their choice' (Vu, 2006: 4) The way in which a desirable employer brand image is crafted is important (Backhaus and Tikoo, 2004).

Employees become quickly disenchanted with workplaces which claim to be one thing but are experienced by their employees as another. Organisations that successfully use employer branding and gain the right to call themselves 'employer of choice' ensure that what they claim to be is what they are. In this way, the brand they use truly reflects the type of organisation they are, and new employees do not become disillusioned because they were attracted by a statement on the employer's web site which proves not to be the factual experience of employees.

Selecting staff

It is said that retention begins with selection. In the past, recruitment and selection processes were designed to ensure person–job match. In fact, it was often said that employees were hired for their ability to perform the job but terminated (fired) for lack of fitness for the job – because their values and those of the organisation did not match.

General selection principles

Robertson and Smith (2001) rated selection methods accuracy. The most effective methods were cognitive ability and integrity tests, and structured interviews, closely followed by work sample and job knowledge tests. Personality tests are less effective but, importantly, may enable reliable assessment of ability to work in a team.

Assessment centres can be used for selection, as a basis for promotion decisions, to identify training and development needs or to select suitable team members to ensure a balanced and productive team. Assessment centres incorporate a range of activities and tests. Cognitive ability and integrity tests combined with structured interviews and work-related activities (a work sample test) and job knowledge tests can provide valuable information for good selection decision making.

Personality tests, or activities which assist in measuring leadership and team skills, when incorporated in an assessment centre, will help identify people suited to working in project teams. As with work sample tests, this information can also be used to identify training needs of current employees, or development required in preparation for promotion to a new role.

Assessment centres offer the opportunity to observe how potential or current employees handle a range of situations that could occur in their actual jobs (Applebaum *et al.*, 1998). Project managers are probably not going to have access to or be trained in the use of the range of tests required to make an effective selection and incorporated in an assessment centre. Increased emphasis on the management of people for organisational success has led to each manager, or supervisor, becoming involved in an increasingly wide range of people-management activities formerly performed by the HR department. Initially, some questioned whether this would signal the end of the HR department as we knew it (Cunningham and Hyman, 1999). HR's role has changed. Involvement in such activities as promotion decisions, performance management, coaching, employee discipline and termination decisions, job design, career planning and development and recruitment and selection by project managers and line managers is now greater than is HR's (Kulik and Bainbridge, 2006). Policies and procedures to guide the conduct of these activities are developed by HR. Managers will want to be directly involved in the interview process and therefore may need to be trained in structured interviewing skills.

Project managers will be more directly involved in recruiting and selecting team members than perhaps in the past, but they should make use of the expertise HR officers possess in this area to assist in making good selection decisions.

Employee referrals can provide a cheap source of job candidates and enable the staffing process to be quickly completed. Current employees will not risk recommending a poor performer or an unreliable person; this would reflect poorly on them and perhaps hinder their personal career progress. Project managers, and other project team members, may recommend to one another people to fill all remaining team roles. However, this seemingly ideal method of recruiting and selecting new employees or team members does have some drawbacks. If all new job candidates have been referred by employees it may result over time in a lack of diversity and creativity. People who think or act differently will not be recommended. Indeed, this can be one of the issues to watch for when using the person–organisation fit to select employees (discussed below). Though employing people who quickly fit into the established culture and do not question any of the norms and practices in the organisation may sound attractive, there is no doubt that some questioning of the *status quo* can be healthy for organisations. Employing people who have had different experiences, worked in different organisations and industries, can lead to more innovative and creative solutions being found to problems, for instance.

Employee referrals can be the quickest and cheapest way of filling vacant positions and of selecting team members, however, these advantages should be weighed against the potential disadvantages of this practice over time.

Person–job and person–organisation fit

The concentration in the past was on ensuring that the applicant could perform the clearly defined role. More recently it has been recognised that today's workplace requires person–organisation fit. The use of employer branding is important here. It can convey a true picture of the organisation and attract the type of person required to fit both the job to be performed and the culture of the organisation. This suggests that it is important to clearly state both the values of the organisation and the role to be performed. But in today's workplace, especially one where project teams are regularly formed and disbanded, this is not always possible. Employees are now selected to work in groups, to form part of different project teams and are expected to frequently change roles (Lievens et al., 2002). Here it is important to select people who are flexible; those who will be willing to take on a range of roles over time. They will, though, need to ft into the organisation's culture.

Selecting team members

As mentioned before, the value proposition of the employee will need to match that of the organisation. Research has found that a closer match between the values of a person and those of the organisation is linked to longer tenure, increased organisational commitment and improved job performance (O'Reilly et al., 1991; Kristof-Brown et al., 2005). Importantly, the preferences of newcomers should match those of their supervisor. Recent research has shown that where concern for people is high for both the employee and their supervisor newcomers will remain, settle in to the organisation and be productive (Van Vianen, 2000). Person–environment fit occurs when the individual's characteristics are compatible with those of their work environment (Kristof-Brown et al., 2005). Four components can be seen as present here: person–job, person–organisation, person–group and person–supervisor fit. Job satisfaction and organisational commitment have been found to be strong for person–job fit and person–organisation fit and to a lesser extent present when person–group fit and person–supervisor fit were present.

Person–group and person–supervisor fit

Selection processes traditionally include an emphasis on person–job fit and more recently on person–organisation fit through the use of personality and related tests. Rarely is the same emphasis placed on person–group and person–supervisor fit; however, these may be areas which need to be included if a project team performance is to be maximised. When selecting people for team settings it has been found that social skills, personality and teamwork knowledge should be assessed. Measures of ability to work effectively with others could be useful for assessing social skills. Personality assessment should include characteristics such as conscientiousness, extroversion, agreeableness and emotional stability. Effective team members are likely to possess knowledge and skills on how to function in a team setting, or to possess team competencies (Morgenson et al., 2005). Research confirmed that the use of structured interviews, and tests to measure the personality characteristics of conscientiousness, extroversion, agreeableness and emotional stability could increase team selection success. In addition, using situational judgement tests to measure teamwork knowledge could also be beneficial (Morgenson et al., 2005).

Using rewards to attract and retain employees

Rewards are strongly linked to retention. Rewards include salary or wages, incentives, benefits and the work environment in general – is it truly rewarding? Is the work environment friendly? Is a 'controlling' or a 'commitment' management style used? Are employees valued? Are their efforts recognised? Is employee health and safety a high priority? Can employees proudly state who they work for? Money alone will not attract and retain good employees, but paying poorly and not valuing employees will lead to increased employee turnover and failure to attract quality applicants.

An understanding of motivation theories and their implications for employee retention is not only of importance to HR professionals; it is important for every manager in an era where an increasing range of HR activities are being devolved to line managers and supervisors. Line managers, project leaders and supervisors now find they need to use their understanding of people and what motivates them to get the best from their direct reports.

Retention

In 1998, turnover in the USA stood at 1.3% per month (Ramlall, 2004). At this rate, an organisation could experience a 100% staff turnover in six and a half years! Knowledge, experience, contacts and relationships are lost when people leave. Vital roles on project teams may no longer be able to be filled by skilled and experienced people. However, deeper analysis may reveal a different picture. Waldman and Arora (2004) advocate focusing on measuring retention, rather than turnover. This involves analysing which employees remain and which employees leave the organisation. If it is the valuable employees who are remaining for several years, even if the turnover rate is higher than desired, the right people are being retained; those who will enable the organisation to achieve its goals. What such analyses usually reveal is areas for improvement within the staffing process. If 30% of new recruits each year are leaving in their first year of employment, but the other 70% are remaining for three or more years, this may indicate problems with the recruitment and selection process especially if they were the group receiving the lowest performance ratings. Why have people who are unable to perform to the required level or who don't wish to work in the area been hired? Attraction, recruitment and selection processes need to be better targeted to avoid hiring unsuitable people.

When key players leave, the effect on the organisation is greater still. Client and working relationships are broken; important customer information and organisational knowledge may be lost. The costs of turnover are great; Fitz-enz (1997) estimated that organisations lose approximately US$1 mn for every 10 managerial and professional employees who leave. The cost of replacing other members of staff, including recruiting and training the replacement and time to full effectiveness is somewhere around one year's wages. Cost alone is great. Loss of knowledge and impact on organisational effectiveness, especially during the replacement period, can result in larger costs in the long term.

Retention is an important issue in an era of skills shortages. Skills shortages can occur even when there are large numbers of people seeking work. Skills shortages relate to the KSAOs which an employee brings to a job. A person without the necessary KSAOs may be able to develop the desired skills, gain the knowledge required or learn to take new and different approaches to, for instance, customer service. However, this is not always the case and it may take too long to reach the required level of expertise.

Retaining staff in project-based industries

We mentioned earlier that retaining high-quality project team employees until project completion can present challenges. The tendency has been for highly skilled staff to commence negotiations for a new role prior to project completion. Indeed, if they have high-level skills they will be sought after by others who are aware that the project is nearing completion. The desire to obtain continuing employment is understandable. Research has shown that professionals willing to relocate to provide the flexible workforce desired in the USA and other parts of the world today do tend to benefit with an increase in income six years after moving of approximately 20% over that which could have been expected if they had remained at their former site (Rodgers and Rodgers, 2000). But this is not the case in all industries.

Relocation for construction professionals is likely to mean moving to a role with a similar rate of pay, the benefit of relocation being continuing employment at the expense of uprooting their family. Relocation will present family and financial pressures for some, so shoring up a position that does not require the family to relocate can be attractive.

Staff who leave before completion of the project, especially those who have been involved in the project from the outset, will lead to loss of tacit knowledge. As with the SAP example provided above, someone without the embedded knowledge will not be able to understand the context of many of the problems facing the project team because they have not experienced them.

Skills shortages, and these exist across a range of specialist areas, lead to power issues. Those in demand will find it easier to locate a new role in order to avoid a period of unemployment between projects. To avoid the negative consequences outlined, the employer may feel the need to offer financial bonuses to those who remain to completion. But how much is enough to retain a high-quality employee whose skills are in high demand? Three months employment and an extra $10,000 for example may not equate to three years employment on a new project. A gap in employment is a strong possibility for the employee who chooses to remain to completion. In addition, with construction projects as opposed to IT and other projects, the need for geographic relocation is more likely to exist.

Experience has seen financial bonuses paid to retain people to completion; however, there may be alternatives that could be considered. The cost of losing a key performer three months prior to completion of a three-year project may be far greater to the employer than paying one month of salary for the high performer to take leave and return for the next major contracted project. Another approach may involve working out a plan with the person, most likely with associated employee development costs, to retain the services of this identified key contributor, thus addressing both the organisation's and the employee's value proposition.

Research in the high-technology manufacturing industry in the USA found that employee development was used both as a retention tool and as a benefit (Benson, 2006). By providing tuition-reimbursement the organisation encouraged employees to undertake formal qualifications. In comparison to on-the-job training and company-run courses, completion of a transportable qualification recognised by other organisations meant that these employees would have 'marketable skills'. Qualifications gained through tuition-reimbursement were viewed as increasing their career advancement opportunities within the organisation and their employability should the company not be able to provide ongoing employment. Employee commitment was found to be more closely linked to tuition-reimbursement programmes if promotion followed within 12 months of completion of studies. Gaining a transportable qualification, one attractive to other employers, increased the likelihood of the employee leaving the organisation if career advancement was not linked to completion of formal qualifications (Benson, 2006). This highlights the need for an integrated approach to workforce planning, employee development, and retention and rewards planning.

Some organisations require tuition fee pay back, with the percentage to be returned to the employer reducing over time until there is no 'debt' after, say, three years. However, contractual 'hand cuffs', as this practice is termed (Associates, 1999), are not commonly used. Baruch (2001) found HR managers did not believe that offering employees 'employability' as part of the new psychological contract would succeed in improving loyalty and commitment and there was some questioning among HR managers of the cost involved. Baruch questioned the feasibility of the 'employability' promise replacing the former loyalty-based relationship. However, evidence already provided in this chapter that generations X and Y don't expect long-term employment may explain why Benson's (2006) more recent research demonstrated that better planning, by linking development programmes to career path planning, may make them more attractive to the workforce of the future.

People management

Human resource practices can be categorised as either 'control' or 'commitment' practices (Wood and de Menezes, 1998), two very different approaches that attempt to shape employee behaviour and attitudes. Control human resource practices are commonly used to increase efficiency and reduce labour costs. This is achieved by introducing strict work rules and procedures. In these situations rewards are commonly linked to measurable outputs (Walton, 1985), similar to piecework plans – employees are paid per item produced or transaction completed.

High commitment HR practices are designed to maximise commitment and employee empowerment with the goal of increasing productivity and effectiveness. This is achieved through a range of linked HR practices. Increasingly, employees are being involved in decision making, jobs are being re-designed to make them challenging and interesting, and rewards systems are being designed to encourage employees to identify with the organisation's goals (Whitener, 2001), for instance, through employee share schemes. The belief here is that employees will work harder to achieve goals they have been involved in setting and where they will directly benefit from the organisation's success. These efforts are all about developing committed employees and mutual trust. High commitment HR practices support retention efforts.

The control or commitment approach can also be applied to ensuring future staffing needs in a project environment. On the one hand, it is possible to build into contracts a requirement that contractors guarantee the supply of specified specialist skilled staff in order to be able to tender for a job. This is the control approach. On the otherhand, it is possible to take the commitment approach and work with preferred suppliers to ensure that they will be able to meet staffing requirements into the future. Retraining of current staff, supporting university students through their studies, indeed attracting them into studies in an identified future high-demand area in partnership with your supplier may yield better results, along with a more co-operative long-term working relationship. Trust, commitment and loyalty are important between partners in projects, just as they are between employers and employees. There is a mutual dependence in each instance.

Rather than seeing people as a resource to be used and then disposed of when needs change, high-commitment HR practices focus on the organisation's long-term needs and future success. That is why employees are viewed here as resources for achieving business goals and should be developed to meet the changing needs of the business. The IT and construction industries commonly employ large numbers of people to complete discrete projects, and on completion of the project may no longer require the services of some or all of those employed for the project. The people employed to form the project team possess specialised skills required for successful project completion. However, it is these skilled employees who expect to work for a range of organisations over their careers (Benson, 2006). Commitment to the one employer is not viewed as possible because of the short-term nature of the projects, or in other industries because employees wish to make their own career decisions, using their high level of skill to locate new roles, thus avoiding the betrayal of trust experienced by their parents (Loughlin and Barling, 2001).

In project-based industries, forming an alliance may reduce this need, with specialist skills being provided from a constant pool available from organisations which work only in that specific area – computer programming, web design, site management, quantity surveying, civil engineering or others. The view that preferred sub-contractors will be able to continue to provide the type and quality of staff required, without discussing future, long-term needs with them and considering together the related supply/demand situation, is one that could lead to disaster.

Whether or not partners or contractors are involved, planning future staffing needs is an important activity and part of overall strategic planning. Brockbank (1999) sees HR activities as belonging to four possible dimensions of competitive advantage.

- *Operational reactive*   Here the basics are implemented and in HR a lot of these will relate to legal compliance. Paying the correct amount, providing safety training and equipment, and so on.

- *Operational proactive*   These activities involve doing the basics better. Finding easier, cheaper, quicker and better ways of doing what must be done.
- *Strategic reactive*   Though often seen as going far enough, strategic reactive indicates that HR people are merely reacting to the strategic direction set by others, rather than helping to set a new and different strategic direction.
- *Strategic proactive*   HR people will propose new and exciting opportunities for competitive advantage here and back them up by ensuring the necessary human resources to deliver are available.

Consider the following scenario in Box 10.1 as an illustration of an engineering and architectural design solutions organisation which wasn't appropriately staffed.

Box 10.1 Scenario

A skills gap continues to exist. The client, for example, the government, says the train system isn't working; new trains, tracks and signalling systems are required. After considerable expense, driverless trains were introduced six months ago but there have been a range of problems and the advantages expected from their introduction have not been gained. Worse still, the public transport users are complaining bitterly about reduced and disrupted services and the government is feeling under pressure from these and other highly vocal groups in the community.

An investigation revealed that driverless trains require state-of-the-art signalling systems which were not installed in preparation for the changeover. The new signalling systems, which will have to be installed quickly now to solve the problems created by the introduction of driverless trains, require highly specialised skills and knowledge to design, implement and operate. To this point, HR have not been involved in any of the discussions. They were involved in developing the redundancy packages for the drivers who were made redundant as a result of the introduction of the driverless trains, but they are no more aware of the causes of the current problems than is the general public.

A proactive strategic approach applied to the procurement system would have avoided this problem in the first place. Instead of involving HR only in a reactive operational role, if HR had been involved in the planning, from the beginning, they would have asked questions and helped to narrow down the total resource base including describing the people required in all areas to support the new trains. This would have required HR to write position descriptions, describing all new roles required to support the driverless train system, outlining in detail all KSAOs. The need to procure the highly specialised skills would have then been identified and HR could consider, in discussions with other members of the strategic planning team, whether these services should be provided by a specialised organisation – outsourced – or whether appropriate employees should be attracted, recruited and inducted in time for the introduction of the new train system.

Working with the supplier of the new trains, instead of being under pressure now to procure the new signalling system and the staff to support it, it would have been possible during the initial procurement stages to identify all related needs. By forming a team to cover all contingencies and deliverables within the supply chain, predicaments such as that experienced here can be avoided. Using the new advanced rail system as an example, it becomes obvious that the project manager has a range of areas of expertise to address as well as a complex supply chain to manage. Experience would indicate that rarely are all complex components of the supply chain strategically thought through to the extent that issues such as a signalling system required supporting operation of the new and expensive train system being addressed. It may have been possible to negotiate with the supplier of driverless trains the best deal possible on the signalling system and for it to be delivered with time for testing so it could be operational when the driverless trains were delivered. By forming a team to cover all contingencies and deliverables within the supply chain during the pre-qualification stage there would have been time to examine which organisations could best develop, deliver and operate the signalling system.

Pulling together a group of people from a range of areas, from outside the organisation when working in an alliance or with other contractors, requires consideration of personality – Will these people work well together? – and combined, does the team contain the knowledge, skills, abilities and other attributes required to achieve success? There will not be time now for the signalling staff to be chosen using all these criteria, and the likely repercussions of this are reduced productivity, decreased service quality and increased turnover.

Rewards

Rewards in a project environment need to attract, motivate and retain top performers at the individual and team level. Recent research has revealed that potential employees are seeking employers who offer a range of benefits, and a higher than average salary won't make up for not addressing some of their wants. Job security is desired (Rabey, 2005) and, for young workers in particular, 'poor quality' employment which lacks a career path or learning opportunities, has been found to be de-motivating (Loughlin and Barling, 2001).

Total rewards

Rewards are part of the value proposition that the employer offers the employee and are made up of *remuneration*, including pay and short- and long-term incentives; *benefits*, including work/life programmes and *careers* encompassing employee development linked to career opportunities (www.mercerhr.com.au)[1] (Gross and Friedman, 2004). For success, total rewards strategies should be aligned with the business strategy. This is especially so when the rewards plan has been 'created to support an organisation's unique human capital strategy' (www.mercerhr.com.au). They must also match the employee's value proposition, so organisations must be able to understand the wants and needs of the type of people who can help them achieve their objectives. Project managers need to consider the combination of knowledge, skills, abilities and other attributes that will be required to deliver their project on time and on budget. Motivating the team to perform becomes the next challenge. For project teams, this may mean considering individual or team, or combined individual and team rewards.

Team or individual rewards

Cacioppe (1999) states that team rewards and recognition may be awarded either on individual behaviour and performance or on whole team performance, with the rewards being equally divided between the members of the team. If all members of a team are to share equally a reward – most commonly financial or exchangeable for money – the assumption is that all have equally contributed. If individuals only are rewarded the assumption is that people have contributed in different ways, or at a different level. This is, perhaps, not always the way things happen in a team. In particular, with long projects, the level of contribution or involvement may vary during different stages of the project depending on the contribution each chosen member has to make to the total project.

More recent research by Benson (2006) has revealed, as already mentioned, that well-managed employee development plans can assist in retaining staff. Employees do appreciate support from their employer to develop their knowledge and skills base and employees are reluctant to leave employers who provide such benefits (Benson, 2006). However, there is a need for these plans to be linked to succession planning and career path planning. Those who are supported to gain non-company-specific qualifications expect to be rewarded by advancing their careers. If the organisation sponsoring their studies does not also develop a career plan for those employees who have had their tuition fees refunded the investment made may be lost.

In a project environment general skills, such as those gained by employees enjoying tuition-reimbursement schemes, may be able to advantage the employer as well as the employee. Project managers and team members may now be well prepared to perform a range of permanent positions within the organisation, albeit only temporarily between projects. The advantage for the organisation is that they have a knowledgeable and experienced employee group to call on to assemble highly talented teams to ensure project success in the future. For individuals it means that their employment is more secure and that they enjoy the challenge of new and different roles, the learning that comes from this and the variety of experiences that these changes in role inevitably provide.

Work design

Employees seek challenging and interesting work (Rodrigues, 2001) that is 'rewarding'. Performing challenging work and different roles enables employees to demonstrate their potential to work in higher-level roles. Employment security, and a clear career path, can be seen as providing a 'future value'; a reason for staying with the organisation. This will need to be linked to opportunities to learn, grow and for career advancement. In a project environment, project placements providing new and different tasks to perform, enabling people to work with experienced successful project managers, and being supported to gain new skills and knowledge can form a large part of the individual's total rewards. For many employees these opportunities for new experiences and to gain new skills and knowledge to advance their career may be seen as more important than higher levels of pay (Gross and Friedman, 2004). This is especially so for early career employees.

Job security

Attraction and retention are not major issues only because skills shortages already exist in some areas and others are predicted in the future. The employment relationship has changed. Job security, a component of the traditional employment relationship, was undermined by downsizing efforts in the late 1980s and early 1990s and a range of mergers and acquisitions which have occurred.

The effects of this trend towards temporary employment and the accompanying feeling of job insecurity are being felt today by organisations. Research has shown that children's and younger adults' understandings of work and employment are shaped by their parent's employment experiences (Loughlin and Barling, 2001). Having formed work attitudes and behaviours that convey a sense of betrayal, younger workers are not committing to their employers in the way their parents did. Having seen their parents, family and friends made redundant after years of faithful service, Generations X and Y are planning their careers around the belief that they cannot count on their employer providing them with long-term employment (Baruch, 2001).

The level of trust and commitment of employees has fallen. As a result, organisations are looking at ways they can re-establish a level of trust with its accompanying commitment and loyalty to combat the negative impact of the current and future skills shortages. This has led to the development of 'high commitment work practices'. The psychological contract, also describes the employee–employer contract and relates to 'the unwritten promises, mutual expectations and obligations that exist between employees and their organisations' (Winter and Jackson, 2006: 422). The psychological contract has long been seen to include the expectation that each employee will exert a level of effort commensurate, in the employee's view, with the total rewards – financial and non-financial – being offered by the organisation for doing so (Nicholson and Johns, 1985).

Project work inevitably involves temporary employment, which can lead to lack of job security. There may be no guarantee of continued employment on completion of the current project. Research has found that limited term employees have reduced levels of commitment to managerial goals, and are likely to demonstrate lower levels of effort and cooperation (Boswell, 2006). This can be because they feel like 'outsiders'. There is some evidence that lack of job security is linked to a reluctance to share knowledge. Here the employee, fearing their employment may not continue, feels that their knowledge is critical to maintaining their value as an employee. In addition, learning that comes from sharing knowledge of mistakes made and which leads to the development of new ways of operating in the future is unlikely to occur when employees feel their job is under threat (Davenport et al., 1997). Rewards and recognition plans will need to be designed so as to encourage willingness amongst all employees to discuss failures and to develop plans to avoid them in the future. We discuss later the possible combination of individual and equally distributed team rewards. Careful design of both components may assist here. Project leaders should be rewarded for encouraging discussions; individuals should be rewarded for contributing examples of failures. A 'no blame' culture will be required for this to succeed.

Development of this new approach to rewards and recognition planning is particularly important because the incidence of temporary employment has increased, in particular in Europe (Brewster et al., 1997), and other developed countries, such as Australia, have also experienced increases in casual employment in recent years (Lowry, 2001). Feelings of insecurity are likely to be accompanied by lower levels of job satisfaction and organisational commitment (De Cuyper and De Witte, 2006) however this did not lead, as might have been expected in the past, to reduced satisfaction in life in general or feelings of personal performance failure. Again, this would appear to support the research quoted throughout this chapter that the next generation of managers, and those currently performing strong support roles – the Generations X and Y employees – have a different outlook on career and the employment relationship in general.

Some basic components of good rewards planning remain. Internal equity, fairness between employees according to level of responsibility and effort, and external equity linked to the current market rate for the profession or trade cannot be ignored. Rewarding individual contribution, and also encouraging team co-operation and performance is required for project success and the type, length and complexity of the project will have to be taken into consideration when developing a rewards plan that will support achievement of the desired outcome.

Sarin and Mahajan (2001) investigated rewards planning processes that support successful new product development projects. Here project teams may be required to work together for a prolonged period of time to produce the desired outcome; in other circumstances, the project may be of considerably shorter duration. They considered the commonly asked question: Should rewards be equally distributed amongst team members according to the level of success of the project or should the position or status of the team member within the organisation, or in relation to their role on the project team, be used as the basis upon which rewards are distributed?

The requirement here is that rewards be allocated in a manner that supports successful completion of the project. However, as already discussed, it is necessary to bear in mind issues such as equity and justice, being viewed as a 'good employer', and ensuring that key talent is attracted to and retained in the organisation. What Sarin and Mahajan's (2001) research revealed was that a range of issues surrounding the project, its length and objectives had to be considered in order to develop a supportive rewards plan. Ease of ability to evaluate individual performance was a factor. Where ease of evaluation was high, position-based rewards could be positively related to the internal dimensions of team performance and therefore deliver a fair means of rewarding performance. In these circumstances, team member satisfaction levels, in relation to allocation of rewards, were high. In situations where ease of individual evaluation was low, distributing rewards equally was found to positively relate to internal dimensions of team performance.

Sarin and Mahajan (2001) also analysed rewards, performance and team member satisfaction in relation to the length of time taken in product development that is relevant here. A process-based reward structure was expected to be positively related to the external dimensions of team performance for long product development cycles. Process rewards were found to relate negatively to time/speed to market. When the process of developing the new product, and stages within this were rewarded rather than for the whole project, or product, delivery speed to market was negatively impacted. It took longer for the team to deliver the product. Outcome-based rewards only marginally improved time to market, or time taken to deliver the project but had a positive effect on performance of long or less complex projects.

What Sarin and Mahajan's (2001) research reveals is that culture is a factor that needs to be considered when developing project team rewards. A position-based differential reward structure was found to be more effective and to support greater team member satisfaction when individual performance could be easily evaluated. This approach recognises that different levels of contribution and risk responsibility are demanded of different members of a project team. Nahavandi and Aranda (1994) have argued team-based approaches to rewards planning must be adapted when they are being used within Western cultural values systems. Here we are working with team members who come from individualistic cultures; individual rewards must, therefore, at least play a role within the total rewards plan. Many organisations have developed highly effective rewards plans which incorporate both individual and group measures against which rewards are allocated. It is necessary to consider culture, desired outcome and expected length of the project when deciding how much emphasis should be placed on each of these two elements when designing rewards plans.

To conclude this section

The change from temporary roles being seen as problematic and leading to project-based organisations having to settle for poorer quality employees or reduced levels of commitment and performance, to a new psychological contract where expectations have shifted may provide considerable advantage for those organisations that regularly form and disband project teams to achieve their goals. Generations X and Y employees don't expect continuing employment to the extent that their parents or grandparents did. Attracting, motivating and retaining key talent may not present the same challenges in the future as they did in the past, as a result of overall changes in the employment relationship and employee expectations. However, complacency will not provide competitive advantage. Understanding the complex relationship between all interacting areas of the employment relationship and ensuring that they are well coordinated will play a vital role in sustainability for project-based organisations in the future.

Rewards and recognition plans must be aligned with the organisation's strategy, project goals and type for a truly successful project outcome.

Chapter conclusion

In this chapter we posed the question: how do we effectively attract and retain valuable and skilled people to be project team members? We discussed the value proposition of key talented staff and how organisations need to align their HR policies with attracting, retaining and maintaining links with past employees who, while moving on to projects outside that organisation, nevertheless may be able to return and be valuable and desirable contributors on future projects.

We also posed the question: how do we present our project as being attractive to our supply chain partner teams? We argue that in addressing the strategic HR issues associated with in-house staff and also applying these principles to strategic supply chain partners can help make an organisation attractive as a partner of prime choice. This, together with ethical and governance as well as learning organisation strategies can reinforce an organisation's attractiveness.

The lesson that we argue can and should be learned from the literature and ideas presented in this chapter is that procurement strategies should address issues relating to attracting talent. This is because a talented workforce is a pivotal part of the resource mix that provides a better chance for project success and business sustainability.

Vignette

Vignette scenario

PeaceOfMind Inc (POM) is a USA/UK jointly-based health care industry service provider that recently embarked upon an ambitious expansion programme. The business model envisaged a series of high-technology health diagnostic services in which referred patients would receive the kinds of expensive and not widely available tests requiring computerised axial tomography (CAT) scans, ultrasound imaging and a range of other diagnostic tools requiring specialised equipment and clinical facilities. A 'cookie-cake cutter' approach to design of the physical facilities would be developed with state-of-the-art IT systems to complement the concept of excellence in service delivery working on the basis of a 24/7 model. Patients would make appointments choosing a phone-in or e-booking system medium that would guarantee and automatically send confirming emails and/or text messages to patients in the event of a greater than 20 minute wait for service. POM began its business with the development of 10 trial sites in major US, Canadian and UK cities by commissioning the physical facilities to be constructed, and adapting existing medical IT systems. These include not only adding some novel features that linked to an existing customer relationship management (CRM) system to be adapted to manage not only the patients and doctors who referred them, but also to manage a growing network of diagnostic specialists whose role was to interpret the images produced by the scanning equipment and be in a position to be on-demand as soon as the scan was taken. The CRM system would match patient, doctor and diagnostic specialists to ensure that the preferred language could be used in reports and communications. This required linking a global network of these specialists so that an appropriate person was available to take a 'job' at a moment's notice. The concept involved using an approach much like the radio taxi operators that contact taxis and direct them to customers; however, it also included access and development of a community of practice (CoP) so that specialists were able to feel that they were in a virtual 'clinic' space, in which they could work, interact with colleagues and develop and maintain their specialised diagnostic skills through the CoP, which also held online conferences. Further, the CoP was to be linked to several universities so that participants could undertake academic courses and engage in mentoring and coaching and undertake research degrees. This programme of projects therefore comprised several types of project. They range from the very physical building project type (building the facilities) to the adaptive IT-type project that took well-known existing software tools but configured them in novel ways. It also included elements of totally ephemeral projects such as establishing the CoP and developing the specialist networks.

The first 10 clinics were to be built concurrently over a 6-month period preceded by a 6-month design finalisation and advanced equipment procurement stage. Much of the software part of the programme had been developed by several of POM's start-up participants. The linking software development component was estimated to take 12 months to complete and the CoP project was already enthusiastically endorsed by an informal CoP that had at its core two of POM's medical specialist directors who were original start-up participants in the company.

It would be possible, for example, to use global time zone differences for a patient in the US mid-west region to have undertaken their diagnostic tests at say 8pm and the results to be accessed via email by a diagnostic analyst with the necessary specialised knowledge in Adelaide, Australia in early afternoon local time. Results could be expeditiously returned with minimum elapsed time being experienced between testing and advising medical specialists and their patients of results. Many of the diagnosis specialists are people with young families. It could suit them to work intensively from home for several hours in the late morning and early afternoon, and to fit this in with child-rearing activities. Another demographic of experts are recently retired 'grey nomads' who have retired from full-time work and are travelling in mobile homes with extensive internet connections so that they could, if recruited, combine targeted part-time work while travelling. A camp site in 'Woop Woop' with inter-net access could become part of a virtual clinic where CoP and diagnostic activities both could be enjoyed. The HR business model envisaged engaging many specialists who otherwise would not be con-sidered as potential employees or contracted specialised service providers.

The plan was expected to pass its proof of concept phase within the second year and a large investment agency had committed to fund expansion from 10 locations to 100 within the next two years and from that, exponential growth was envisaged.

This ambitious plan required a massive recruitment and selection exercise if it were to fulfil its promise. POM's start-up owners had branded the company name based on the customer value proposition of getting a rapid and accurate response to diagnose their medical condition, as well as providing those working for POM a work–life balance that fits their family and social commitment patterns, to not only encourage them to participate in this bold venture but to also make their work more interesting and rewarding. The CoP was planned to address linking far-flung individuals into a virtual clinic as well as provide them with on-the-job opportunities for professional and research development. Part of POM's mission was to develop a strong relationship with a construction contracting organisation that would guarantee a work–life balance plan for its on-site and head office staff when working on POM clinic projects. They had already engaged in discussion with six global construction organisations that had enthusiastically embraced relationship contracting and POM was presently drafting the pre-qualification criteria to choose three organisations that could work with POM for the first decade in developing the clinics in a projected 1,000 cities across the globe.

Issues to ponder

1   Discuss three issues raised here that appears to radically challenge the way that the procurement process addresses attracting talent;
2   POM envisages a relationship contracting approach to construction of the clinics. Discuss three advantages and disadvantages to this approach in terms of obtaining a competitive solution to a procurement quandary of achieving best value;

3   There are at least three types of project that comprise this programme. Compare and contrast the 'talent' market with likely availability and demand for each type. You are unlikely to be familiar with more than one of these so you should not be afraid to speculate what you think may be valid. Through group discussion, compare perceptions of the scope and scale of the HR attracting talent issues raised;

4   The POM business plan could be envisaged as revolutionary. Discuss three key issues that you have gleaned from this chapter and the cited literature that you think will be the most challenging for POM to deal with and

5   What would rank as the three most important work–life balance issues that POM have attempted to address in their business plan. How do you propose that these can be implemented?

Note

1 Mercer Human Resource Consulting Pty Ltd, 'Aligning Total Rewards Strategy with Business Success', 31 July 2006, accessed from <www.mercerhr.com.au>. Also published in *Human Resources*, the Human Resources Institute of New Zealand's magazine, April/May 2006.

References

Allen, T. D. (2001). 'Family-Supportive Work Environments: The Role of Organizational Perceptions'. *Journal of Vocational Behavior.* 58(3): 414–435.

Allen, T. D., Herst, D. E., Bruck, C. S. and Sutton, M. (2000). 'Consequences Associated with Work-to-family Conflict: A Review and Agenda for Future Research'. *Journal of Occupational Health Psychology.* 5: 278–308.

Anon (2005). Modular approach helps the industry deal with shrinking workforce. *The Australian.* Sydney: 43.

Applebaum, S. H., Harel, V. and Shapiro, B. (1998). 'The Developmental Assessment Centre: The Next Generation'. *Career Development International.* 3(1): 5–12.

Arthur, M. M. (2003). 'Share Price Reactions to Work-Family Initiatives: An Institutional Perspective'. *Academy of Management Journal.* 46(4): 497–505.

Associates, H. (1999). *Survey Findings: Design and Administration of Educational Reimbursement Programs.* Lincolnshire, IL: Hewitt Associates.

Australian Bureau of Statistics (2004a). Australian Social Trends. Canberra: Australian Government printing Service, ABS Catalogue No. 4102.

Australian Bureau of Statistics (2004b). Labour Force Australia, Spreadsheets. Canberra: Australian Government printing Service, ABS Catalogue No. 6202.0.55.001.

Backhaus, K. and Tikoo, S. (2004). 'Conceptualizing and Researching Employer Branding'. *Career Development International.* 9(5): 501–517.

Baruch, Y. (2001). 'Employability: A Substitute for Loyalty?' *Human Resource Development International.* 4(4): 543–566.

Batt, R. and Valcour, P. M. (2003). 'Human Resources Practices and Predictors of Work-family Outcomes and Employee Turnover'. *Industrial Relations.* 42: 189–220.

Behson, S. J. (2005). 'The Relative Contribution of Formal and Informal Organizational Work-Family Support'. *Journal of Vocational Behavior.* 66(3): 487–500.

Benson, G. S. (2006). 'Employee Development, Commitment and Intention to Turnover: A Test of Employability™ Policies in Action'. *Human Resource Management Journal.* 16(2): 173–192.

Bittman, M., Hoffman, S. and Thompson, D. (2004). Men's Uptake of Family-Friendly Employment Provisions, Policy Research Paper Canberra, Australian Government Department of Family and Community Services, 22: 24.

Boswell, W. (2006). 'Aligning Employees with the Organization's Strategic Objectives: Out of "Line of Sight," Out of Mind'. *International Journal of Human Resource Management* 17(9): 1489–1511.

Boyar, S. L., Maertz Jr, C. P., Pearson, A. W. and Keough, S. (2003). 'Work-Family Conflict: A Model Of Linkages Between Work and Family Domain Variables and Turnover Intentions'. *Journal of Managerial Issues.* 15(2): 175.

Brewster, C., Mayne, L. and Tregaskis, O. (1997). 'Flexible Working in Europe'. *Journal of World Business.* 32(2): 133–151.

Brockbank, W. (1999). 'If HR Were Really Strategically Proactive: Present and Future Directions in HR's Contribution to Competitive Advantage'. *Human Resource Management.* 38(4): 337–352.

Burke, R. J. and Mikkelsen, A. (2006). 'Burnout Among Norwegian Police Officers: Potential Antecedents and Consequences'. *International Journal of Stress Management.* 13(1): 64–83.

Butler, A., Gasser, M. and Smart, L. (2004). 'A Social-Cognitive Perspective on Using Family-Friendly Benefits'. *Journal of Vocational Behavior.* 65(1): 57–70.

Cacioppe, R. (1999). 'Using Team-individual Reward and Recognition Strategies to Drive Organizational Success'. *Leadership and Organization Development Journal.* 20 (6): 322–331.

Casper, W. J. and Buffardi, L. C. (2004). 'Work-life Benefits and Job Pursuit Intentions: The Role of Anticipated Organizational Support'. *Journal of Vocational Behavior.* 65(3): 391–410.

Chan, L. L. M., Shaffer, M. A. and Snape, E. (2004). 'In Search of Sustained Competitive Advantage: The Impact of Organizational Culture, Competitive Strategy and Human Resource Management Practices on Firm Performance'. *International Journal of Human Resource Management.* 15(1): 17–35.

Cunningham, I. and Hyman, J. (1999). 'Devolving Human Resource Responsibilities to the Line: Beginning of the End or a New Beginning for Personnel'. *Personnel Review.* 28(1/2): 9–27.

Davenport, R. H., De Long, D. W. and Beers, M. C. (1997). Managing the Knowledge of the Organization, Working Paper. Ernst & Young LLP, Center for Business Innovation 24.

Davis-Blake, A., Gibson, G. E. J., Dickson, K. E. and Mentel, B. (2001). Workforce Demographics Among Engineering Professionals, a Crisis Ahead? A Report of the Center for Construction Industry Studies, The University of Texas at Austin, Under the Guidance of the Owner/Contractor Organizational Changes Thrust Team Austin, Texas October 2001, Austin Texas, Consruction Indiustry Institute (CII) and The University of Texas at Austin: 54.

De Cuyper, N. and De Witte, H. (2006). 'The Impact of Job Insecurity and Contract Type on Attitudes, Well-Being and Behavioural Reports: A Psychological Contact Perspective'. *Journal of Occupational and Organizational Psychology*. **79**: 395–409.

DeFillippi, R. J. and Arthur, M. B. (1998). 'Paradox in Project-Based Enterprise: The Case of Film Making'. *California Management Review*. **40**(2): 125–139.

Dessler, G., Griffiths, J. and Lloyd-Walker, B. M. (2007). *Human Resource Management*. 3rd Edn. Frenchs Forest, NSW: Australia Pearson Education Australia.

DiVanna, J. A. and Rogers, J. V. (2005). *People: The New Asset on the Balance Sheet*. New York: Palgrave Macmillan.

Dolphin, R. R. (2004). 'Corporate Reputation – A Value-Creating Strategy'. *Corporate Governance: International Journal of Business in Society*. **4**(3): 77–92.

Eaton, S. C. (2003). 'If You Can Use Them: Flexibility Policies, Organizational Commitment and Perceived Performance'. *Industrial Relations*. **42**(2): 145–167.

Ensworth, P. (2001). *The Accidental Project Manager: Surviving the Transition from Techie to Manager*. New York: John Wiley & Sons.

Fitz-enz, J. (1997). 'It's Costly to Lose Good employees'. *Workforce*. **76**(8): 50–51.

Francis, V. and Lingard, H. (2004). A Quantitative Study of Work-life Experiences in the Public and Private Sectors of the Australian Construction Industry, Final Report. Brisbane, Construction Industry Institute Australia: 142.

Francis, V., Lingard, H. and Gibson, A. (2006). A Qualitative Study of Work-life Experiences in the Ppublic and Private Sectors of the Australian Construction Industry, Final Report. Brisbane, Construction Industry Institute Australia: 142.

Frank, F. D. and Taylor, C. R. (2004). 'Talent Management: Trends that Will Shape the Future'. *Human Resource Planning*. **27**(1): 33–41.

Frone, M. R. (2000). 'Work-family Conflict and Employee Psychiatric Disorders: The National Comorbidity Survey'. *Journal of Applied Psychology*. **85**(6): 888–895.

Gorey, K. M., Rice, R. W. and Brice, G. C. (1992). 'The Prevalence of Elder Care Responsibilities Among the Work Force Population'. *Research on Aging*. **14**: 399–418.

Grant-Vallone, E. J. and Donaldson, S. I. (2001). 'Consequences of Work-family Conflict on Employee Well-being Over Time'. *Work & Stress*. **15**(3): 214–226.

Greenhaus, J. H. and Beutell, N. J. (1985). 'Sources of Conflict Between Work and Family Roles'. *Academy of Management Review*. **10**(1): 76–88.

Greenhaus, J. H., Parasuraman, S. and Collins, K. M. (2001). 'Career Involvement and Family Involvement as Moderators of Relationships Between Work-family Conflict and Withdrawal from a Profession'. *Journal of Occupational Health Psychology*. **6**: 91–100.

Gross, S. E. and Friedman, H. M. (2004). 'Creating an Effective Total Reward Strategy: Holistic Approach Better Supports Business Success'. *Benefits Quarterly*. **20**(3): 7–12.

Grover, S. L. and Crooker, K. J. (1995). 'Who Appreciates Family-Responsive Human Resource Policies: The Impact of Family-Friendly Policies on the Organizational Attachment of Parents and Non-parents'. *Personnel Psychology*. 48(2): 271–288.

Grzywacz, J. G. and Marks, N. F. (2000). 'Family, Work, Work-family Spillover, and Problem Drinking During Midlife'. *Journal of Marriage and the Family*. 62(2): 336–348.

Halman, J. I. M. and Braks, B. F. M. (1999). 'Project Alliancing in the Offshore Industry'. *International Journal of Project Management*. 17(2): 71–76.

Hammer, T. H., Saksvik, P. Ø., Nytrø, K., Torvatn, H. and Bayazit, M. (2004). 'Expanding the Psychosocial Work Environment: Workplace Norms and Work-family Conflict as Correlates of Stress and Health'. *Journal of Occupational Health Psychology*. 9(1): 83–97.

Handy, C. (1997). *The Hungry Spirit*. London: Random House.

Hartman, F., Ashrafi, R. and Jergeas, G. (1998). 'Project Management in the Live Entertainment Industry: What is Different?' *International Journal of Project Management*. 16(5): 269–281.

Holland, J. L. (1985). Making Vocational Choices : A Theory of Vocational Personalities and Work Environments. 2nd Edn. Englewood Cliff, NJ: Prentice-Hall.

Hull, D. and Read, V. (2003). Simply the Best Workplaces in Australia, Acirrt Working Papers, Working Paper. Sydney, University of Sydney, acirrt, www.acirrt.com, 88: 41.

Hutchinson, A. and Gallagher, J. (2003). Project Alliances: An Overview, Melbourne, Alchimie Pty Ltd, Phillips Fox Lawyers: 33.

Jacobs, J. A. and Gerson, K. (2004). *The Time Divide: Work, Family, and Gender Inequality (The Family and Public Policy)*. Boston: Harvard University Press.

Kahn, R. L., Wolfe, D. M., Quinn, R. P., Snoek, J. D. and Rosenthal, R. A. (1964). *Organizational Stress: Studies in Role Conflict and Ambiguity*. New York: John Wiley.

Karasek, R. A. and Theorell, T. (1990). *Healthy Work: Stress, Productivity and the Reconstruction of Working Life*. New York: Basic Books.

Kaufman, B. E. (1999). 'Evolution and Current Status of University HR Programs'. *Human Resource Management*. 38(2): 103–110.

Konrad, A. M. and Mangel, R. (2000). 'The Impact of Work-life Programs on Firm Productivity'. *Strategic Management Journal*. 21(12): 1225–1237.

Kristof-Brown, A., Zimmerman, R. D. and Johnson, E. R. (2005). 'Consequences of Individual's Fit at Work: A Meta Analysis of Person-job, Person-organization, Person-group, and Person-supervisor Fit'. *Personnel Psychology*. 58: 281–342.

Kulik, C. and Bainbridge, H. (2006). 'HR and the Line: The Distribution of HR Activities in Australian Organisations'. *Asia Pacific Journal of Human Resources*. 44(2): 240–256.

Lievens, F., van Dam, K. and Anderson, K. (2002). 'Recent Trends and Challenges in Personnel Selection'. *Personnel Review*. 31(5): 580–601.

Lingard, H. and Francis, V. (2004). 'The Work-life Experiences of Office and Site-based Employees in the Australian Construction Industry'. *Construction Management & Economics*. 22(9): 991–1002.

Lingard, H. and Francis, V. (2005a). 'The Decline of the "Traditional" Family: Work-life Benefits as a Means of Promoting a Diverse Workforce in the Construction Industry of Australia'. *Construction Management and Economics*. 23(10): 1045–1057.

Lingard, H. and Francis, V. (2005b). 'Does Work-family Conflict Mediate the Relationship Between Job Schedule Demands and Burnout in Male Construction Professionals and Managers?' *Construction Management & Economics*. 23(7): 733–745.

Lingard, H. and Lin, J. (2003). 'Managing Motherhood in the Australian Construction Industry'. *Australian Journal of Construction Economics and Building*. 3(1): 15–24.

Lingard, H., Brown, K., Bradley, L., Bailey, C. and Townsend, K. (2007). 'Improving Employees' Work-life Balance in the Construction Industry: A Project Alliance Case Study'. *American Society of Civil Engineers*. Forthcoming.

Lipman, I. (2002). 'The Sears Lectureship in Business Ethics at Bentley College: "Business Ethics in the 21st Century."' *Business and Society Review*. 107(3): 381–389.

Loughlin, C. and Barling, J. (2001). 'Young Workers' Work Values, Attitudes, and Behaviours'. *Journal of Occupational & Organizational Psychology*. 74(4): 543–558.

Lowry, D. (2001). 'The Casual Management of Casual Work: Casual Workers' Perceptions of HRM Practices in the Highly Casualised Firm'. *Asia Pacific Journal of Human Resource Management*. 39(1): 42–62.

Maslach, C., Jackson, S. E. and Leiter, M. P. (1996). *Maslach Burnout Inventory Manual*. 3rd Edn. Palo Alto, CA: Consulting Psychologists Press.

Maslach, C., Schaufeli, W. B. and Leiter, M. P. (2001). 'Job Burnout'. *Annual Review of Psychology*. 52(1): 397–422.

Meyer, J. P. and Allen, N. J. (1991). 'A Three-Component Conceptualization of Organizational Commitment'. *Human Resource Management Review*. 1(1): 61–89.

Moen, P. and Sweet, S. (2004). 'From "Work-Family" to "Flexible Careers": A Life Course Reframing'. *Community, Work, & Family*. 7: 209–226.

Morgenson, F. P., Reider, M. H. and Campion, M. A. (2005). 'Selecting Individuals in Team Settings: The Importance of Social Skills, Personality Characteristics, and Teamwork Knowledge'. *Personnel Psychology*. 58: 583–611.

Nahavandi, A. and Aranda, E. (1994). 'Restructuring Teams for the Re-Engineered Organization.' *Academy of Management Executive*. 8(4): 58–68.

Netemeyer, R. G., Boles, J. S. and McMurrian, R. (1996). 'Development and Validation of Work-family Conflict and Family-work Conflict Scales'. *Journal of Applied Psychology*. 81: 400–410.

Nicholson, N. and Johns, G. (1985). 'The Absence Culture and Psychological Contract: Who's in Control of Absence?' *Academy of Management Review*. 10(3): 397–407.

O'Driscoll, M. P., Poelmans, S., Kalliath, T., Allen, T. D., Cooper, C. L. and Sanchez, J. L. (2003). 'Family-responsive Interventions, Perceived Organizational and Supervisor Support, Work-family Conflict and Psychological Strain'. *International Journal of Stress Management*. 10: 326–344.

O'Reilly, C. A., Chatman, J. and Caldwell, D. R. (1991). 'People and Organizational Culture: A Profile Comparison Approach to Assessing Person-organisation Fit'. *Academy of Management Journal.* **34**: 487–511.

Parker, S. and Wall, T. (1998). *Job and Work Design: Organizing Work to Promote Well-being and Effectiveness.* Thousand Oaks, CA: Sage Publications Inc.

Perlow, L. A. (1998). 'Boundary Control: The Social Ordering of Work and Family Time in a High-tech Corporation'. *Administrative Science Quarterly.* **43**(2): 328–358.

Pfeffer, J. (1998). 'The Real Keys to High Performance'. *Leader to Leader.* **8**: 23–29.

Pringle, C. D. and Kroll, M. J. (1997). 'Why Trafalgar was Won Before it was Fought: Lessons from Resource-based Theory'. *Academy of Management Executive.* **11**(4): 73–89.

Rabey, G. (2005). 'The Power Within'. *Industrial and Commercial Training.* **37**(2): 106–110.

Raidén, A. B., Dainty, A. R. J. and Neale, R. H. (2006). 'Balancing Employee Needs, Project Requirements and Organisational Priorities in Team Deployment'. *Construction Management and Economics.* **24**(8): 883–895.

Ramlall, S. (2004). 'A Review of Employee Motivation Theories and their Implications for Employee Retention Within Organizations'. *The Journal of American Academy of Business, Cambridge.* **5**(1/2): 52–63.

Robertson, I. and Smith, M. (2001). 'Personnel Selection'. *Journal of Occupational and Organizational Psychology.* **74**: 441–472.

Rodgers, J. R. and Rodgers, J. L. (2000). 'The Effect of Geographic Mobility on Male Labor-Force Participants in the United States'. *Journal of Labor Research.* **211**: 117–132.

Rodrigues, C. A. (2001). 'Fayol's 14 Principles of Management Then and Now: A Framework for Managing Today's Organizations Effectively'. *Management Decision.* **39**(10): 880–889.

Ruona, W. E. A. and Gibson, S. K. (2004). 'The Making of Twenty-first-century HR: An Analysis of the Convergence of HRM, HRD, and OD'. *Human Resource Management.* **43**(1): 49–66.

Rushton, K. (2002). 'Business Ethics: A Sustainable Approach'. *Business Ethics, A European Review.* **111**(2): 137–139.

Russell, G. and Bowman, L. (2000). Work and Family: Current Thinking, Research and Practice, Report. Canberra, Department of Family and Community Services, Commonwealth of Australia, ISBN 0 642 43249 X: 58.

Sarin, S. and Mahajan, V. (2001). 'The Effect of Reward Structures on the Performance of Cross-functional Product Development Teams'. *Journal of Marketing.* **65**(2): 35–53.

Secret, M. (2000). 'Identifying the Family, Job, and Workplace Characteristics of Employees Who Use Work-family Benefits'. *Family Relations.* **49**(2): 217–225.

Shellenbarger, S. (1998). 'Those Lists Ranking the Best Places to Work for are Rising in Influence'. *Wall Street Journal.* New York.

Thompson, C. A., Beauvais, L. L. and Lyness, K. S. (1999). 'When Work-Family Benefits Are Not Enough: The Influence of Work-family Culture on Benefit Utilization, Organizational Attachment, and Work-family Conflict'. *Journal of Vocational Behavior.* **54**(3): 392–415.

Ulrich, D. and Brockbank, W. (2005). *The HR Value Proposition*. Boston, MA: Harvard Business School Press.

van Marrewijk, M. (2004). 'The Social Dimension of Organizations: Recent Experiences with Great Place to Work Assessment Practices'. *Journal of Business Ethics*, 55(2): 135–146.

Van Vianen, A. E. M. (2000). 'Person-organization Fit: The Match Between Newcomers' and Recruiters' Preferences for Organizational Cultures'. *Personnel Psychology*. 53: 113–149.

Veigaa, J. F., Baldridge, D. C. and Eddleston, K. A. (2004). 'Toward Understanding Employee Reluctance to Participate in Family-friendly Programs'. *Human Resource Management Review*. 14(3): 337–351.

Vu, U. (2006). Ethics, Branding Top Issues of Interest for HR. *Canadian HR Reporter*.

Waldman, J. D. and Arora, S. (2004). 'Measuring Retention Rather than Turnover: A Different Complementary HR Calculus'. *Human Resource Planning*. 27(3): 6–9.

Walker, D. H. T. (2005). *Having a Knowledge Competitive Advantage (K-Adv) A Social Capital Perspective*. Information and Knowledge Management in a Global Economy CIB W102, Lisbon, 19–20 May, Franciso L. Ribeiro, Peter D. E. Love, Colin H. Davidson, Charles O. Egbu and B. Dimitrijevic, DECivil, 1: 13–31.

Walker, D. H. T., Hampson, K. and Peters, R. (2002). 'Project Alliancing vs Project Partnering: A Case Study of the Australian National Museum Project'. *Supply Chain Management*. 7(2): 83–91.

Walton, R. E. (1985). Toward A Strategy of Eliciting Employee Commitment Based on Policies of Mutuality. In *Human Resource Management, Trends and Challenges*. Walton R. E. and P. R. Lawrence, Eds. Boston, MA: Harvard Business School Press: 35–65.

Waterman, R. H. J., Waterman, J. A. and Collard, B. A. (1994). 'Toward a Career-Resilient Workforce'. *Harvard Business Review*. 72(4): 87–96.

Way, S. A. (2002). 'High Performance Work Systems and Intermediate Indicators of Firm Performance Within the US Small Business Sector'. *Journal of Management*. 28(6): 765–785.

Wheatley, M. (1997). 'Goodbye, Command and Control'. *Leader to Leader*. 5: 21–28.

Whitener, E. M. (2001). 'Do "High Commitment" Human Resource Practices Affect Employee Commitment? A Cross-level Analysis Using Hierarchical Linear Modeling'. *Journal of Management*. 27(5): 515–535.

Winter, R. and Jackson, B. (2006). 'State of the Psychological Contract: Manager and Employee Perspectives Within an Australian Credit Union'. *Employee Relations*. 28(5): 421–434.

Wood, S. and de Menezes, L. (1998). 'High Commitment Management in the U.K.: Evidence from the Workplace Industrial Relations Survey, and Employers' Manpower and Skills Practices Survey'. *Human Relations*. 51(4): 485–515.

Wood, S. and Wall, T. D. (2002). Human Resource Management and Business Performance. In *Psychology at Work*. Warr P., Ed. London: Penguin: 351–374.

Chapter 11

Case study – developing a centre of excellence (CoE)

Chris Cartwright and Derek H. T. Walker

Introduction

Chapters 1 through 10 provided a solid and broad PM perspective on project procurement practices. These aspects are further discussed and focused upon through case studies presented in the next four chapters that provide detailed discussion reflecting relevant theory and practice – focusing upon a single organisation and a single aspect. In this case the organisation was Ericsson P/L and the focus of discussion is on the procurement process (mainly internally) of the skills, resources and expertise necessary that resulted in developing a successful PM centre of excellence (CoE).

This CoE development task, which links well to chapters in this book, was not outsourced to consultants but was organically grown from within the organisation (links with Chapter 1). It was a highly 'experimental' project type and while a 'tender' was not called for, iterations of a highly rigorous business plan were required to gain top management support and adequate resourcing – the business case was scrutinised and evaluated in depth (links with Chapter 1). The project also involved careful identification and management of key stakeholders together with consideration of Ericsson as part of a supply chain delivering value to customers (links with Chapter 3). While the CoE did not explicitly address ethical issues dealt with in Chapter 4, it did require that corporate governance measures be designed-into the way that the CoE would operate. In terms of strategy, it was emergent with an entrepreneurial flavour and a transformational impact (links with Chapter 5). The CoE was appreciated and evaluated using a notional CMM with some development of BSC and tangible–intangible mapping being explicated in a formal way. The CoE extensively developed a portal tool for e-business and so it is of immediate relevance to Chapter 7. Similarly, the initiative was of itself an organisational learning exercise. This closely links to Chapters 8 and 9 because this required an organisational culture appreciation to enact transformational change. A CoE is aimed to develop key staff as well as upskill staff generally and so links closely to Chapter 10.

This chapter is structured as follows. The next section will outline the context and background history of the CoE's development – its *raison d'être*. This is followed by a discussion of the underpinning theory that the case study illustrates and tests – the 'what' issues. The following section explains the 'how' and 'why' issues. The last section provides an evaluation and discussion of implications for PM procurement practice.

Context and history

This case study investigates how the local arm of the international telecommunications company[1] developed a project management office (PMO) which was rated by Ericsson's global head office in October 2005 as 'best in class'. The rating was accomplished using a measurement instrument, the Corporate Practices Questionnaire (CPQ) from Human Systems Ltd, the same process was conducted across 52 countries within the organisation. The PMO was established in 1999 at a time when the whole Information Technology and Telecommunication (IT&T) industry was in a state of turmoil. The led to disruptive change and risk events (such as Y2K and an explosion in (IT&T) company share prices) that triggered expectations for the need for fundamental change to respond to identified new business opportunities. Digital electronics technology was also continuing to drive convergence between computers and telecommunication. Each side of the information technology (IT) and computerised multimedia digital communication divide was positioning itself for the future.

There was also fundamental change occurring at a political level. Free market economic policies adopted by the Australian government resulted in the sale of 49% as the beginning of a long drawn-out privatisation process of Telstra P/L (Australia's government-owned telecom). This meant that the government could no longer use its ownership position as a lever for economic development of local industry, and Telstra was no longer forced to specify local content as a criterion for equipment purchasing decisions.

Many countries in the Asia-Pacific region were also using their tax regimes to attract local manufacturing – 10-year tax-free windows were being negotiated for companies establishing a new factory in, for example, Malaysia's Multi Media Corridor. The need to compete with economic models used by the computer industry was forcing a major reorganisation on Ericsson. It was planned to close more than 30 of its 50 manufacturing facilities globally. The global manufacturing model, much of it outsourced, meant that the Australian factory, though recognised as being one of the best in the country, was to be gradually phased out. This closure meant a complete change in the way the local company did business. Until then, a large automated factory delivered product directly to Telstra which was subsequently installed and commissioned by Telstra personnel. The changes being driven by senior management in Telstra meant that they were also looking at how they could more effectively meet the needs of their customers and shareholders. Telstra needed equipment to be procured with design services, installation, commissioning and even in some cases operation. The opportunity became obvious for a telecommunication product provider to become a services provider. The term 'total solution' became a part of Ericsson's vocabulary and the need became apparent for a provider to manage the end-to-end activity. Project management was seen as the mechanism to effectively manage the end-to-end delivery.

The transition from a product-based company to a service-based company was a major cultural change (which continues six years later); the Australian and New Zealand Ericsson company (ANZA) was leading this global organisation down a services-focused path. Senior management set targets to achieve a certain percentage of income to be derived from services – typically ANZA achieved double the percentage of the group's average. In 2004, a major milestone was achieved when ANZA achieved 50% of sales income from services. These services included consulting, education, design, installation, integration, commissioning, operation and project management (PM). The global management team nominated three key issues for the future as part of the cultural change needed – leadership, sales competence and PM competence. The local CEO nominated PM as one of three key enablers in their local strategy.

In 1996, a movement started within the organisation to establish a PMO and consequently a number of business cases were regularly put to the management team. The breakthrough came in 1999 when the Director of Services agreed to fund a PMO for 12 months to test the feasibility of the concept. The rationale included the following

- *Because it is what we do* – by 1999 the factory was closing and more than 70% of the income came from delivering projects;
- *Primary customer interface* – once products and services were sold, the customer saw the project manager as the face of the organisation;
- *It provides our primary opportunity to improve profitability and customer satisfaction* – cost control and good stakeholder management, fundamental parts of project management, drive profitability and customer satisfaction;
- *Project management facilitates risk management* – organisational and financial risk can be managed in a more systematic way using project management. Deciding how the organisation selects opportunities that become projects was built into the new sales process;
- *Project management facilitates knowledge management* – project closure and the collection and dissemination of lessons learned are key parts of the project management methodology used;
- *It ensures governance* – consistently using a project methodology provides certainty and visibility for organisational and financial governance;

- *Project management drives process improvements in order to enhance organisational effectiveness and efficiency* – as the project management delivers a project horizontally across the organisation they can see the disconnects and inefficiencies; initially they were tasked with highlighting these and
- It *must be leveraged to increase sales opportunities* – the project management team is in a unique position within the customer's organisation, they are seen more positively than the sales team. They may see opportunities that the sales team may not be aware of, or they may be able to include a change process to add value to the project.

What is a PMO and why have one?

Chapter 5 discussed aspects of what a PMO might look like. In the early 1980s Naisbitt (1982) predicted the transition from an Industrial age to an Information age would occur through 10 major trends – the 'megatrends' as he called them. Haldane (1999) focuses on one, *High Tech/High Touch* as it applies today in the discipline of project management, making the point that a projectised workforce drives PMO growth. A particular management structure is needed if, as is currently the case in many companies, project delivery outcomes present a risk to a major proportion of an organisation's finances. Most organisations in the 21st century rely on some form of matrix structure to manage the functional and project dimensions (Pettigrew and Fenton, 2000). With an accelerating shift from manufacturing to services delivery, the type of manager that focuses on cost control is being regarded as lacking an ability to also focus on strategic business development issues. The formation of a PMO can fill this void and ensure that both commercial and strategic project success become major organisational drivers. The PMO resides at the cutting edge of the organisational management matrix.

The term PMO is used to describe many different types of organisation. In most large single projects there will be a level of administration within the project to ensure effective delivery; this may be called a PMO, and as a result there is a view that a project office is responsible for mainly administrative tasks. KPMG, in their PM survey describe a PMO as having a broader mandate including many internal and external projects. As an organisation matures there will be a changing role for the project office from an administrative focus, through people development, then a business focus, followed by a strategic role. The definition of KPMG put forward for a PMO in the future is 'A strategic function responsible for coordinating, prioritising, planning, overseeing, and monitoring an organisation's projects to achieve business strategy and benefits' (KPMG, 2003: 3). By 2005 this was certainly the role that the ANZA PMO was required to fulfil. Whether one uses the terminology project office, PMO, programme management office, enterprise

PMO or portfolio management office, it describes the management function that oversights an organisation's projects and is responsible for project management performance. Miranda (2003) makes the point that the support a PMO provides may range from acting as a repository for PM information through to being totally accountable for project performance.

The Gartner Group (Light and Berg, 2000) puts forward three roles for the maturing PMO:

1   *The PMO as a repository* – custodian of the project methodology and not involved in the decision-making process;
2   *The PMO as a coach* – provides guidance on projects, performs project reviews on request, may establish and support project planning, monitors and reports on projects but does not order corrective action and
3   *The PMO as manager* – operating as an agent of senior management, manages the project portfolio, manages the master resource plan, reviews project proposals and is accountable for the portfolio.

While a mature PMO may take a management role, it must also continue to provide the roles of knowledge repository and coach – while many activities may be added to the role, few if any will be deleted. Kerzner (2003) describes three slightly different PMO types:

1   *Functional* – mainly responsible for managing a resource pool;
2   *Customer groups* – focused upon projects in common customer groups to better manage customer relationships and
3   *Corporate* – focused on corporate and strategic issues.

The merger of Hewlett Packard and Compaq provided an opportunity to select the best in class PMO from the two organisations, Levine (2005) summarises four main focus areas where the PMO is defined as having the broadest impact:

1   *Project management expertise* – raising the level of project management competence and achieving competitive advantage with an industry-leading success rate;
2   *Best-in-class methodology* – implementing a standard methodology using best practices to ensure repeatable success;
3   *State-of-the-art enterprise project management system* – adopting Ericsson Project Management Institute (EPMI) tools to provide employees with a world-class system that enables knowledge-sharing and visibility of all work on one common database and
4   *Portfolio management* – business strategy aligned and interlocked through active reporting, tracking and analysis of projects and programmes.

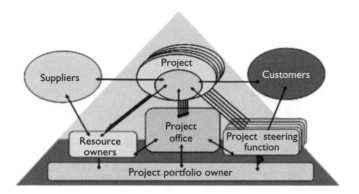

Figure 11.1 Ericsson EPMI model.

Bergmann (2002) presented a PMO governance model in 2002 at a conference in Stockholm as a standard for the Ericsson group. Figure 11.1 illustrates the organisational structure and how the PMO links the organisation to delivered projects. Bergman is an EPMI, and one of the founders of the Ericsson PM Project Model for PROPS. The method, developed in 1989, is now into its fourth revision.

The defined model placed the PMO at the centre of the services world, all projects having oversight with PMO links to senior management and the portfolio. Each project is able to draw on suppliers or internal resources to deliver the required outcomes. Project governance is ensured through a gating[2] process buried in the PROPS methodology. The project steering function consists of a project sponsor owning each project and following the gating process. Only one individual owns accountability and ultimate responsibility for the complete project portfolio, this may be the CEO who may delegate the accountability to the manager of the PMO.

The PMO form that Ericsson has evolved into now is illustrated in Figure 11.2; it includes the following tasks:

* *New assignment handling* – acting as the central desk for projects, providing estimates where requested, providing structure and substance to the bid and initiation process; ensuring a governance process is in place, followed for each sales opportunity;
* *PM staffing and competence development* – as capability owner for project management, ensuring there are sufficient suitably competent project managers available on an as-needs basis. This was supported by both internal and external certification processes;

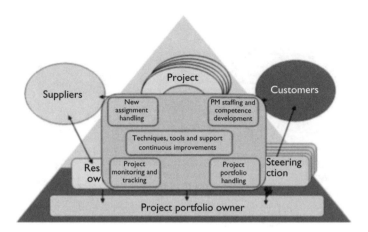

Figure 11.2 More advanced PMO model.

- *Methods, tools and support* – providing access and training in best of breed methods, tools and processes, providing help and training where required;
- *Continuous improvement* – monitoring processes for quality, suitability and adherence, and providing feedback to management on any opportunities for improvement;
- *Project monitoring and tracking* – Providing accurate meaningful reports to management on progress for outcomes, schedule, cost, margin and risk on projects, programmes and the complete portfolio and
- *Project portfolio handling* – Supporting senior management to achieve their strategy by managing organisational resources, providing synergies between projects and proactively monitoring portfolio performance.

The purpose of having a PMO is to make available existing PM skill sets across the organisation and to grow PM maturity. As the PMO develops its own maturity in providing greater strategic PM influence and developing PM expertise, tools and talent, it increases the efficiency of an organisation's PM activities and also increases its effectiveness by providing expert advice to strategically select and prioritise projects to be undertaken. These are in line with a shift from skill-development towards reflection, knowledge transfer and sensemaking that is in line with the concept of a centre of excellence (Walker and Christenson, 2005). Further, the Ericsson case study example matches emerging trends in advanced PMOs that have been reported upon in North America (Rad and Levin, 2002) and in some global organisations operating in Europe (Andersen *et al.*, 2006) to an extent that when compared, the Ericsson example is indeed at the cutting edge of advanced PMOs. If we accept that the model above reasonably represents a CoE concept then it is interesting to trace the journey that led Ericsson Australia to this destination.

The journey to best-in-class

There were three major reorganisations of the ANZA PMO during the period 1999–2005 that, looking back, followed the Gartner model. They were to provide the functions of the Project Manager Support Office, the Organisation Project Support Office and the Business Delivery PMO.

The establishment of the PMO in 1999 followed a push from across the organisation to provide some structure behind the growth of PM. There had been, for a couple of years, a business case developed to establish a PMO but the senior management still saw this as another level of administration. The organisation was a strong functional organisation that either did not see the need, or may have felt threatened, by the strengthening of the PM function. In the second half of the 1990s two major projects had failed and this further reinforced the need for action. There was a general belief within the executive team that the projects failed because of poor PM practices. In 1999 the Director of Services signed off on the business case and agreed to fund the PMO for 12 months.

Project manager support office

Figure 11.3 illustrates the impact that the PMO exerted upon the organisation. To achieve project success there were two streams that needed to function effectively: the project stream to provide competent project managers who performed well and the organisational stream that needed to select suitable projects and then provide the required level of governance. Every presentation

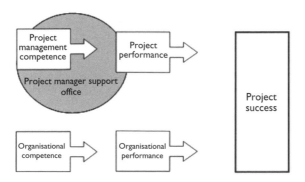

Figure 11.3 PM support office model – Stage 1.

by the manager of the PMO started with this graphic reinforcing of what needed to be addressed.

The Project Manager Support Office was established to be responsible for the following activities:

1   *Own the project management process* – EPMI, the department responsible globally for PM processes, had developed their model for projects PROPS and this was used intermittently across ANZA; the PMO institutionalised the use through further training and coaching.

2   *Set benchmarks and standards* – PROPS nominated particular documents to be used and processes to be followed. Templates were developed and the discipline was imposed to ensure that project managers followed this process.

3   *Benchmarking* – the organisation needed to understand and track the level of competence and process maturity in project management. The decision was made to join a benchmarking ring conducted by Human Systems Ltd. Dr Terry Cooke-Davies had been running a benchmarking ring in Europe for a number of years and was looking to establish Human Systems in the Asia-Pacific region. Ericsson was a founding member of PACRIM 1, a benchmarking ring managed by Dr Lyn Crawford, representing Human Systems, from the University of Technology in Sydney. The main tool that Human Systems use to measure maturity is the CPQ, a forerunner of the Project Management Institute's (PMI) Organisational Project Management Maturity Model (OPM3). The CPQ measures the organisation's approach to project management processes and then the deployment of that approach. ANZA PMO approached EPMI to adopt the tool across the Ericsson group and it was adopted in 2002. The tool was modified to meet Ericsson's needs and renamed the Project Environment Maturity Assessment (PEMA). This was the tool used for assessment in 2005. The CPQ was conducted at ANZA in 1999 and showed a level of 1.5–2.0 across most PM processes. There was also a major discrepancy between the approach and the deployment scores, showing we were not consistently applying our own processes.

4   *Centralise and develop project managers* – a critical decision was taken to centralise all project managers within the PMO and manage them as a key resource. Previously project managers were distributed throughout the organisation and support and development was very uneven. Bringing the project managers together not only allowed Ericsson to develop a team spirit and facilitated imposing discipline and providing the required level of support. Continuous skill development was one of the foundations for improvement. The PMO approached the Royal Melbourne Institute of Technology (RMIT) to provide structured PM training modules, and a number of the team used the opportunity to

complete formal PM qualifications. The organisation was undergoing a change towards a performance-based culture through a process called Competence and Performance equals Reward (CaPeR). The PMO developed a set of behaviourally based competencies for project managers to support the CaPeR process. Competency mapping was used to understand individual and organisational development needs. This information was used to source training, to allocate project managers to particular projects and to develop reward strategies. Role descriptions were put in place for Project Managers, Senior Project managers and Principal Project Managers.

5   *Own the profession* – there was an understanding that the organisation needed to become more professional in project management; across the world PM was starting to develop as a profession. The PMO assumed the role of Project Central, for information on projects, for the supply of project managers and became the PM organisational home.

6   *Certification of project managers* – the PMI had established their Project Manager Professional PMP® examination-based certification programme and, as it was a globally recognised programme, the decision was taken to encourage project managers within the PMO to become certified as PMP®s. In 1999 the first four PMs became certified and by 2005 more than 50 members of the organisation had become certified. As part of the reward strategy all costs including annual membership fees were paid by the organisation.

7   *Must do process* – the sales departments established their targets for the following year each September, part of this process being to study the market place and nominate which major opportunities they needed to be successful in to achieve their sale budgets; this created the '*Must Win list*'. The PMO then took responsibility to deliver each opportunity once a contract was signed. The PMO coined the term '*Must Do list*', the terminology linked the PMO with the business and resulted in invitations for the PMO Manager to attend the sales Must Win meetings; hence the PMO had a clearer view of the sales funnel. The term Must Do became part of the organisation jargon. The CEO would begin his quarterly leadership meetings with an overhead on financials and the second overhead was the traffic lights for the Must Dos. A monthly Must Do meeting was held where the CEO reviewed the Must Do projects that had red traffic lights. It was during some of these meetings that many of the functional managers resented some of the stones being lifted in front of the CEO. It was also the first time that the CEO and many of the management team saw the project managers running what were quite major projects.

While not an activity, one of the ongoing requirements for the PMO at this time was to continually justify its position within the organisation.

The establishment of a PMO was seen by many middle managers as a threat to their power. There were many other agendas and power plays that make up the day-to-day turmoil of everyday business life, and the manager of the PMO needs to be politically aware and savvy to understand what the current issues and pitfalls are. One of the strategies from the PMO was to consistently communicate the same message to the rest of the organisation.

Organisation project support office

During the first three years of the PMO there was continual questioning from the management team: 'Do we really need a project office?' This was especially true once the initial version of portfolio management was established and when the PMO staff (in the language of the organisation), 'started lifting some of the stones in the organisation'. The PMO highlighted issues that some managers would have preferred remained hidden. It now had a set of Balanced Scorecard (BSC) measures that linked directly to the company BSC. The triple constraint of budget, time and customer satisfaction were all improving; in 2002 the targets were

Projects on Budget	80%
Projects on time	80%
Customer satisfaction	80%

Figure 11.4 illustrates the growing impact that the PMO exerted upon the organisation to push further into project performance and impinge upon organisational competence.

While the responsibilities now assigned to the PMO started to expand, none were removed; the PMO still needed to ensure that it had a team of competent, motivated and available project managers consistently applying established

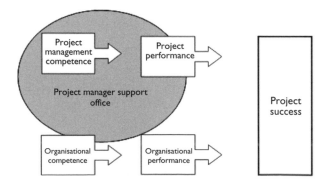

Figure 11.4 PM support office model – Stage 2.

global processes. The required activities now included the following:

- *Drive adherence to process through reviews and facilitation* – each project was now required to run a customer and internal satisfaction survey and a formal wash-up meeting as part of the conclusion phase. There was also a requirement that a peer review was conducted early during the execution phase to ensure the plan had sufficient rigour, and constraints assumptions and risks were understood. The PMO was also asked to conduct *ad hoc* reviews or health-checks when Sponsors felt there was a need. At times the results of these reviews were used to establish blame and finger-point rather than use them as a fundamental plank for improvement.
- *On-screen report to management* – during 2002 the PMO started to take more responsibility for portfolio management. While the Must Do process had enabled the organisation to restrict problems with out-of-control projects it was only an early version of portfolio management. An on-screen report was developed for the portfolio based on the Must Do traffic lights and the process was renamed. The report was named Key Projects.
- *Key Projects* report – the on-screen version was available to the whole global organisation with a level of security for sensitive information. A project sponsor was able to drill down for complete financial and delivery details. Figure 11.5 illustrates a sample of the portal report information provided.

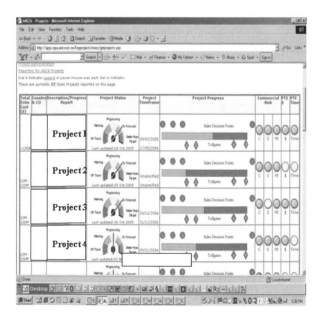

Figure 11.5 Key projects dashboard.

- *Establish management of projects* – EPMI referred to 'Management of Projects' rather than 'Portfolio Management' to define the management of the organisation's total load of projects. When the PMO took on the role of portfolio manager it gained the responsibility to ensure that the correct processes were followed during the sales processes by holding the various decisions for a project. This ensured that only contracts that were deliverable and met the target for margin were signed. The second responsibility was to ensure the sponsor used the governance built into the PROPS model; toll gates were held and decisions documented.
- *Performance manage PMs* – initially the PMO provided competent, motivated, professional project managers. The PMO took on the role of performance managing the PM team to support the organisation by taking direct responsibility for project outcomes – not just providing support to the PMs.
- *Manage forecast load* – the sales funnel for projects is normally uneven and relatively opaque. The PMO in managing forecast load needed to understand and plan for future projects by reviewing probabilities and scenarios for PM support needs. The organisation needed support not only in the delivery of projects but also in the selling process, where quick realistic estimates for cost, time and risk allowed flexibility in responding to customer requests. In the sales environment speed is critical; the time line tends to be today, not next week; planning for this type of response is difficult, and it normally tended to be an add-on activity for a project manager already managing a project.
- *Capability owner for project management* – the PMO now took on the role of capability owner – no longer just being responsible for just PM competence. The role of capability owner entailed not only proving the correct quality of project manager but also the correct quantity. Flexibility in capability level was achieved by using a base of fully employed project managers, supplemented by a small number of contractors.
- *Cross-functional reporting of disconnects* – the demise of the quality organisation during the various restructures had left no accountability for continuous improvement. The re-branding of quality as 'business excellence' provided an opportunity to address the gap. The PMs as part of the horizontal matrix saw where there were disconnects in the delivery capability and they were asked to highlight, as part of their performance management activities and progress reporting, where they were. This caused some friction within the functional organisation.

There was still a need to ensure that the PMO was seen by all stakeholders to be adding value to them to justify its position in organisation. It was not until 2003 that there appeared to be an understanding by all senior managers that the PMO was a necessary business expense and after that time there was no request to revisit the PMO business case.

Late in 2002, as part of the ongoing benchmarking activities the ANZA PMO conducted a second maturity measurement using the Human Systems CPQ. The results showed there was a major improvement since the first measurement in 1999 with the approach score averaging over 4.0 and the deployment score at 3.3. Around this time the global organisation decided to use Human Systems and the CPQ as the basis for an audit process of PMOs across the global organisation. The decision was taken that all PM processes should be a global responsibility and that they should be applied locally, the mantra became *Global Approach with Local Deployment*. The global organisation took accountability for the approach score and the local companies were given accountability for the deployment score. The CPQ tool was re-engineered to include some measures that the global team needed but the CPQ was still intact to ensure integrity of the current data. The ability to benchmark with external organisations was seen as a major advantage. The new tool was named the PEMA, a tool that measures how a local company sells and delivers the Network Rollout product suite.

The business delivery PMO

In early 2004, there was another restructuring of the organisation and the opportunity was taken to add to the responsibilities of the PMO. The project scorecard was now looking much better and the opportunity was taken to raise the bar on the targets:

Projects on budget	90% (overall total within budget)
Projects on time	100%
Customer satisfaction	80%

The PMO now took on a role that tended towards that illustrated in Figure 11.6.

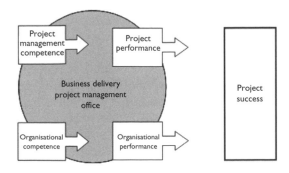

Figure 11.6 PM support office model – Stage 3.

The PMO was expected to provide all current services plus the following:

- *Accountable to support the sales team* – the PM was now seen to be an integral part of the bid team. The PMO needed to provide business savvy commercially competent project managers.
- *Accountable for estimates* – the PM now needs to sign off on the estimate and is accountable to deliver the project to this estimate. The estimate will be used to drive the pricing structure and become the basis for benefit management.
- *Manage order desk and end-to-end delivery process* – the PMO is now totally accountable to deliver projects; previously the sales organisation saw that as a need to oversight the delivery to safeguard their customer relationship. The new order was for the sales team to focus on selling and for the services team, especially the project managers, to focus on delivery.
- *One version of the truth* – one of the major links between the project and the hosting organisation is the financial performance of the project. Most organisations use a large ERP system, for example, SAP/R3, to manage and report finances. The information within the system is reported to the market as profit and/or loss. A project was planned and the planned cost was also entered in the system. From then, project managers used the Estimate at Completion (EAC) and the Estimate to Completion (ETC) to manage and report on their project. Many financial systems have problems reporting estimates so there may be differing views of the current budget numbers. There may appear to be many versions of the truth. At one regular executive meeting the PMO manager reported two sets of figures: one, the figures within the financial system, and the other the total of the EAC figures from current projects from the PMO database. There was a major discrepancy; the executive team stated that the numbers within SAP were the ones reported to the market and that they must be correct. The financial team supporting SAP was moved into the PMO to provide better support to the PMs and align the project's and organisation's financial numbers. The mantra became *'one version of the truth and SAP is King'*.
- *Train project sponsors* – each project has, according to the PROPS methodology, an internal senior manager who owns the project, in a business sense. The senior manager, the sponsor of the project, uses the tollgate governance process to ensure the defined business benefits are delivered. By 2004 there was a belief that the role of the sponsor was not adding the value it should. Many sponsors were not fulfilling their role and providing the required support to project teams. The PMO

was asked to provide education in the selection of sponsors and their roles and responsibilities.

- *Provide business support* – The PMO was now seen as able to add value to the bid process and overall governance. There were requests to facilitate risk management sessions, planning sessions, audits, bid management, wash ups and lessons learned sessions.
- *Cross-functional process improvement* – the PMs were now asked to take an active role in removing disconnects within the organisation. They had targets set for process improvements as part of their CaPeR scorecard which drove their reward structure.
- *Accountable to deliver the agreed margin/benefits* – the PMO became totally responsible for delivering the margin agreed for the project when the contract was signed. One of the PMs from the PMO was involved in the bid to provide estimates, high-level plans and an understanding of associated risks. Margin delivery was included as part of each individual PM's performance management process.

During the period 1999–2005 there was anecdotal evidence both locally and globally that the ANZA PMO was doing innovative things and was among the best in the group. There was a team providing benchmarking services within the corporate headquarters that was assessing companies to establish what their level of deployment maturity was and then helping them with improvement plans. This team had asked Human Systems to measure the maturity of the global approach and had been scored at a maturity level of 4.2. The audit team was asked to assess the ANZA PMO maturity and visited the company in September of 2005 to conduct the PEMA[3] assessment. The assessors conducting the audit made the following points:

- PEMA is a method for measuring project performance maturity in an organisation, by interviewing individuals in the organisation.
- The organisation is the target of the assessment, not the individuals.
- Answers shall reflect the level of deployment of project management practices, not the intended approach.
- The result of a PEMA is delivered as a report in a feedback meeting.
- The PEMA result and the feedback shall be used as a basis for improvement work in the organisation.

The feedback from the audit team was that ANZA scored 4.12 for deployment, the highest score across the 51 assessments conducted.

'Our general impression of the level of Organisational maturity at ANZA, is that the organisation is very mature, and contributes well to Operational Excellence. We have observed several good initiatives to coordinate and foster good Sales & Delivery Management. We encourage the MU management team to continue to support these initiatives.' (Hans Lock, lead auditor).

Evaluation – lessons learned from the journey

1   It has been said many times, but it still holds true, that a major change programme MUST be driven from the top. In the case of the ANZA PMO there was consistent support from the CEO and the management team. Whenever support, including financial support, was asked for it was given and each manager was seen to 'walk the talk'. During the period 1999–2005 there were three CEOs; each was totally committed to improving the maturity of the PMO.

2   The decision to centralise the team of project managers was critical to the success of the ANZA PMO. The need for discipline, profession-alism, up-skilling, teambuilding and general motivation was made much easier through co-location and the physical bringing together of the team. Costs for each project manager were charged back to each user as they sourced PMs, which appealed to the users as they no longer had to pay for a PM if they were not directly employed on a project. Access to immediate PM support was available due to the common team.

3   The PMO is a constantly emerging function; there is never a time to sit back and watch. The PMO must meet the current needs of the organisa-tion and these needs change daily. Stakeholder management is critical, ensuring there is an understanding of each stakeholder's needs and how well they are being met. While there were three major reorganisations of the PMO during the period addressed there was ongoing balancing and alignment of competence and quantity of the project management capability. Many times signing one contract causes a ripple effect as a particular PM is assigned to a new project.

4   The PMO must be seen to be adding value and be responsive; successes must be communicated and celebrated. In the beginning the PMO was forcing itself on the organisation whether they wanted it or not, as the services provided were seen to add value. Then the organisation started to approach the PMO and ask for services and at times suggest other services that could be offered. The PMO also ensured that the services they offered were not available from other sources. In the early stages, many of the units selling products and services did much of the upfront estimating and planning themselves – when they asked for support it had to be provided quickly and professionally, ensuring that the unit

avoids falling back into a self-help mode. Reinforcement from the management team, asking where estimates came from, if the PMO had signed off on the plan and who conducted the risk analysis, ensured the profile and their commitment to the PMO. The PMO must make it easier for the selling unit to get support so they can then spend more time on relationship management which is a critical part of successful selling.

5   In today's business climate it is critical that an organisation has access to an accurate set of financial data. The USA Sarbanes-Oxley laws[4] now provide serious penalties if organisations are not providing accurate information to the marketplace. The adoption of the correct version of the truth supports this direction. Once managers realise they are at risk they become much more interested in project governance.

6   Bring key stakeholders and the organisation along with you. In the project management profession we have created a new jargon and ways of thinking and working. Talking to senior managers about CPI and EAC may result in blank faces; it is your responsibility to educate the whole organisation in project management. The PMO needs to be in step and one step ahead at the same time – in step with the organisation to ensure it is effectively meeting its current needs and one step ahead so that you are anticipating and forming future needs.

7   As with any change project it is the people who have to change. There was a team of people who passionately believed in what they were doing; there were times when it was rocky and some very good people were lost. The PMO was able to keep a core of PMs who grew with the PMO in professionalism, process, knowledge, stature and confidence as the ANZA PMO matured into a successful PMO.

Chapter summary

The purpose of this chapter is to provide a practical case study example of a project procurement decision and how it links to theory discussed in other chapters in this book. This chapter's focus centres on project governance and a decision to undertake a project using in-house resources. The rationale for using in-house resources for the CoE development was an emergent strategy in improving project governance. It made sense to rely upon and develop in-house expertise for two reasons – thus the 'make' rather than 'buy' procurement decision was appropriate. First, this was experimental and lessons learned needed to be internalised. Second, the CoE project was itself an organisational cultural-change and business transformation project – thus it was a typical emergent strategy project.

This chapter provides a useful case study on developing governance tools and processes with evolution of a PMO into a CoE, thus it also contributes to PM theory and practice.

Table 11.1 Summary of aspects relating to this book

Chapter – perspective	Comments and discussion
Chapter 1 – Value proposition and make-or-buy decision	This used an in-house project delivery team in developing the CoE over several years using a 'make' decision. Expected value was learning and business/cultural transformation to high customer focus and service delivery
Chapter 2 – Project type and procurement choice	The CoE project was akin to a research and development change management project, highly complex and evolving with high relational and integrated design and delivery
Chapter 3 – Stakeholder management	No formal stakeholder identification management processes were followed. However, close personal contact and interaction with senior levels of management was maintained. Operational departments were perceived as being the initial key customers of the CoE that in turn serviced paying customers, clients and supply chain partners
Chapter 4 – Governance	The whole thrust of the initiative was to improve organisational governance of PM services within the framework of well-established organisational ethical standards and norms
Chapter 5 – Business strategy	An emergent strategy of evolving the CoE through a maturity process. High focus on project portfolio and programme management to authorise and manage projects. CoE seen as integral with business strategy
Chapter 6 – Performance measurement	A BSC approach with extensive use of a 'traffic lights' communication tool as part of the CoE developed tools. Mapping specific tangible with intangible outcomes, was not specifically developed, however, an internal view of capability maturity (based on the PMI OPM3 emerging concept) used to judge CoE development progress
Chapter 7 – E-business	The CoE substantially used an in-house developed e-portal as its platform for managing and monitoring projects
Chapter 8 – Innovation and organisational learning	A key CoE purpose was to promote learning and PM excellence. The CoE served as a training ground as well as facilitated mentoring and knowledge transfer throughout the business
Chapter 9 – Cultural dimensions	The CoE 'project' was itself a business process change management exercise. As a global organisation, it necessitated cross-cultural teams as well as understanding organisational culture from the perspectives of various business units. The CoE performed an increasing cross-organisational portfolio management role for project prioritisation and authorisation
Chapter 10 – Selecting and growing talented people	The CoE project promoted staff development and building strategic competencies. It enabled strategic HRM for the organisation's migration from a predominantly manufacturing firm orientation towards a services delivery organisation. The CoE promoted strategic growth, skills development and business excellence

Notes

1 Ericsson Australia P/L.
2 For more of project or product stage gates see Cooper (2005).
3 Project Environment Maturity Assessment (PEMA).
4 See for detail of this act URL http://www.soxlaw.com/index.htm

References

Andersen, B., Hendriksen, B. and Wenche, A. (2006). 'Project Management Office Establishment Best Practices'. *Project Perspectives – Annual Publication of International Project Management Association.* **XXVIII**(1): 30–35.

Bergmann, I. (2002). The EPMI project Office Model. *Ericsson Project Management Institute.* Stockholm: Ericsson.

Cooper, R. G. (2005). *Product Leadership: Pathways to Profitable Innovation.* New York: Basic Books.

Haldane, D. (1999). *The Future of Project Management.* Newtown Square, PA: Project Management Institute Research.

Kerzner, H. (2003). 'Strategic Planning for a Project Office'. *Project Management Journal.* **34**(2): 13–25.

KPMG (2003). Programme Management Survey – Why Keep Punishing Your Bottom Line? General Report. Singapore: KPMG International Asia-Pacific: 20.

Levine, H. A. (2005). *Project Portfolio Management – A Practical Guide to Selecting Projects, Managing Portfolios, and Maximizing Benefits.* San Francisco: Jossey-Bass.

Light, M. and Berg, T. (2000). The Project Office: Teams, Processes and Tools, The Gartner Group, R-11-1530: 30.

Miranda, E. and NetLibrary Inc. (2003). *Running the Successful Hi-Tech Project Office* at http://www.netLibrary.com/urlapi.asp?action=summary&v=1&bookid=87747.

Naisbitt, J. (1982). *Megatrends: Ten New Directions Transforming Our Lives.* New York; Warner Books.

Pettigrew, A. and Fenton, E. M., Eds. (2000). *The Innovating Organization.* Series The Innovating Organization. Thousand Oaks, CA: Sage.

Rad, P. F. and Levin, G. (2002). *The Advanced Project Management Office: A Comprehensive Look at Function and Implementation.* Boca Raton, FL: St Lucie Press imprint of CRC Press.

Walker, D. H. T. and Christenson, D. (2005). 'Knowledge Wisdom and Networks: A Project Management Centre of Excellence Example'. *The Learning Organization.* **12**(3): 275–291.

Chapter 12

Trust, commitment and mutual goals in Australian construction industry project alliances

Peter Rex Davis and Derek H. T. Walker

Chapter introduction

Chapter 2 introduced the concept of project alliances. Issues relating to trust and commitment, which are relevant to this chapter, were also raised and discussed in Chapter 3 and aspects of organisational culture were presented in Chapter 9. This chapter provides findings relating to project alliancing on construction projects in Australia, based upon interviews with 49 senior participants involved with 134 relationship-style projects (average value AU$150 million) that was originally published in Peter Davis' PhD thesis (Davis, 2006). The prime focus is centred upon how trust and commitment is enhanced under the project alliance procurement approach during the project team selection stage.

This chapter is structured as follows. The next section will outline the context and background of project alliances in a little more detail than elsewhere in this book. This is followed by a brief outline of the research methodology to provide an understanding to the reader of the rigour and depth of the study. Following this a discussion on the findings and an evaluation and discussion on implications for PM procurement practice is provided.

Context and theoretical background

This chapter discusses project alliances; however, it is relevant to review some procurement choices that lead towards a project alliance.

The model (Figure 12.1) shows a continuum of transactional to relationship-based procurement choices and illustrates some of the more common procurement choices with the most transactional choice, tendered lowest price, at the extreme edge of being non-relational and highly transactional. Joint ventures, at the mid-point in the procurement choices model, represent a strategic choice where organisations collaborate for mutual benefit. They may do this for strategic reasons as discussed in Chapter 5 or to enhance their competitive advantage. Walker and Johannes (2003a),

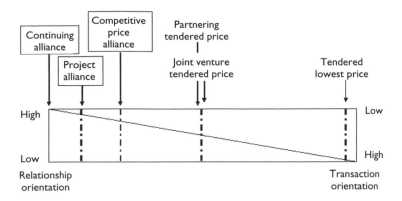

Figure 12.1 Relationship-based procurement choices.

reporting upon a study of nine top tier international construction contractors in Hong Kong,[1] highlight risk sharing, jointly offering complementary services, and brand identification as motivation behind JV formation. Other JVs may be motivated by a need to learn from their JV partner to stimulate organisational learning and/or innovation, as discussed by Walker and Johannes (2003b) and in Chapter 8 of this book. The JV relationship is indicated in Figure 12.1 as partly transactional because JV partners usually expect to benefit from an exchange of value in their joint project delivery enterprise. Doz and Hamel (1998) argue that many JVs are formed as a precursor to a strategic merger or acquisition, or perhaps as a ploy on the part of one partner to infiltrate the other JV partner's customer base with a view to exploiting the relationship at some future time.

Figure 12.1 also shows partnering at a similar point along the transactional and relational continuum. Partnering was defined earlier in Chapter 2. It is highly relational in its intent with partners signing up to a partnering agreement, developing a partnering charter and subscribing to an agreed well-structured dispute resolution mechanism that allows parties to settle misunderstandings, contentious issues and resolve claims for contract claims. Doing so in a responsive and effective manner helps reduce transaction costs and retains joint enthusiasm for achieving common goals. However, Walker and Hampson (2003a,c) argue that partnering can lead to one partner making profit while another partner suffers a loss on a project. This situation is remedied in a project alliance where the relationship is more encompassing.

Generally speaking, an alliance is a group of organisations working together in a cooperative arrangement with an aim to reduce overall costs, share project risk and reward and increase profits (Allen, 1995; Das and Teng, 1998a; KPMG, 1998). Parties have a focus on relationships (Pascale and

Sanders, 1997) that build on trust in construction business agreements (Kubal, 1994). An alliance can link teams more closely than transactional procurement approaches, or indeed partnering. The entities involved seek to align objectives and collectively develop an appropriate project scope from an early stage. An alliance links organisational performance of participating firms within the framework of a legally enforceable agreement (KPMG, 1998). Parties involved in an alliance agreement seek to align objectives and become closely involved in the development of the project scope and risk contingency budgets. They plan projects together from the outset and operate in a system of interdependence and understanding their partners' business drivers (Kubal, 1994; Allen, 1995; Haimes, 1995; Pascale and Sanders, 1997; Boyd and Browning, 1998). Competitive advantage stems from relationships bonded by integrity and trust. Once established, these bonds prove to be very strong and resilient to external forces. These relationships weave a net that will prevent competitors' entry.

Alliance parties jointly set aside profit margins and agree how gains may be shared before starting the project. These profit margins are then placed 'at risk' for all parties in the alliance before the project commencement and are released only at the end of the project. The 'profit at risk' approach demands greater active joint commitment. This is because, if agreed project metrics are not met, then all parties suffer a loss. Similarly, they jointly gain if performance metrics are exceeded using an agreed reward structure (Walker and Hampson, 2003c). The key point is that rewards or sanctions are *jointly risked*, based on project results not individual organisation's performance, so this provides the motivation for each party to do its best to help others in the alliance; in other words: *they all sink or swim together*.

Three factors must be present for an alliance to be successful:

1   The first is alignment of objectives (Kubal, 1994; Allen, 1995; Pascale and Sanders, 1997; Boyd and Browning, 1998). To realise this participants attempt to align objectives through scope development, targeting relationships that build trust (Kubal, 1994; Pascale and Sanders, 1997).
2   The second is win–win attitudes (Kubal, 1994; Hampson and Kwok, 1997; Tomer, 1998). Relationship contracting (RC) research is punctuated with phrases that include: 'win–win outlook', 'common goal attainment' and a 'search for synergy'. These terms, whilst not explicit in contract documents developed for alliance projects, do form an implicit underpinning theme (Hollingsworth, 1988; Hutchinson and Gallagher, 2003). This RC language is at odds with transactional procurement discourse, although benchmark documents that have captured a change in transactional contracting strategies advocate the use of RC strategies (Latham, 1994; DETR, 1998). These reports, when compared with more recent texts that discuss relationship-based procurement, show where enhanced value is provided to project participants in an RC environment (Keniger and Walker, 2003).

3   The final factor for a successful alliance is risk allocation and commercial incentives (Scott, 1993; Fellows, 1998; KPMG, 1998). By cooperating, the participants aim to reduce overall project costs, share project risk and reward and increase mutual profits (Allen, 1995; Das and Teng, 1998b; KPMG, 1998; Hutchinson and Gallagher, 2003; Ross, 2003; Whiteley, 2004). Successful alliances are not based on low-bid tendering. Awkward issues, such as price and change to the scope, are accounted for by participants at the earliest opportunity (Kubal, 1994; Allen, 1995; Pascale and Sanders, 1997; KPMG, 1998; Walker and Hampson, 2003c).

These three factors are founded on trust, cooperative rather than adversarial relations, collaboration rather than competition, problem solving and innovation rather than sanctions or contractual penalties (Boyd and Browning, 1998). Most importantly, an understanding of relationship development is required (KPMG, 1998; Thompson and Sanders, 1998).

A continuing alliance that remains beyond a single project is the most relational of all, as illustrated in Figure 12.1, and would apply where a contractor agrees to work with a client (or a client plus other project team members) on an agreed and continued basis. This may be advantageous when the client seeks consistency in delivery methods and approaches across projects or where the project cost may be a minor part of the procurement decision. In some industry sectors, when building a factory for a pharmaceuticals company or some electronic components suppliers, for example, speed of construction or the quality of mechanical and electrical services components may be of a much higher-order importance. Some chain stores or restaurants also choose to form longer cross-project alliances where their project 'signature' style is consistent or where joint learning leading to quick fit-out times is an essential procurement choice criterion.

The above provides a framework to test how a number of project alliances in Australia build trust and commitment before a project commences. Our results are presented in the section that follows – an insight into the research methodology.

Research methodology and studied alliance projects' context

The findings that are discussed in this chapter relate to a recent PhD study (Davis, 2006). This study employed an earlier phase involving an extensive survey of marketing and tendering practices employed by construction industry players in Australia. This part of the study established a theoretical framework with which to test, using qualitative research techniques, how a number of alliance projects were executed and performed in terms of relationship-based characteristics or dimensions.

The purpose for presenting this part of the chapter is to indicate the level of rigour achieved and to highlight the validity of the data-gathering process and analysis techniques used. As this is a book chapter and not a thesis, most readers will not be interested in the details of the methodological aspects – readers with this interest may refer to the PhD Thesis.[2] The scope of a book chapter is naturally constrained compared to that of a PhD thesis so we decided to focus on two key aspects of an alliance relationship-based procurement choice. These two aspects, at the early stage, are alliance formation and its subsequent relationship development.

Forty-nine project managers were interviewed using a semi-structured survey instrument (questionnaire). Data gathered from these interview transcripts were analysed using QSR N6 software. Particular industry practitioners were chosen because they were specifically representative of typical consortia associated with alliance projects. The sample of people interviewed in the PhD research project had considerable experience from which they were able to speak with authority. Care was taken to ensure respondent stratification being balanced across the industry, geographic location of projects across Australia and the interviewees' number of years experience in the industry. Interviewees' roles in industry and the projects that they were involved with were typical and representative of contemporary alliance/relationship projects. The industry stakeholders comprised consulting engineers and project managers, contractors and client representatives. Personnel from key infrastructure and *nation building* projects were selected as projects worthy of consideration. Projects included North Side Storage Tunnel, Sydney; Woodman Point (WA21), Perth; Wahroonga Dam, Queensland and the National Museum of Australia (NMA) in Canberra.

The questionnaire was designed to extract information about relationships that respondents have with their associates in the forming of an alliance. Only open-ended format questions were used in the interview. The survey instrument had five sections. The list of sections below identifies the main grouping questions pertinent to this chapter:

1   Background – context

    a   What were the project and respondent characteristics?

2   Establishment of the alliance

    a   What aspects of the alliance establishment were easy to manage and what aspects were hard to manage?
    b   What was your perception of potential benefits and shortcomings associated with the alliance establishment process?

3   Relationship building

    a   How was a *win–win* mentality developed?

A four-step process was devised as illustrated in Figure 12.2. Each of the four steps included a number of interrelated stages that enabled the

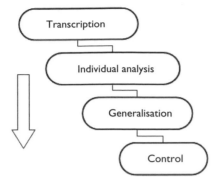

Figure 12.2 Four-step data analysis.

researcher to understand the complexity of the relatively unstructured raw data that had been gleaned from the interviews.

Interviews varied in length – the shortest was 15 minutes and the longest just over 1 hour; generally they lasted for approximately 40 minutes. Interviews were conducted on a one-to-one basis, and designed to stimulate conversation and break down any barriers between the interviewer and interviewee. Interviewees were allowed to talk freely without interruption or intervention to deliver a clear picture of their perspective. Interviews were mainly conducted in a respondent's head office and occasionally site offices were used. Many of the interviewees were interstate, outside of the researcher's home base (Perth, West Australia). Interstate interviews were all carried out over the phone at predetermined times to suit interviewees. All of the interviews apart from one were digitally recorded using an ipod(tm) digital recording instrument. Transcripts were made, checked and converted to a verbatim word document for manipulation and importation into QSR N6, a software program designed to enable reduction of data arising from qualitative interviews.

Discussion of research findings

The following outlines some results from the PhD thesis that are relevant to this chapter.

Selection of alliance partners

Typically, alliance partner selection takes place on the basis of alliance consortia providing evidence that potential partners can provide the best set of resources, skills and capabilities to deliver best-value for a project. There are some differences, however, between a pure alliance model and a Target

Out-turn Cost (TOC) or Cost Competitive Alliance. The latter is said to provide better cost and time certainty to public sector clients. According to Cowan and Davis (2005) the key elements of the project delivery phase are the same in both variants and represent similar risk patterns and project management structures. However, it is the tension associated with delivery of price where a departure lies.

In a pure alliance, for example the NMA, tension occurs between the client and a single non-client consortium after the preferred consortium selection has been made and work proceeds to develop a TOC. This can lead to time and cost uncertainty. This uncertainty may be overcome in the private sector through commercial drivers and business hurdles that stop a project from proceeding if determined parameters are exceeded. In the public sector, a business case may dictate that a project is almost certain to proceed – for Australian examples of this see Cowan and Davis (2005). This actuality may instil a lack of confidence in the commerciality or competitiveness of the TOC.

Indeed Whiteley and Henneveld (2006) suggest that a cost competitive selection process is regularly used to overcome an industry perception that unless there is market tension the target price will be inflated.

In a cost competitive alliance alternative a single client team develops a competitive TOC with two consortia working in parallel and each proponent submits their concept design and associated TOC at the end of the development period. The client team will then select the best value offer, taking into account price and non-price criteria (Cowan and Davis, 2005). The competitive nature of the process redirects price tension from the client and focuses it back onto the non-client participants who are in competition until their project alliance agreement is accepted. This may have adverse trust and commitment implications as discussed later.

The NMA project provides a good example of a pure project alliance and illustrates alliance selection processes. The successful NMA alliance team was not judged on an estimated tender cost criterion. The assumption of the selection panel, who recommended the successful consortium, was that a superb team, acting as a focused and coherent group with a vision to provide an excellent end product, would deliver best-value for the project – at cost, time and quality that is at least equivalent to global best-practice benchmarks. Details of the selection process can be found in Walker and Hampson (2003c) and follows the process described by several alliances facilitators (Hutchinson and Gallagher, 2003; Ross, 2003).

Initially, an open call for expressions of interest was made on the NMA project with an industry-briefing forum being conducted to explain the concept and detail evidence required of consortia to support their submission. Written proposals were received and the selection panel short-listed several leading proponents based in part on their technology innovation, resource capability and their understanding of the alliance concept.

The short-listed candidates were interviewed and took part in an extensive two-day workshop with each of the short-listed consortia tested on evidence submitted in their proposals. In addition, they worked together with panel members to fully flesh-out exactly how the project could proceed. This involved reference to substantive detail regarding target cost and time budgets. Discussion to refine these aspects, together with improved team-behaviour and communication, developed joint problem solving and built trusting and committed relationships.

The preferred alliance consortium was selected and this group continued with a facilitated workshop (with a facilitator who would continue after the workshop in a 'coaching role'[3] to help maintain the relationship between alliance parties) to establish and agree on performance metrics, risk/reward agreements and further develop the project budget (termed from then on as TOC). The successful alliance consortium was then confirmed and the alliance agreement executed and the budgets were subsequently finalised.

An important difference between this approach and other procurement choices (discussed in Chapter 2 of this book) is that the project budget was refined only *after* the alliance was formed so that this group was committed to an out-turn cost. Another fascinating innovation was that profit margins were agreed to be fixed at the participants' average profit over the past six years. Alliance participants agreed to a probity process of open book access to verify the average profit margin over the previous six years as well as inspection when requested of their accounts during the project execution. The margins varied between the alliance members because they differed by organisation – the alliance comprised design consultants, a main contractor and specialised service contractors, and each of these had different profit margin structures. Profit margin was quarantined to be placed 'at risk' and agreed to be paid on successful completion of the project (based upon the metrics agreed upon during the selection phase). A reward/penalty structure related to improving the TOC was based upon agreed ratios. The defining element of this process was that it delivered commitment by involved parties and clarified many of the up-front issues often left unclear on other procurement forms. For example, there was a high level of trust generated because risks were clearly devolved to those best to bear them and processes and procedures for working collaboratively were established and agreed upon. The NMA approach was similar to that for many of the project alliances studied in the (Davis, 2006) PhD study.

The selection process could therefore be perceived as a risky business and according to many of the respondents in the research study, it required careful planning, investigation and forethought.

The risk associated with the selection of consortium partners required that they evaluate trust and monitor its development to the best of the available methods at the time. They would do this mindful of their limited information concerning the individuals associated with proposed stakeholder organisations. In the majority of situations their preference would be to work with individuals and organisations with whom they had prior favourable experience. Previous experience would reduce risk and enable a more tangible assessment of trust and commitment. The established model of alliance relationship development added a burden to the contractors and was indicated to be unfamiliar territory to them. This aspect once again moved them into a risk area, and trust and commitment were evaluated in a workshop environment with appropriate facilitation or coaching to reduce risk-moves. Examples of risk-moves include a large concession that requires reciprocation, a proposal for a compromise, a unilateral action of tension reduction or a candid statement about one's motives and priorities (Dwyer *et al.*, 1987). The workshop environment, albeit a *manufactured* environment that used management games/selection themes, enabled the participants to assess and fast-track experience over a reduced time frame and achieve a suitable tangible outcome.

In the PhD research (Davis, 2006), the primary themes of the selection process were described as trust, commitment and mutual goals. Trust enabled contractors to differentiate themselves or be selected in a different way from the more traditional and familiar price-alone selection. Early development of trust engendered harmony within the stakeholder group. This harmony was found to continue into the implementation phase of the project. Commitment to others in the alliance workshops through resources allocation, for example, generated attitudes that reduced risk for all parties by moving risk to those parties that could best manage that risk. Risk was intelligently allocated rather than merely being shifted or passed on to others with less power to resist accepting that risk. It enabled a client to have a better understanding of the process and complexity of the project they were generating in more detail than would ordinarily happen. Mutual goals were described as a common understanding or focused alignment of expectations in a communicative way. The clients used the term 'Progressive engagement' to describe goal development that took place in the early stages of an alliance project.

The data supported the conclusion that alliancing facilitated much broader appreciation of risk, complexity and capabilities of parties to manage risk. It also facilitated better planning and joint problem solving. This front-end capacity and marshalling of skills where and when they can best be deployed set the project team on a potentially higher performance trajectory path as discussed in Chapter 1 Project phases and performance figure.

Workshops for relationship building and development

The relationships characteristics that the respondents were striving toward were likened to personal relationships. As indicated in the above section, respondents would endeavour to identify suitable partners whom they could work with and trust. Individuals were important, and their respective organisations tended to be placed in the background as a secondary criterion. Organisational identity was a selection criterion, but individuals from favoured organisations were scrutinised in more depth as it was their knowledge that was in demand. This links with the trust models presented in Chapter 3, particularly trust in phases. Caution was counselled with respect to personnel selection and advice was provided that selection of individuals should be based on *best-for-project* as opposed to selecting people for their personal affability or friendship. Although study interviewees indicated that these traits are often an outcome of the relationship workshops and ensuing projects, a commitment to alliance principles and required skills in specific areas were said to be major selection factors. Commitment and capability were described as the '*big ones*' when referring to selection criteria. Skills displayed by an individual would need to complement existing expertise that was a priori in the existing client or non-client group making the selection.

In terms of relationship development, participants would have to show their selectors that they were credible – in other words: were the statements they made or actions that they took in the relationship development workshops believable? Effective experience and competence were strong indicators of credibility. Often these were based on their experience with each other on past projects, whether relationship or transactional type projects. Seemingly the respondents were prepared to put faith in the word of those whom they were dealing with in the relationship development workshops, on the basis that promises made would be delivered. Clients would look for openness and a willingness to share. Rational checks, for example calling for CVs and commercial checks were used by some respondents to supplement the less subjective measures of credibility. A preparedness to provide *open books* was described as showing credibility. Some of the respondents were clearly far less formal with rational checks than others. This aspect reinforces the validity of the trust models presented in Chapter 3, particularly the 'trust and distrust' model.

The relationship development workshops provided a context for the relationships to develop. Its structure provided a framework on which the relationship would be built. The actual process of the relationship development would provide, in itself, value to the project. Often, however, the value created in the process would not be captured in a tangible way through, for example, reflecting and reporting upon benefits derived. This failure caused several respondents, mainly clients, to miss the potential *value* of these workshops and presume that the relationship development afforded little net gain to the overall project. Some argue that this positively differentiates a pure alliance over a cost-competitive alliance.

Several underlying attributes of the relationship building workshops were identified. They included, from a contractor's perspective: trust; development of bonds and tests of the relationship. Because the non-client team were assessed as a collective entity in any alliance selection exercise, it was important that they were able to show a bond between them that would translate into a team that was prepared to integrate with the client participants. Testing these bonds with technical issues was part of the relationship development workshops. From the respondent's perspective, an integrated team was the goal of the relationship development workshop; a team working together as if part of the same organisation, solving problems and having preparedness to confront *hard* issues. Collaboration provided focus and alignment to the team that was to influence the project outcome. Trust and commitment was said to be a test of collaboration with an example being given concerning collaborative decision making. The sentiment was expressed that only when trust and commitment were manifest would positive decision making be possible.

Coaches were indicated to have an enormous impact on the relationship development process and used to deliver a fair workshop outcome that was balanced and objective. Both client and contractor respondents made use of coaches and/or facilitators.

The role of trust

Trust was an important concept raised by respondents. Trust enabled contractors to differentiate themselves or be selected in a different way from the more traditional and familiar price-alone selection. Early development of trust engendered harmony amongst the participants. More than a third of contractors-recorded comments related to trust building. Respondents linked trust with respect in relationship development. Relationship development was designed to deconstruct formality using a mixture of activities that put participants through a combination of scenarios to assess their reactions, for example, whether they displayed empathetic behaviours in a particular circumstance or not.

Contextualising this, several respondents commented that it was easy for them to respond from the heart in the alliance development environment, suggesting that mutual understanding was a key to their relationship. Respondents entered the alliance development with open minds and proceeded with the assumption that they could trust others and hoped for reciprocation.

Research results indicated that trust was being assessed in parallel with commitment. Commitment to others in alliance development (through resources allocation, for example) generated attitudes that reduced risk for all parties. It enabled a client to have a more detailed understanding of the process of the project they were generating than would ordinarily happen. The non-client team (the contractor and supply chain consortium) were assessed as a collective entity, so it was important that they were able to demonstrate bonding that would translate into their team being prepared to integrate with other project team participants (i.e. design and client representative teams). Testing bonds with technical scenarios was applied as part of the relationship development workshops.

From the respondent's perspective an *integrated team* was the goal of the relationship development workshop; a team working together as if part of the same organisation, solving problems and being prepared to confront hard-to-solve issues. Earlier in the workshop, when teasing out underlying issues, clients were noted to assess their personal feelings of trust towards contractor personnel. All respondents indicated that they perceived increasing levels of trust throughout the alliance development duration. When asked if they had any special way of measuring or determining evidence of trust, several clients indicated that they often use performance measurement systems. The performance measurement systems successfully measured such things as communication.

Effective communication was said to be indicative of relationships and relationship building and was accordingly proposed as an appropriate determinant closely aligned with trust development. Interestingly, some of the client groups had formalised documentation that provided policy and guidance, setting out procedures for measuring attributes of trust and trust-building behaviours. Increasing levels of trust and commitment had a dual purpose. It tested the ability of participants to work in a cooperative and collaborative environment. In addition, it built relationships; these relationships established rapport between participants before the start of the implementation phase of the project – in effect by the time the actual contract was awarded there would be an effective relationship. In this context an important link was made between experience and trust.

Respondents perceived that they should endeavour to generate experience of participants as quickly as possible. This was important, as respondents were conscious that only through experience were they able to generate the requisite levels of trust for the continuance of the alliance project. Trust and commitment were said to be a test of collaboration; an example given concerned collaborative decision making. The sentiment was expressed that only when trust and commitment were evident would positive decision making aligned to mutual goals be possible.

Organisational development

Organisational development of the alliance was a high priority requiring senior (executive-level) staff to be present at the relationship development workshops. Organisational commitments made in meetings could therefore be relied upon; statements by senior staff made or actions that they took in the relationship development workshops were understood as credible. It was emphasised that failing to honour these commitments would be badly received, affecting an organisation's future work. It appeared that this sanction would be sufficient penalty from the respondent's perspective.

Proven effective experience and competence were strong indicators of organisational credibility. Often these indicators were based on experience on past projects; whether alliance or more traditional construction projects. Seemingly the respondents were prepared to put faith in the word of those that they were dealing with in the relationship development workshops, on the basis that promises made would be delivered. Consensus from the respondents found that reaching agreement prior to establishing a contractual relationship was a definitive test of commitment. This emphasised the defining attributes that differentiate a pure alliance over a target outturn cost alliance. Examples of commitments were given and several innovative points were raised by the respondents. One respondent noted a list of commercial outcomes that established an organisational commitment to life skills, including training and community relations. These outcomes were afforded to individuals outside the usual recognised boundaries of alliance personnel.

Organisational development clearly focused the minds of the collective participants. All activities that took place were targeted towards strategies and processes that would benefit their project. Mutual goals were described as a common understanding or focused alignment of expectations in a communicative way. The clients used the term 'Progressive engagement' to describe the goal development that took place in the early stages of an alliance project. Collaboration provided focus and alignment to the team that was to influence the project outcome.

Participants argued that trust and openness encouraged participant focus on problem solving and developing innovation that would generate new knowledge in the alliance project organisation. More than two-thirds of contractors commented on the nature of the alliance management team (AMT) that was formed as an outcome of the relationship development. Analysis of interview data indicated that the alliance team would create a life of its own, capable of transcending personal and company behaviours. The AMT became integrated with organisational fit from within the newly established hybrid team. The team defined their organisational boundary rather than accept boundaries of individual organisations joining the alliance.

Respondents were, however, concerned about organisational interaction. Major concerns centred on the population of intra-organisational and inter-organisational relationships, together with policies on placing suitable personnel into the alliance (people-resource allocation issues). It was clear from responses that intra-organisational alliance relations impacted upon inter-organisational relations and relationships. This was described as a consequence of the relationship building and selection exercises that were carried out in the alliance developments. Thinking of innovative ideas in an open fashion was one of the ways that the contractor participants enunciated this sentiment.

Participants often become an integrated team by the end of the alliance development period. This integration became so strong that several respondents suggested that it would appear to outsiders that they all came from the same organisation, sharing assumptions values and team artefacts such as a project vision and agreed plans and procedures.

The process of developing relationships draws together many facets from initial strategy, through commitment to durability and finally enduring relationship quality. These three phases are not mutually exclusive and each represents a collection of iterative macro processes. The boundaries referred to in the phases of relationship development are indistinct and dependent on supply chain activities; they may change with individual relationships: for example Araujo *et al.* (1999) propose four interfaces from standardised, through specified and translational to interactive. These serve to balance the costs and benefits of establishing and maintaining relationship interfaces. Boundary penetration, the degree each organisation partner penetrates the other organisations in a joint action, is regular and serves to reduce partitioning of participants (Heide and John, 1990). Each phase interacts with subsequent phases through seamless boundaries that are punctuated with incremental investments.

The three-stage alliance relationship development model illustrated in Figure 12.3 draws on key papers and 49 interviews focusing on relationship development (RD) and is based on the relevant literature

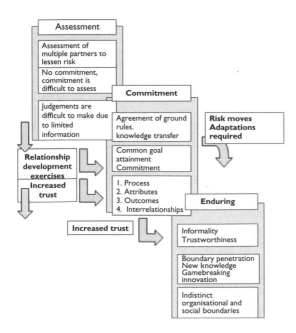

Figure 12.3 A three-stage alliance relationship model.

(Ford *et al.*, 1985; Dwyer *et al.*, 1987; Wilson, 1995; Pascale and Sanders, 1997; Ford *et al.*, 1998; Donaldson and O'Tool, 2001; Walker and Hampson, 2003b). In the first stage objective judgement would be difficult due to the intangible nature of the construction service and the client would be looking for indicators that would reduce risk associated with decision making (Day and Barksdale, 1992; Patterson, 1995). Many strategies were identified by the respondents, essentially a client would be asking the question: 'Can you do this for me?' The question would be seeking information concerning resources and technology together with managerial expertise (Ford, 1982; Wilson, 1995; Ford *et al.*, 1998). The thrust of questioning in the first instance should be directed towards a firm. Analysis of individuals would follow in subsequent stages. In the first stage there would be limited commitment and if commitment was evident it would be difficult to recognise and evidence.

Trust should be initially developed with the use of relationship development exercises as the relationship evolves and intersects with the second phase. This has been shown to fast-track commitment (Davis, 2006). The commitment in itself would lead towards trust and enhance trust maintenance, albeit relatively limited at this stage in the model. At the conclusion of the first stage, which will permeate the second stage boundary rather than happen at any pre-determined time, there would be an option for either party to end the relationship. This action follows both the literature and the respondents' comments. It is noteworthy to recognise that a predisposition towards upstream relationship variables of trust has been identified as stronger than downstream and this fact should be accounted for in the workshops (Davis, 2006). At this stage the parties often exhibit guarded information transfer and limited trust. Trust testing may take the form of an offer to reconfigure capital items of plant or IT equipment to meet the needs of a client. It is suggested that workshops should take place in the second stage of the relationship development model in line with current practice in alliance procurement and as described by the respondents (Hutchinson and Gallagher, 2003; Ross, 2003).

In the second stage illustrated in Figure 12.3 questions that the parties to the workshop would be endeavouring to answer would include: 'What are our common goals?' and 'How can we do this project together?' they would also be looking to establish ground rules. The ground rules would not only be required for the workshop meetings but also for the project to be carried out. Parties would be thinking about risk. Open discussion should revolve around adaptations required to deliver best-for-project innovation. These preceding points lead to deeper trust between the parties and embed commitments based on propositions about future actions. Trust would be growing through this stage, with commitments being increased. The process of the relationship development workshop, its underlying attributes, the measured goal and commitment outcomes, together with organisational interrelationship issues are fundamental to success. These are addressed in the alliance relationship development model.

Trust and its links to commitment were identified as important attributes of a relationship marketing approach to relationship development; this association was not missed by the respondents. They should not be overlooked in the development of an alliance relationship. Effective communication and information sharing were shown to increase trust and trust development. Goal setting and outlining deliverables were indicators of commitment, aligning these commitments with commercial objectives were also provided as examples where trust development in this stage would enhance the relationship. The reciprocal nature of trust was explored and practitioners in a construction relationship development programme should be cognisant that often the reciprocal nature of trust is displayed through knowledge transfer in what has been described as communities of practice (COPs). Nahapiet and Ghoshal (1998), Wenger (1999) and Wenger and Snyder (2000) describe COPs as groups of people linked together through their interests in an environment in which sharing and exchanging knowledge is a primary goal.

The third stage of relationship development was referred to in the relationship marketing literature as enduring and variable. As indicated in the literature, the relationship would be becoming close to fruition. Indistinct organisational and social boundaries would become less evident as boundary penetration of actors creates an informal hybrid team that is to be in place for the duration of the project (Wilson, 1995). Cohen and Levinthal (1990) use the term *absorptive capacity* to describe the acceptance of new knowledge which includes a propensity towards openness and tolerance of mistakes. In this context having a history of gamebreaking ideas suggest absorptive capacity is important to this stage of the alliance relationship development model. People who have an ability to note an opportunity of an idea transferred from one discipline to another are useful to the relationship at his stage. For example, in one of the projects discussed by the respondents several were able to cite an example of technology transfer of a significant nature that was indeed gamebreaking. This competency together with others cited by Maskell (1998) including reliability and trustworthiness cannot be bought and are not readily available to competition in the market place. They augment the competitiveness of an organisation through their continual reuse in the enduring relationship. In the last stage, enduring, strong ties that encompass technical, social and knowledge areas are forged. Regular evaluations still remain in place to ensure the quality of the relationship.

Implications for alliance procurement practice

We argued in this chapter that the relationship development process is crucial to a successful relationship contract/alliance. Evidence presented from the research (Davis, 2006) suggests that there is a recognisable structure to relationship development that is underpinned with specific themes which should be considered when managing the relationship development process (see Figure 12.3). Trust and commitment have been explicitly addressed in this chapter and are recognised as important implicit elements. We stress that relationship-based procurement is dependent upon and is reinforced by joint learning from joint problem-solving activities established in the relationship development workshops.

Respondents in the study, when asked about relationship benefits associated with alliance projects, were positive. They expressed belief that the selection process that included relationship development workshops added significantly to the benefits that they all accrued from the project. Benefits were not limited to process benefits but flowed into product or relationship benefits. The alliance development provided a context for the relationships to develop; it provided a framework or scaffolding from which the relationship would be built upon. The actual process of the relationship development in itself would provide value to the project. Often, however, value created in the process would not be captured in a tangible way; through reporting, for example. This failure caused several respondents – mainly clients – to underrate the value and presume that the relationship development afforded little gain to the overall project. Maskell (1998) refers to this action as the codification of tacit knowledge, in as much as it may remain tacit whilst it is available only to an individual. It is only when information is shared with others who grasp its significance, as they did in the workshops, that it becomes codified. The allocation and management of risk in alliance development follows the work of relationship development by Dwyer *et al.* (1987), Ford (1982) and Wilson (1995). Finally, the benefits ascribed to relationship marketing from the literature were wholly supported; Han and Wilson (1993) and Wilson and Mummalaneni (1986), for example, cite examples where relationships between long-term buyer–suppliers save inspection costs and consequently provide benefits in quality and reduced lead-in times. The primary benefits that all the respondents in the study spoke about were trust, commitment and the development of mutual goals that were engendered at an early opportunity and reinforced throughout the project development stage.

From the respondents' perspectives it appears that the entire process of relationship development hinged around commitment, trust, satisfaction and mutual outcomes. From the respondents' collective perceptions it was all important and entirely reciprocal, regardless of upstream or downstream engagement. Connections that evolved in the team drove the process and enhanced the outcome of the project, and as indicated above added substantially to the value that was derived from the project. The implications for the above on project procurement using project alliances are that joint learning and problem solving is a vital vehicle for developing the trust and commitment required for project alliancing to work. The issue of a

TOC-based selection or a selection on perceived skills, aptitudes, behaviours and abilities to work collaboratively is a complex one. While many clients with stakeholders who expect some sort of test based upon tender-type price may feel pressured to attempt an alliance selection process with project TOC as a criterion, we caution that the study evidence suggests that this may compromise the achievable levels of trust and commitment; unwanted consequences may result from this.

Table 12.1 Summary of aspects relating to this book

Chapter – perspective	Comments and discussion
Chapter 1 – Value proposition and make-or-buy decision	This study reviewed relationship-based procurement options that considered 'value' in a wider context'. While projects were outsourced, a best-value proposition drove the procurement choice
Chapter 2 – Project type and procurement choice	Projects fell into typical tangible construction project types though intangible value was expected as a project outcome
Chapter 3 – Stakeholder management	The project alliance concept is highly focused upon supply chain advantages gained from greater synergy and better client–contractor interactions that reduce transaction costs
Chapter 4 – Governance	Project alliances have highly developed probity and governance structures
Chapter 5 – Business strategy	Project alliances are one-off and emergent; however, these often lead to a more structured arrangement so insights into the strategy schools can help us understand why some project alliances remain as project-specific while others develop into programme arrangements. The dynamic capabilities and competitive advantage strategic management aspects are highly relevant to content in this chapter
Chapter 6 – Performance measurement	None of the non-traditional aspects of this chapter were specifically addressed; however, KPIs and metrics are an important part of project alliance probity
Chapter 7 – E-business	No direct reference to this chapter, though in the NMA case study the availability of a web portal that connected alliance partners was an important selection issue. Communication and problem solving is an important factor, so groupware provides vital facilitating tools
Chapter 8 – Innovation and organisational learning	One of the main drivers for alliances is the expectation that it can facilitate organisational learning and innovation
Chapter 9 – Cultural dimensions	Trust and commitment that meld organisational identity from a series of individual organisations into one alliance requires cultural alignment, this is highly relevant to this chapter
Chapter 10 – Selecting and growing talented people	Achieving alliance objectives requires selection and retention of high-value talent with skills and behaviours that require careful selection and retention strategies. This is of concern to alliance partners

Chapter summary

The purpose of this chapter was to present leading-edge research findings of an emerging form of project procurement in the construction industry-project alliancing. This chapter drew extensively upon relationship marketing literature and fills a gap elsewhere in the book where it was not appropriate to include ideas from that body of knowledge. Further, in this chapter we were able to not only present advanced conceptual models but also, through reporting on the Davis (2006) PhD study, present practical first-hand experience of how this form works in practice. Implications for practice were also offered.

This chapter argues that developing trust and commitment can effectively deliver superior project outcomes but that some of the outcomes are intangible (team relationships, knowledge transfer, improved decision making, possible waste reductions).

Table 12.1 indicates how this chapter fits into the first 10 chapters.

This chapter provides a summary of many useful alliance case studies, and so it represents a considerable contribution to the project procurement literature.

Notes

1 See Johannes (2004) for more details follow the link for Johannes thesis on http://dhtw.tce.rmit.edu.au/dwressup.htm, accessed 17 January 2007.
2 See http://dhtw.tce.rmit.edu.au/dwressup.htm follow the link to P. Davis PhD thesis.
3 For details about a facilitator's role in alliance projects and coaching levels provided contact Andrew Hutchinson through www.alchimie.com.au/ welcome and Jim Ross at www.pci-aus.com/ or Evans and Peck at www.evanspeck.com.au/

References

Allen, A. L. (1995). Oil and Gas Greenfield Project Alliances Theory and Practice. Report Presented as Part of the Requirements of the Award for the Degree of Master of Project Management. Perth, Curtin University of Technology.
Araujo, L., Dubois, A. and Gadde, L.-E. (1999). 'Managing Interfaces with Suppliers'. *Industrial Marketing Management.* **28**(5): 497–506.
Boyd, L. and Browning, P. (1998). 'Facilitating Partnering in Local Government'. *Local Government Management: LGM.* **32**(3): 10–13.
Cohen, W. M. and Levinthal, D. (1990). 'Absorptive Capacity: A New Perspective on Learning and Innovation'. *Administrative Science Quarterly.* **35**(1): 128–152.
Cowan, B. and Davis, J. (2005). Development of the 'Competitive TOC' Alliance – A Client Initiative at http://www.alliancenetwork.com.au/pdfs/Paper%20on%20Competitive%20Alliances%20Ver%204.pdf, 08.12.2006.
Das, T. K. and Teng, B.-S. (1998a). 'Between Trust and Control: Developing Confidence in Partner Cooperation in Alliances'. *Academy of Management Review.* **23**(3): 491–512.

Das, T. K. and Teng, B.-S. (1998b). 'Resource and Risk Management in the Strategic Alliance Making Process'. *Journal of Management.* **24**(1): 21–41.

Davis, P. R. (2006). The Application of Relationship Marketing to Construction. *PhD, School of Economics, Finance and Marketing.* Melbourne: RMIT University.

Day, E. and Barksdale, H. C. (1992). 'How Firms Select Professional Services'. *Industrial Marketing Management.* **21**(2): 85–91.

DETR (1998). Rethinking Construction, Report. London, Department of the Environment, Transport and the Regions.

Donaldson, B. and O'Tool, T. (2001). *Strategic Marketing Relationships: From Strategy to Implementation.* Sussex: John Wiley and Sons Ltd.

Doz, Y. L. and Hamel, G. (1998). *Alliance Advantage – The Art of Creating Value Through Partnering.* Boston: Harvard Business School Press.

Dwyer, F. R., Schurr, P. H. and Oh, S. (1987). 'Developing Buyer-Seller Relationships'. *Journal of Marketing.* **51**(2): 11–27.

Fellows, R. F. (1998). *Construction Projects: The Culture of Joint Venturing.* Proceedings of the CIB World Building Congress 1998, Gavle, Sweden, 7–12 June 1998, Fahlstedt K., 3: 1505–1512.

Ford, D. (1982). The Development of Buyer Seller Relationships in Industrial Markets. *International Marketing and Purchasing of Industrial Goods: An Interaction Approach.* Hankansson H. Uppsala: John Wiles & Sons.

Ford, D., Gadde, L.-E. and IMP Project Group. (1998). *Managing Business Relationships,* Chichester; New York: Wiley.

Ford, D., Hakansson, H. and Johanson, J. (1985). 'How Do Companies Interact'. *Industrial Marketing and Purchasing.* **1**(1): 26–40.

Haimes, G. A. (1995). Partnering and Alliancing in Engineering Construction; Report Presented as Part of the Requirements of the Award for the Degree of Master of Project Management. Perth, Curtin University of Technology, Perth. W. A.

Hampson, K. and Kwok, T. (1997). 'Strategic Alliances in Building Construction: A Tender Evaluation Tool for the Public Sector'. *Journal of Construction Procurement.* **3**(1): 28–41.

Han, S.-L. and Wilson, D. D. S. (1993). 'Buyer-Supplier Relationships Today'. *International Marketing Management.* **22**: 331–338.

Heide, J. B. and John, G. (1990). 'Alliances in Industrial Purchasing: The Determinants of Joint Action in Buyer-Seller Relationships.' *Journal of Marketing Research.* **27**(1): 24–36.

Hollingsworth, D. S. (1988). 'Building Successful Global Partnerships'. *The Journal of Business Strategy.* **9**(3): 12–15.

Hutchinson, A. and Gallagher, J. (2003). Project Alliances: An Overview, Melbourne, Alchimie Pty Ltd, Phillips Fox Lawyers: 33.

Keniger, M. and Walker, D. H. T. (2003). Developing a Quality Culture – Project Alliancing Versus Business as Usual. In *Procurement Strategies: A Relationship Based Approach.* Walker D. H. T. and K. D. Hampson, Eds. Oxford: Blackwell Publishing: Chapter 8, 204–235.

KPMG (1998). Project alliances in the Construction Industry, Literature Review. Sydney: NSW Department of Public Works & Services, 7855-PWS98–0809-R-Alliance.

Kubal, M. (1994). *Engineered Quality in Construction: Partnering and TQM.* New York: McGraw-Hill.

Latham, M. (1994). Constructing the Team, Final Report of the Government/Industry Review of Procurement and Contractual Arrangements in the UK Construction Industry. London: HMSO.

Maskell, P. (1998). Globalisation and Industrial Competitiveness: The Process and Consequences of Ubiquitification. In *Making Connections: Technological Learning and Regional Economic Change*. Malecki E. J. and P. Oinas, Eds. Brookfield, Ashgate: Aldershot, Hants: 35–59.

Nahapiet, J. and Ghoshal, S. (1998). 'Social Capital, Intellectual Capital, and the Organizational Advantage'. *Academy of Management Review*. 23(2): 242–266.

Pascale, S. and Sanders, S. (1997). *Supplier Selection and Partnering Alignment: A Prerequisite for Project Management Success for the Year 2000*. The 28th Annual Project Management Institute 1997 Seminars and Symposium, Project Management Institute, Project Management Institute: 19–26.

Patterson, P. G. (1995). 'Choice Criteria in Final Selection of a Management Consultancy Service.' *Journal of Professional Services Marketing*. 11(2): 177–187.

Ross, J. (2003). *Introduction to Project Alliancing*. Alliance Contracting Conference, Sydney, 30 April 2003, Project Control International Pty Ltd.

Scott, B. (1993). *Partnering and Alliance Contracts – A Company Viewpoint*. Perth, Australia: Curtin University.

Thompson, P. J. and Sanders, S. R. (1998). 'Partnering Continuum'. *Journal of Management in Engineering – American Society of Civil Engineers/Engineering Management Division*. 14(5): 73–78.

Tomer, J. F. (1998). 'Beyond Transaction Markets Toward Relationship: Marketing in the Human Firm: A Socio-Economic Model'. *The Journal of Socio-Economics*. 27 (2): 207–228.

Walker, D. H. T. and Hampson, K. D. (2003a). Developing Cross-Team Relationships. In *Procurement Strategies: A Relationship Based Approach*. Walker, D. H. T. and K. D. Hampson, Eds. Oxford: Blackwell Publishing: Chapter 7, 169–203.

Walker, D. H. T. and Hampson, K. D. (2003b). *Procurement Strategies: A Relationship Based Approach*. Oxford: Blackwell Publishing.

Walker, D. H. T. and Hampson, K. D. (2003c). Project Alliance Member Organisation Selection. In *Procurement Strategies: A Relationship Based Approach*. Walker D. H. T. and K. D. Hampson, Eds. OxfordL: Blackwell Publishing: Chapter 4, 74–102.

Walker, D. H. T. and Johannes, D. S. (2003a). 'Construction Industry Joint Venture Behaviour in Hong Kong – Designed for Collaborative Results?' *International Journal of Project Management, Elsevier Science. UK*. 21(1): 39–50.

Walker, D. H. T. and Johannes, D. S. (2003b). 'Preparing for Organisational Learning by HK Infrastructure Project Joint Ventures Organisations'. *The Learning Organization, MCB University Press. UK*. 10(2): 106–117.

Wenger, E. C. (1999). 'Communities of Practice: The Key to Knowledge Strategy'. *The Journal of the Institute for Knowledge Management*. 1(Fall): 48–63.

Wenger, E. C. and Snyder, W. M. (2000). 'Communities of Practice: The Organizational Frontier'. *Harvard Business Review*. 78(1): 139–145.

Whiteley, D. (2004). Project Alliancing an Investigation into the Benefits and Attributes of the Project Alliance Contracting Strategy. Masters, *Faculty of the Built Environment*. Perth, Australia: Curtin University.

Whiteley, D. and Henneveld, M. (2006). *A Practical Guide to Achieving Outstanding Performance from Alliance Contracting*. LexisNexis Commercial Litigation & Construction Law WA Hyatt Regency Perth WA, 20 September 2006, Development L. P., LexisNexis Professional Development: 1–22.

Wilson, D. (1995). 'The Integrated Model of Buyer-Seller Relationships'. *Journal of the Academy of Marketing Science*. **23**(4): 335–345.

Wilson, D. and Mummalaneni, V. (1986). 'Bonding and Commitment in Buyer-Seller Relationships: A Preliminary Conceptualisation'. *Industrial Marketing and Purchasing*. **1**(3): 45–58.

Chapter 13

Innovation management in project alliances

Steve Rowlinson and Derek H.T. Walker

Chapter introduction

Chapter 2 introduced the concept of project alliances. Issues relating to trust and commitment, which are relevant to this chapter, were also raised and discussed in Chapter 3, and aspects of organisational culture were presented in Chapter 9. Chapter 12 presented PhD findings from a range of project alliances and focused on alliance formation at the front-end stage of a project. This chapter provides findings relating to how innovation was enacted by an innovative organisation in a specific project alliancing case study in Australia. This links with Chapter 8 and models discussed in that chapter on how organisations can become innovative. This chapter's focus is centred on how alliancing can foster, support and enhance innovation. As part of this focus, the work environment is identified as critical and so this links with Chapter 10 in which the quest for talent is clearly acknowledged.

The specific focus of this chapter is the investigation of how innovation was enacted within a specific alliance project in Queensland, Australia – the Wivenhoe alliance. In order to properly understand the nature of innovation a model for innovation management in organisations is presented. The chapter then goes on to explain the nature of the Wivenhoe alliance and how innovation was promoted within it. The case is then reviewed and the lessons learned and the model used at Wivenhoe compared with theory.

Context and theoretical background

Innovation is not limited to the production of new products. Van de Ven (1986: 590) defines innovation as 'the development and implementation of new ideas by people who over time engage in transactions with others within an institutional order'. His definition focuses on four basic factors (new ideas, people, transactions and institutional context). He notes that four basic problems confront most general managers in encouraging innovation to flourish and become part of the organisation's culture. These are (1) a human problem of managing attention, (2) a process problem in managing new ideas into good currency, (3) a structural problem of managing part–whole relationships, and (4) a strategic problem of institutional leadership.

Innovation and knowledge sharing

Theory about the innovation process links in well with ideas on 'sticky knowledge' as proposed by Szulanski (1996) and how innovation is diffused. This is discussed further in the context of learning organisations in Chapter 8 of this book. The models presented by Peansupap (2004) in his PhD thesis and published elsewhere (Peansupap and Walker, 2005a,b,c; 2006a,b) provide deep insights and models of how ICT was diffused within highly computer-literate construction organisations. This helps inform us about factors affecting the challenges that organisations face when managing innovation. Further, the model presented by Maqsood (2006; Maqsood *et al.*, 2007) in Chapter 8 of this book illustrates the critical role played by people (individuals, groups and an organisations) working in harmony to align innovation potential with its realisation using their knowledge and skills. This was shown as using 'push' and 'pull' forces to drive innovative organisations towards becoming learning organisations. Procurement choices can influence the strength and direction of the innovation trajectory.

Innovation can be applied to organisation structures and to individual work practices. It can also be applied to business strategy and to business and operational relationships. Figure 13.1 illustrates such a view of innovation that combines elements of factors affecting ICT diffusion from Peansupap (2004) and (Maqsood, 2006) with ideas presented by Prather (2000: 18) and also Bucic and Gudergan (2002: 13). This also links to the environmental factor of the organisation's response to gaining a knowledge advantage (Walker, 2005) as explained more fully in Chapter 6 of this book.

Many models exist for implementing innovation within an organisation. The model adapted from Prather (2000: 18) presents a simple but useful view of innovation within an organisation and the management of that innovation. At the core of the model is the innovative organisation – the goal and direction the organisation intends to move in. Prather identified three areas which contribute to the creation and maintenance of this innovative culture. The areas of education, application and environment overlap to provide three key elements of innovation management: the innovation climate; shared problem-solving processes; individual thinking, tools and practices. The models discussed in Chapter 8 of this book based on Peansupap (2004) and Maqsood (2006) provide a useful way of extending Prather's model by indicating the knowledge and innovation push-and-pull forces that must be understood and harnessed to create a learning organisation. Using this simple model the approach adopted on the Wivenhoe alliance to influence innovation through its procurement approach will be analysed within this case study.

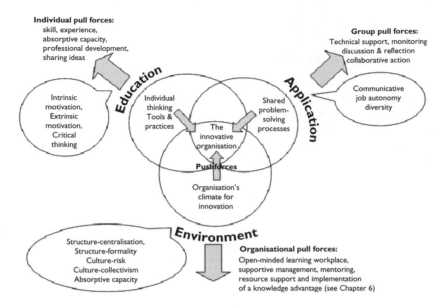

Individual pull forces:
skill, experience,
absorptive capacity,
professional development,
sharing ideas

Group pull forces:
Technical support, monitoring
discussion & reflection
collaborative action

Education

Application

Intrinsic
motivation,
Extrinsic
motivation,
Critical
thinking

Individual
thinking
Tools &
practices

The
innovative
organisation

Shared
problem-
solving
processes

Communicative
job autonomy
diversity

Push forces

Organisation's
climate for
innovation

Environment

Structure-centralisation,
Structure-formality
Culture-risk
Culture-collectivism
Absorptive capacity

Organisational pull forces:
Open-minded learning workplace,
supportive management, mentoring,
resource support and implementation
of a knowledge advantage (see Chapter 6)

Figure 13.1 Developing an innovative organisation.

Education is an important element of the innovation process. Innovations originate with individuals, not with groups nor organisations. Thus education and training techniques such as brainstorming and lateral thinking are part of the process of developing the individual, and individuals in groups, in order to stimulate, promote and nurture innovation within the organisation. Education is pulled by existing skills, experience, absorptive capacity (Cohen and Levinthal, 1990), professional development and ability and desire to exchange ideas.

Application refers to the way in which the organisation is structured in order to handle new ideas and to see them through to implementation. Groups apply knowledge and are pulled towards innovation by supporting forces such as technical support, mentoring, discussion forums and time and ability to reflect and take part in collaborative action, where ideas are socialised, reframed and developed. This also requires a set of procedures and protocols which allow ideas to be evaluated, developed and then implemented. Indeed, this process continues through to the measurement of the impact of the innovation and its subsequent export to other organisations and other parts of its own organisation. For an example of a structured approach to this the reader might refer to the Grass Roots Innovation Team (GRIT) at 3M and follow through how the process is developed in that renowned, innovative organisation.[1]

Finally, the environment within which the project and the innovations take place is a key element in determining the successful implementation of an innovation strategy. The culture of organisations is formed by leadership, attitudes to success and failure, the facilitation of communication and sharing of information across the organisation and the role of the bureaucracy within the organisation. This includes an open-minded workplace, leadership that champions questioning and supports reflection and adaptation, and mentoring. Chapter 6 discusses some tools that can be built into a project procurement governance system to encourage the effective pursuit of an organisation's knowledge advantage (Walker *et al.*, 2004; Walker, 2005; Walker *et al.*, 2005). Thus, all three areas of education, application and environment combine to provide the infrastructure in which innovation can take place.

The interesting element of this case study is that, in many ways, the approach to procurement was an innovative approach and also the project team had innovation as a major goal within the project itself. Hence, this can be seen as an example of meta-innovation. Thus, in this case study we will examine both the innovative approach to procurement and the way in which innovation within the project was managed.

There are a number of elements and attitudes (which could be considered important dimensions of the organisational climate for promoting innovation) that create an organisational climate that supports groups and individuals to freely share ideas and collaborate to deliver innovation. Together, these elements shape the organisational climate which nurtures innovation. Hence, one would expect an innovative organisation to adopt a set of values which reflected those dimensions listed above. However, creating a set of values is a somewhat intangible issue (see Chapter 9 of this book).

In order to ensure that these values are achieved and maintained there is a need for tangible support which enables these values. These supports can be described as awards, as taboos and repetitive reinforcement (Prather, 2000: 21). With the inculcation of values and the provision of these supports there is the propensity for actors within the organisation to trust one another and so collaborative and cooperative relationships are established, a key goal of alliance teams. This is particularly important in an organisation such as the Wivenhoe alliance where many of the project team members are from different organisations. Thus, in this instance, awards were given to team members for actively supporting the alliance's values. Taboos are seen as actions which have a negative impact on innovation and which are managed by subtly or overtly ensuring that these actions do not interfere with the innovation process. The repetitive reinforcement in this instance took place through the use of a project facilitator, an innovation manager, an alliance coach and various other means discussed later.

Alliancing

The concept of alliancing is introduced in Chapter 2, and Chapter 12 also provides detailed discussion on the selection of alliance partners. Alliance projects are initiated in an attempt to overcome the adversarial relationships which are common within the construction industry. The concept is based upon open book accounting and project participants having an attitude of best-for-project in the decision-making process. All alliance participants have an involvement in determining the project scope and risk and reward allocation. As a consequence, this approach leads to simplified approaches to dispute resolution by dealing with issues as they arise rather than involving the project team in the common paper chase of requests for information and claims for variations and changes.

There is often confusion on the differences between partnering and alliancing in the construction industry. The most meaningful distinction between partnering and alliancing is described by Walker and Hampson (2003a: 53) as: '...with partnering, aims and goals are agreed upon and dispute resolution and escalation plans are established, but partners still retain independence and may individually suffer or gain from the relationship. With alliancing the alliance parties form a cohesive entity that jointly shares risks and rewards to an agreed formula...'

Under such a definition it can be seen that key project team members come together to form a pseudo company for the duration of the project. This enables the benefits of working as one company to achieve a common goal to be realised. There are particular circumstances when alliances may function as an ideal model for procurement:

- Concept design has been undertaken but detailed design is yet to be completed thus leaving the projects in a state of 'living design' which is conducive to a flexible approach and so enabling value management.
- Total project's cost has been estimated but the budget has yet to be set – project alliance objectives go beyond an emphasis on price, to include the ability to innovate and to manage relationships both within and between alliance participants, including stakeholders.
- The alliance partners brought together are chosen such that they complement one another's strengths and weaknesses.
- There is a high level of stakeholder and community awareness of the project and its benefits and issues.
- The time frame is recognised as being a constraint.

The key principle of the alliancing approach is that the personnel assigned to the project are the best people from each company in terms of the needs of the alliance (see Chapter 10 for more details on the quest for talent). Team members are expected to be leaders in their field and hence require little monitoring or guidance – they also tend to welcome taking the initiative.

However, in order for the alliance to function as a team certain prerequisites and other transaction costs have to be implemented:

- The key focus for the alliance is to form a strong alliance culture within the team. Unresolved internal conflicts will doom a project to failure, so great importance is placed on team building to assist with the development of the alliance culture. The generation of trust and friendship within the team is a key determinant of this.
- Team building should not only commence at the very outset of the alliance but also continue until completion of the project and the disbanding of the alliance. Team-building activities that build trust and commitment can include games, and Outward Bound-type activities and such things as emotional intelligence programmes.
- Knowledge sharing is an important aspect of the alliance team. Open and free communication within a trusting environment is a key component of collaborative design. In essence, there must be no secrets as hidden agendas within the team.
- Importantly, the alliance deed which member companies sign at the outset of the alliance contains a no-litigation clause. The aim of this is to ensure amicable, speedy resolution of disputes within the team rather than an escalation to the courts.

The alliance participants collaboratively agree the target cost for the project and then set about developing the design and also, at the outset, establish arrangements for sharing risk and reward arising from the project. Typically, as was reported in the case of the National Museum of Australia, project profit is placed at risk to ensure that cost targets are achieved. However, as an incentive, cost savings are shared between the alliance participants under some form of 'gain-sharing' formula which is designed to encourage innovation, maximise collaboration and focus production on efficiency (Hutchinson and Gallagher, 2003; Walker and Hampson, 2003b).

There have been a number of examples of alliances in Australia over recent years (Davis, 2006), such as The Ampolex Wandoo alliance (construction of an oil platform and field development in Western Australia), WMC East Spur alliance (a gas field development off Western Australia), Brisbane Water Enviro alliance (upgrade of three water treatment plants in Brisbane) Sydney's Northside Storage tunnel project and the National Museum of Australia. In general, such projects have been reported to have exceptionally high levels of collaboration, cooperation and communication, leading, generally, to technical and managerial problems being resolved effectively and efficiently in the best interests of the project. This process is facilitated by participants putting project interest above individual company interest. However, alliances have been criticised in some instances for precluding local contractors from the bidding process, For example, the Report by the

House of Representatives Standing Committee on Industry, Science and Technology, 'A sea of indifference – Australian industry participation in the North West Shelf project' states: 'Through (alliancing) partners in a project have a special interest in supplying key components and services' and 'construction of platform modules was undertaken...without competitive tenders being called. Australian industry was denied any chance of undertaking this work' (Standing Committee on Industry, 1998: 63).

Underpinning theory

Alliancing is categorised by scholars into two main types, strategic alliancing and project alliancing. The most commonly adopted definition of strategic alliances is based upon two related ideas. First, the act of establishing inter-organisational relations that engages in collaborative behaviour. Second, these acts have a specific purpose which has a long-term strategic consequence in responding to opportunities and challenges that face an organisation – whether that is for an organisation currently competing globally or positioning itself to take asvantage of future opportunities (Doz and Hamel, 1998). The inter-organisational relationships can be grouped into three broad classifications of services: cross-company consortium (service), opportunistic and stakeholder alliances (Howarth et al., 1995).

A strategic alliance is also seen as an inter-organisational arrangement which usually exists between two or more companies, which extends beyond a specific project and whose parties would expect ongoing, mutually beneficial business (Doz and Hamel, 1998). According to Bronder and Pritzl (1992: 412), a strategic alliance exists when the value chain between at least two organisations with compatible goal structures are combined for the purpose of sustaining and achieving significant competitive advantages. A strategic alliance can provide access to resources such as capital, information, technology, management expertise, markets, customers, distribution channels, land and labour. Such resources may not be available to an organisation acting alone. Greater access to resources allows an organisation to reduce its level of uncertainty in a demanding and turbulent environment. Strategic alliances also enable organisations to speed up the market-entry process and increase their responsiveness to consumer markets, as indicated in Table 13.1 – adapted from Howarth et al. (1995).

Yet, no successful strategic alliances can be developed without trust. Trust in a strategic alliance also includes the concept of reciprocity, which implies a long-term focus, the acceptance that obligations are mutual, and room for adjustment if one partner is suddenly placed in a compromising position (Howarth et al., 1995). As occurs with partnering and relational contracting, trust between strategic alliance partners is important because it creates an opportunity and willingness for further alignment, reduces the need for partners to continually monitor one another's behaviour, reduces

Table 13.1 Alliance-type characteristics

Alliance type	Characteristics
Service alliances	• Requires the lowest level of interdependence between partners with the smallest amount of changes and the lowest level of joint commitment • Provide economies of scale • Provide the ability for the partners to undertake large-scale projects with a limited purpose • Difficulties arise due to the diversity of the interests and goals of the partners • Result in a loss of commitment from members
Opportunistic alliances, for example, joint venture	• Provide access to the resources of the partner organisations • Motivated by the existence and recognition of a market opportunity • Partners might exploit one another's resources and then move on to pursue the opportunity alone
Stakeholder alliances, for example, suppliers, customers, employees	• The closest link between member organisations of all • Seek to build strong, long-term relationships • Assist in achieving the organisational goals by major stakeholders

the need for formal controls, and reduces the tensions created by short-term inequities. It allows partners to focus on long-term business development as well as reducing cost and time. The characteristics of successful strategic alliances as well as successful business relationships proposed by Hampson and Kwok (1997) – trust, commitment, interdependence, cooperation, communication and joint problem solving – reflect a similar theme.

A successful alliance also requires creativity. It has been shown in the past that alliances that have failed are typically the second alliance that a group of companies undertake together. The problem arises when the individual team members who were on the first alliance insist on using the same practices by repeating them since they worked on the previous alliance, despite the fact that they might not be appropriate for the second. The new team members do not understand why the practices are adopted and do not feel any sense of ownership. Since both strategic and project alliances are tailor-made mechanisms, such alliancing would fail due to the lack of creativity by the team members from the first project team and the new team members do not feel committed to work.

The main difference between project alliances and strategic alliances is project alliances have a defined end, which is most commonly the practical completion date of a project. The parties are brought together for a specific project or outcome and a project alliancing agreement is also legally enforceable. Hutchinson and Gallagher (2003) describe a project alliance as a project delivery strategy where several participants join together to share risks and outcomes on a project and where sponsor and commercial participants' objectives (client's objectives) are aligned to

- maximise performance;
- proactively manage risk;
- reduce cost and
- achieve outstanding results in sponsor key project objectives.

They define project alliance as: '...an integrated high performance team selected on a best person for the job basis; sharing all project risks with incentives to achieve game-breaking performance in pre-aligned project objectives; within a framework of no fault, no blame and no dispute; characterised by uncompromising commitments to trust, collaboration, innovation and mutual support; all in order to achieve outstanding results' (Hutchinson and Gallagher, 2003: 8).

Love and Gunasekaran (1999) stated that alliances can be either collaborative or cooperative (Hamel *et al.*, 1989; Bronder and Pritzl, 1992; Huysman *et al.*, 1994) based on core competences. Kwok (1998) describes project alliances as a cooperative arrangement between two or more organisations that forms part of their overall strategy, and contributes to achieving their major goals and objectives for a particular project. Hamel *et al.* (1989) suggest that organisations that enter into collaborative alliances (short term) are aware that their partners are capable of disarming them. Parties to these alliances have clear objectives and understand that their partner's objectives will affect their success. Yet, collaboration does not always provide an opportunity to internalise a partner's skills. Love and Gunasekaran (1999) suggest that a 'psychological barrier' may exist between alliance partners caused by the fear that their partner(s) may out-learn or deskill them. Wood and Gray (1991) state that organisations typically enter collaborative relations to reduce environmental complexity and to gain more control over environmental factors. Such collaboration may cause new dependencies to be created, which may increase environmental complexity and turbulence. They argue that increases in complexity may increase transaction costs, the need to manage bilateral and multilateral relations and the need to develop new skills.

Cooperative alliances (long term) encourage alliance partners to commit their resources to the relationship to gain mutual learning (Love and Gunasekaran, 1999). There is a lower level of competition and as a result, partners may feel more committed to work together and exchange their knowledge and resources. Ketelhöhn (1993) suggests that cooperative strategic alliances can create a competitive advantage. Organisations that rely on cooperation have been found to obtain lower costs for as long as they maintain trust internally and externally – among employees and members of their network.

There are many studies, see for instance Lingard and Francis (2004; 2005; 2007) and Chapter 10 of this book, which indicate that work–life conflict has a damaging effect on job satisfaction, organisational commitment, productivity, employee turnover and absenteeism. For the individual, this is associated with burnout, mental health issues and dysfunction within the family setting. Indeed, the construction industry is well known for its long and inflexible work hours, particularly for those based on-site or working on projects. Despite the considerable resistance within the industry to the adoption of new ways of scheduling work, the Wivenhoe alliance attempted to create a high-performing team by innovating in terms of work–life balance and reducing work hours. The additional effect of this change in balance was aimed at encouraging the retention of staff and enhancing recruitment in the future.

Case study details

The following will provide background additional information for the reader to better understand its context and relevance to this book.

The Wivenhoe alliance

South East Queensland Water (SEQ Water) owns and operates Wivenhoe, Somerset and North Pine Dams in Queensland, Australia and is the sponsor organisation in an alliance with

- Leighton contractors – the contractor;
- Coffey Geosciences – specialist consultants, geotechnical and hydrogeology;
- MWH – specialist consultants, water;
- Department of Commerce (New South Wales) – providers of finance.

The alliance was formed with the intention of upgrading the existing Wivenhoe Dam to ensure that it could cope with extreme rainfall and 'any conceivable flood' in the future. The $70m Australian project involved construction of a 165 m wide additional spillway, a five-span traffic bridge, post-tensioning of the existing spillway and associated works, 80 km from Brisbane. The upgrade was completed six months ahead of schedule and ten percent below budget. The alliance commenced by undertaking a number of studies dealing with design options, hydrology, construction, environment, geotechnics and stakeholders (including the local communities) prior to developing the design and setting a target cost for the project.

In terms of benefits arising from the alliance formation, MHW claim that the approach leads to innovative solutions,[2] its web page states that 'The Wivenhoe Alliance has demonstrated how collaborative projects promote innovation. Significant savings were achieved through an innovative way of disposing of earthworks soil. Rather than transporting the soil to an agreed disposal area, the Alliance designed and constructed some low hills in close proximity to the dam and construction site. The hills were landscaped and planted to create environmental refuge areas and to match the surrounding landforms. The result was aligned with the client's environmental objectives for the project as well as achieving beneficial cost savings.'

The key issue to come out of the alliance as far as this case study is concerned is the facilitation and management of innovation within the alliance. Innovation took place in a number of fields, such as project learning, value management, sustainability, design innovation, construction innovation, process and product innovation, working conditions and work–life balance. Each of these areas of innovation will be reviewed below but, in order to put this into context, the underpinning theory behind alliances is briefly expounded below.

Project learning system

The contractor in the alliance had for a long time intended to set up an organisation-wide project learning system. The idea behind the system is that workers, estimators and directors on subsequent projects can make reference to a database of knowledge which will make their current decision making more efficient and effective, based on past experiences within the organisation. The aim is to ensure that the same mistakes are not repeated and that good ideas are kept and repeated in practice. The nature and culture developed within the alliance was such that individuals and the team were willing to take risks with new ideas and new technology. Consequently, all bought into the idea of the project learning system and participants felt obliged to make their contributions to the system. Hence, ideas behind things that went well, or didn't go well, were harnessed and stored in a database set up to capture this knowledge. As far as these participating organisations were concerned this was their first experience of such a project learning system.

Value management

In a similar manner, participation in the value management process was facilitated by the culture of the alliance. All participants bought into the idea of 'best for project' and so were willing to reveal in a timely manner issues which affected the adequacy or efficiency of the project's components and which might, in a different procurement system, have lain hidden until their consequences became apparent. Participants willingly and openly discussed problems which they felt were beyond their own individual capacity to solve, and the alliance team addressed these issues in a fashion which indicated a joint responsibility being taken by all for the outcome. Thus, although the value-management process cannot be considered to be an innovation, the way in which the value-management process was undertaken was considered revolutionary by the participants. Specifically, the open and free communication engendered and the attitude of 'no-blame' led to an atmosphere which facilitated the successful value management of the project. In addition, innovations and ideas which contributed to this value management process were treasured and were actually recorded in an innovation register as illustrated in Figure 13.2.

Sustainability

The Wivenhoe Alliance worked with Queensland University of Technology (QUT) in Australia to complete a Sustainability Assessment of the project, adopting a triple-bottom-line approach, that is, Economic, Environmental, Social sustainability and also including Health and Safety as a key indicator. Again, the alliance approach led to an open discussion of sustainability issues which transcended the individual partners' specific interests in that sustainability was seen to be a project cost and benefit and not a benefit to one participant at a cost to the other.

APPENDIX 4 - INNOVATION REGISTER

WIVENHOE ALLIANCE — PAA Phase — **REGISTER** ***Innovations & Ideas***

Innovation/ Idea			Action			SAVINGS				
Date	*Innovation*	*By*	*Who*	*When*	*Result*	*Status*	*Envelope*	*Cost Saving*	*Target Cost Saving*	*Banked Cost Saving*
20/03/2005	Substitute DICL with plastic equivalent for services	J Long / C Dill	B Maher	6/04/2005		Approved	80,000	100,000	100,000	100,000
15/03/2005	Redesign of uplift slab with ground anchors	G Lam	B Maher	1/04/2005	Beign designed	Approved	40,000	60,000	60,000	0
11/03/2005	Hard Landscaping optimisation (delete rotunda)	SEQWater	AMT	6/04/2005	VM Raised and circulated	Pending	40,000	60,000	60,000	0
24/01/2005	Project Learning System	K Litfin	N Atkin	30/04/2005		Open	0	0	0	0
1/12/2004	VM - Redesign of crash barrier x-section on Western side	G Lam				Realised	10,000	20,000	20,000	20,000
26/11/2004	Fuse Plug material optimisation	M Rosulka	M Rosulka	10/12/2004	Design changed. VM - 26/11/2004	Open	400,000	500,000	450,000	300,000
24/11/2004	Extent of Divider Wall stopped at the top of the ogee crest steel formwork (DSN 418)	J Bengin	W Crowther	24/11/2004		Realised	2,000	5,000	3,000	3,000
22/11/2004	Existing spillway stormwater management	M Dixon	M Dixon	22/11/2004		Implemented	0		0	0
22/11/2004	Bioremediation Pit	N Woods / M Dixon	N Woods / M Dixon	22/11/2004		Implemented	0		0	0
22/11/2004	Design Drawing Register improvement	N Atkin	N Atkin	3/11/2004	less time updating spreadsheet	Implemented	0		0	0
22/11/2004	Crash Barrier Cut off Trench	S Perrett / B Peters	K Litfin			Realised	200,000	300,000	250,000	250,000
22/11/2004	Crack monitoring with stencils instead of microscope	M Gurgus	M Gurgus	22/11/2004	Save time of monitoring	Implemented	0		0	0
22/11/2004	Cobble Beach Rock/sand filter on U/S side	S Macnish	S Macnish			Open	150,000	250,000	200,000	0
1/11/2004	VM - Raise height of wavewall to increase MFL	W Crowther / J Long	J Long		Investigating	Open	0	0	0	0

Figure 13.2 Example of the innovation register.

Design innovation

Design innovation was driven and focused by the 'Best for Project' philosophy. The design team was challenged to innovate and value manage every aspect of the project and the construction team encouraged to challenge any aspect of design that could be modified for the benefit of the project, financially or otherwise. The factors of safety which the design team conventionally used were challenged by the rest of the alliance, and the design team were then forced to come down from their pedestal and become a complete team member. Again, the alliance culture and the drive to innovate which this engendered was a major factor in allowing the team to achieve these goals.

An example of a design innovation is illustrated in Box 13.1 below.

Box 13.1 Innovation example

Wivenhoe Dam is being upgraded to be able to withstand the Probable Maximum Flood (PMF), which is essentially a 1 in 100,000 year event. If we did have such an event, whilst all the radial gates on the existing spillway are open, the water would still hit the gates as it passed through the existing spillway. This would cause the gates to bounce up and down, and there is a risk that due to the gates being held open by cables, the cables could break if the gates bounced too high or too much.

The initial solution was to install a locking pin on the gate, such that when it is opened to the fully-opened position, it will be locked open, hence the gate will not bounce and potentially fail. It was unanimous that there must be a better solution, but no one knew what it was. It was agreed that a Value Management workshop should be convened to help identify a solution; two days were allocated to this task.

On the first day, the issues were highlighted and the end result was discussed. It was at this point that everyone commenced brainstorming. We then came together and shared out ideas. The end result was that an innovative solution was achieved. By looking outside the box and being forced to look at the problem from a direct angle, a better solution was achieved.

The solution was to install a wide plate upstream of the Radial gate, the plate would act as a penstock and force the water down to a level such that when it passed the gates it was at least 1 metre below the bottom of the gate. This removed the need for the locking pin because there would be no bouncing of the gate. The extra benefit is that additional works that were previously required are now not needed due to the Baffle Structure being installed.

During the subsequent design of the Baffle Structure, construction modifications were added to allow the structure to become its own permanent scaffolding and access platform. It also has legacy value to the owner, as it will provide a safe access platform for the maintenance workers of the dam. The solution is a non-mechanical one, that is, very little maintenance will be required. The Baffle Structure solution saved the project in excess of $750,000. This was possible only due to forward planning, by having enough time to value manage the issue, design a new solution and then install it.

Construction innovation

Construction innovation was seen to be not a standalone issue. There was an implicit recognition that the design process had a major effect on construction innovation and construction effectiveness. Hence, it was seen to be a whole team task to develop innovations in construction. In order to facilitate this, effective planning is vitally important. The team needs time to think through the construction consequences of design decisions and search for better alternatives. Such an approach can be effective only if the planning and design and construction innovation process take place early in the project. Hence, it was recognised that effective planning at the outset was an important facilitation device for the innovations in construction which did take place.

Collaboration example

An example of the collaborative approach to innovation and the willingness to take risks was a design innovation that occurred on a crash barrier upgrade on the existing dam wall. The design was challenged by the construction team and in conjunction with the design team, a new and evolutional type of design was developed. The design was implemented and construction commenced. The design underwent a design review (this occurs on a monthly basis) and the design was found to be too risk-adverse for the reviewers and the design was altered. The financial result was substantial extra cost due to the re-work involved. On a normal project, this would have been classed as a variation and extra costs would have been incurred, but due to the make-up of the alliance and culture of the team members, there was no negative feeling or resentment and the costs were absorbed into the project. The mindset was 'although the designers cost us money this time, they have saved us time, money and effort in the past and will save more in the future'. Thus, even through failure the alliance culture was reinforced.

Process/product innovation

Innovations are not limited to design and construction, but can occur in any area of the project. Process and product innovations occur every day and are usually innovations that occur in secondary roles as opposed to direct roles. An example of this was an improvement to the design drawing register process which made the location of the most up-to-date drawing so much easier. The benefit from such an innovation is intangible – one cannot assess the costs saved from avoiding working from the wrong drawings but such savings had undoubtedly accrued. Such an innovation would not have likely occurred on a traditional project as there is not the same 'best for project' attitude.

Change in working practices

A compressed work week was introduced part way through the construction phase of the dam. The site was initially operating on a 6-day, 58-hour week comprising five 10-hour days plus an 8-hour day on Saturday. In March, 2005 there was a move to a 5-day week with an 11.5 hour-day. This change in working practices was studied by Lingard et al. (2007) (see Chapter 10 for more details), and they reported that employees' well-being and satisfaction with work–life balance were generally high and that perceptions of work–life conflict were favourably low. Indeed, knock-on effects for work, such as productivity gains and loyalty to the organisation, as well as non-work activities such as more scope for family and domestic duties were seen to be part of the improvements brought about by the change in working practices. There was also evidence that the changes led to improved employee retention. Thus, if one considers changing working practices as an innovation one can then assume that this innovation within the alliance has allowed for the improvement in employee satisfaction and will perhaps have a subsequent effect on work effectiveness and innovation. However, the direct link has yet to be proven.

Indeed, the fact that this was an alliance project was seen by many interviewees to have facilitated the implementation of the shortened working week and led to a change in attitudes. An individual response, as reported by Lingard et al. (2007), bears testimony to this: 'I think this job is an exception, not the rule. I think alliancing is conducive to being able to do this because we can make more money by being smarter and changing the designs and being efficient, whereas the "old school" way of making money in construction is going faster and harder and longer – squeezing as much as you can out of resources over a finite period of time.'

In addition, Lingard et al. go on to conclude that 'the compressed work week was very successful in improving employees' work–life balance. Further, the project's time and cost performance suggest that the change to a five-day week did not hinder the attainment of objectives in other key result areas (KRAs) of the project. Alliance employees were very satisfied with the compressed work week and reported a number of benefits, including increased physical and psychological well-being, greater motivation, improved productivity, increased job commitment and increased involvement in home/family activities. These results suggest that strategies designed to improve employees' work–life balance may be key components of high performance work systems in the Australian construction industry. Further, the collaborative nature of project alliances, in particular the sharing of risks and rewards and the focus on "best for project" decision-making, appear to provide a supportive work environment in which innovative work–life balance initiatives can be implemented.'

Capture and measurement of innovation

In order to ensure that innovation is actually taking place it is necessary to capture and measure the innovations and the innovation process is. Such an approach was adopted on the web by the use of innovation tracking tools which included

- Innovation register (see Figure 13.2)

    - During the Wivenhoe alliance, innovations were monitored and recorded using a specifically designed innovation register which captured the innovations as they occurred and tracked their progress through to completion under the general headings of ideas, actions and savings. However, it was recognised that not all savings could be measured in dollar terms and yet the idea could still be innovative. The concept of benefit and best for project are important in this respect and need to be recognised in order to maintain the 'alliance culture'.

- Value-management register

    - In a similar manner a value-management register was compiled which monitored both the value-management meetings and events and the value-management outcomes. This provides both a useful source of ideas for future projects and provides a mechanism for enabling a cost analysis of the value-management process if required.

- Project learning system

    - The project learning system provides a database of all the good and bad ideas that have occurred on projects and the net effect of these ideas. In this instance, the project learning system was a prototype and was deemed to be a success. Indeed, such a success that the contractor has established it on a state-wide basis for all projects.

- Completion report

  - As a mechanism for formally recording the process and progress of the alliance a completion report was compiled with the intention of capturing all those events and issues which occurred throughout the course of the project. Within this report there is a discussion of the techniques used, the innovations which developed and those which actually did not come to fruition under tools utilised on the project.

Making sense of the case study

So, how can one rationalise what has taken place in the Wivenhoe alliance in terms of innovation and innovation management? One way of looking at the process which took place is to plot on a grid innovation process in one direction and innovation culture in the other direction, and then locate the Wivenhoe alliance and other organisations on this grid. A typical firm, such as a supermarket, which has no culture of innovation and no processes to stimulate innovation might rank poorly on this model and be found in the bottom left quadrant. In contrast, 3M Company, which has a culture that inspires innovation and which has a set series of processes to ensure innovation takes place, might be located at the top right-hand quadrant of this model. When looking at the Wivenhoe alliance, the case study has shown that it does have a culture which promotes and nurtures innovation. It also has processes to generate ideas, get them through the evaluation phase and having them implemented. However, the alliance process is very much an *ad hoc* one compared with that of 3M. There appears to be no strategic or underlying innovation management plan for the alliance. Perhaps this is the nature of the alliance concept rather than merely the Wivenhoe alliance itself; the nature of the alliance is that it has defined beginning and end dates and so there is no scope for or drive towards strategic, long-term views. Hence, given these shortcomings one would place the Wivenhoe alliance in a less strategic position than the 3M Company.

Case study discussion

The case study indicates that the opportunity for innovation in alliance projects is very high compared with conventional hard dollar contracts. However, the very nature of the alliance is that it is of a finite duration and so the ability to structure innovation effectively is somewhat limited. Thus, the alliance approach to procurement has inbuilt within its culture the necessary characteristics to promote innovation but, in all instances, this innovation process needs to be managed. The case study indicates how this innovation process can be encouraged and indeed monitored and measured.

Of great interest is the changing working practices implemented in this alliance. The fact that both performance and work–life balance issues appeared to improve when the working week was reduced is quite impressive. Indeed, for most of the respondents to indicate that such an approach was not possible within the hard-dollar contract is very enlightening. It would seem that the alliance approach has benefits in terms of both the project performance and individual achievements and working practices. Thus, Alliancing might well hold the key to the cultural change required within the industry to improve performance universally.

There are many factors relating specifically to alliances that can be related back to Prather's (2000) model of innovation management as a combination of education, application and environment and its adaptation by us as illustrated in Figure 13.1, combining ideas from Bucic and Gudergan (2002). These factors describe the reasons why innovation is, generally speaking, more successful in alliances than in traditional firms and projects. All of these factors can be seen in the Wivenhoe alliance.

Chapter summary

The purpose of this chapter was to provide a case study example of an alliance project that had successfully demonstrated innovation through alliance practices. We also sought to discuss the theoretical basis for the observed innovative practices and outcomes and to link these to other chapters in this book so that readers can extend or revise what they have grasped from our work.

The chapter started with some context relating to the case study as well as background concepts and theory. We introduced an extension to the innovation diffusion and knowledge transfer models in Chapter 8 by showing how education, application and environment are linked to drive an innovative project alliance organisation that can generate, use and transfer knowledge at the individual, group and organisational level. We then provided some alliance theory and discussed where the case study fitted in with what theory suggests should happen in practice. The evidence presented suggests that a well-functioning alliance can spark innovation and illustrated how the alliance delivered useful innovations where more traditional project procurement approaches may stifle such innovation. Our Table 13.2 provided a summary of lessons from the case study linked to the model presented in Figure 13.1.

In line with our format for the case study chapters we present Table 13.3 that indicates how this chapter fits into the first 10 chapters.

This chapter provides a summary of many useful alliance case studies and so it represents a considerable contribution to the project procurement literature.

Table 13.2 Summary notes of factors influencing innovation

Factors leading to alliance innovation	Notes
Education – individual-level factors Intrinsic motivation Extrinsic motivation Critical thinking Leading to the creative process	In the area of education, or individual-level factors, there is clearly a greater intrinsic motivation of staff. The 'best for project' mentality means that team members will be thinking of ways to challenge the *status quo* and improve the project's performance. There is also a greater extrinsic motivation in alliances. By actively seeking more efficient and cost-effective methods, the alliance increases its potential for financial reward. An increased appreciation of critical thinking and reflecting on past actions and existing knowledge in the present context means that team members will constantly be coming up with new ideas to solve current problems
Application – group-level factors Communicative job autonomy Diversity leading to the learning process	In the group-level factors, or application area, there is a greater level of communicative interaction on the Wivenhoe alliance than is seen in most projects. The informal and personalised communication style encourages lateral thinking, acceptance of criticism and willingness to consult others. The level of job autonomy of staff members encourages decentralisation and gives freedom to the individual and empowerment. Finally, the diversity of team members means that people from a range of backgrounds and experiences are brought together for the duration of the project
Environment – alliance-level factors Structure centralisation Structure formality Culture of risk taking Culture of collectivism Absorptive capacity The knowledge stocks	At the alliance level, or the environment-level factors described by Prather (2000), there are many factors that encourage innovation. The centralised structure, while it can be restrictive in a normal firm, can actually encourage innovation in alliances. The fact that all of the decision makers are present on site means that the approval process for new ideas can happen quickly and easily. A formal structure can prohibit innovation in traditional firms, but the Wivenhoe alliance overcomes this by working in cross-functional teams. The fact that alliances have a risk-orientated culture contributes greatly to innovation. They accept mistakes and allow the freedom to fail more readily than traditional firms. Alliances are typically used for more 'high risk' projects and those where the scope isn't well defined. Since alliance members know they will get paid for the work they do, and that only the profit margin is at risk, they are generally less risk-averse than projects undertaken with hard-dollar contracts. The final two alliance-level factors that lead to innovation described by Bucic and Gudergan (2002) are a culture of collectivism and absorptive capacity. The collectivist culture can be seen at the Wivenhoe alliance in everyone's 'best for project' mentality. Everyone is willing to cooperate to make the project successful, and they put the goals of the project ahead of their own personal goals. The absorptive capacity of an alliance refers to its ability to recognise valuable information and make use of it. This can clearly be seen at the Wivenhoe alliance, both through their innovation register and Project Learning System. They have clear procedures for the recording and sharing of good ideas. All of the individual-group- and alliance-level factors come together, and through well-defined procedures, contribute to alliance innovation

Table 13.3 Summary of aspects relating to this book

Chapter – perspective	Comments and discussion
Chapter 1 – Value proposition and make-or-buy decision	This study reviewed how value was delivered from innovation and how the alliance structure supported this
Chapter 2 – Project type and procurement choice	The project fell into typical tangible construction project types though intangible value was expected as a project outcome through learning and improved processes for future infrastructure work by the client
Chapter 3 – Stakeholder management	The project alliance was highly focused upon internal team stakeholders and providing examples of innovation for the client organisation to replicate and adapt in future
Chapter 4 – Governance	Project alliances have highly-developed probity and governance structures. This alliance was particularly focused on triple-bottom-line outcomes
Chapter 5 – Business strategy	This project alliance's characteristics can be made sense of by reviewing the dynamic capabilities and competitive advantage strategic management aspects relevant to content in this chapter
Chapter 6 – Performance measurement	This chapter links in well with the knowledge advantage discussed as part of the CMM section
Chapter 7 – E-business	Examples of an ICT innovation register (Figure 13.2) and several information management registers are provided
Chapter 8 – Innovation and organisational learning	One of the main drivers for alliances is the expectation that it can facilitate organisational learning and innovation. This is of particular relevance to this chapter
Chapter 9 – Cultural dimensions	Trust and commitment that melds organisational identity from a series of individual organisations into one alliance requires cultural alignment. This is highly relevant to this chapter
Chapter 10 – Selecting and growing talented people	Achieving alliance objectives requires selection and retention of high-value talent with skills and behaviours that require careful selection and retention strategies. This is of high concern to alliance partners

Notes

1 A useful web link to 3M to find out more about a range of 3M discoveries and innovations can be found at URL http://solutions.3m.com/wps/portal/3M/en_US/our/company/information/history/century-innovation/ accessed 1 September 2007.
2 See www.mwhglobal.com.au/alliance-delivery accessed 17 January 2007.

References

Bronder, C. and Pritzl, R. (1992). 'Developing Strategic Alliances: A Conceptual Framework for Successful Co-operation'. *European Management Journal*. 10(4): 412–421.

Bucic, T. and Gudergan, S. (2002). *The Innovation Process in Alliances*. The Third European Conference on Organizational Knowledge, Learning and Capabilities, Athens, Greece, 5–6 April, Mylonopoulos H. T. N., Athens Laboratory of Business Administration: 1–39.

Cohen, W. M. and Levinthal, D. (1990). 'Absorptive Capacity: A New Perspective on Learning and Innovation'. *Administrative Science Quarterly*. 35(1): 128–152.

Davis, P. R. (2006). The Application of Relationship Marketing to Construction. PhD, *School of Economics, Finance and Marketing*. Melbourne: RMIT University.

Doz, Y. L. and Hamel, G. (1998). *Alliance Advantage – The Art of Creating Value Through Partnering*. Boston: Harvard Business School Press.

Hamel, G., Doz, Y. L. and Prahlad, C. K. (1989). 'Collaborate With Your Competitors – And Win'. *Harvard Business Review*. 67(1): 133–139.

Hampson, K. and Kwok, T. (1997). 'Strategic Alliances in Building Construction: A Tender Evaluation Tool for the Public Sector'. *Journal of Construction Procurement*. 3(1): 28–41.

Howarth, C. S., Gillin, M. and Bailey, J. (1995). *Strategic Alliances: Resource-Sharing Strategies for Smart Companies*. Melbourne: Pitman Publishing.

Hutchinson, A. and Gallagher, J. (2003). Project Alliances: An Overview. Melbourne: Alchimie Pty Ltd and Phillips Fox Lawyers: 33.

Huysman, M. H., Fischer, S. J. and Heng, M. S. (1994). 'An Organizational Learning Perspective on Information Systems Planning'. *The Journal of Strategic Information Systems*. 3(3): 165–177.

Ketelhöhn, W. (1993). 'What Do We Mean by Cooperative Advantage?' *European Management Journal*. 11(1): 30–37.

Kwok, T. (1998). Strategic Alliances in Construction: A Study of Contracting Relationships and Competitive Advantage in Public Sector Building Works. PhD, *Faculty of the Built Environment*. Brisbane: Queensland University of Technology.

Lingard, H. and Francis, V. (2004). 'TheWork-life Experiences of Office and Site-based Employees in the Australian Construction Industry'. *Construction Management & Economics*. 22(9): 991–1002.

Lingard, H. and Francis, V. (2005). 'The Decline of the "Traditional" Family: Work-life Benefits As a Means of Promoting a Diverse Workforce in the Construction Industry of Australia'. *Construction Management and Economics*. 23(10): 1045–1057.

Lingard, H. and Francis, V. (2007). ' "Negative Interference" between Australian Construction Professionals' Work and Family Roles: Evidence of an Asymmetrical Relationship'. *Engineering, Constrcution and Architectural Management*. 14(1): 79–93.

Lingard, H., Brown, K., Bradley, L., Bailey, C. and Townsend, K. (2007). 'Improving Employees' Work-life Balance in the Construction Industry: A Project Alliance Case Study'. *American Society of Civil Engineers*. Forthcoming.

Love, P. E. D. and Gunasekaran, A. (1999). 'Learning Alliances: A Customer-supplier Focus for Continuous Improvement in Manufacturing'. *Industrial & Commercial Training*. 31(3): 88–96.

Maqsood, T. (2006). The Role of Knowledge Management in Supporting Innovation and Learning in Construction. PhD, *School of Business Information Technology*. Melbourne: RMIT University.

Maqsood, T., Walker, D. H. T. and Finegan, A. D. (2007). 'Facilitating Knowledge Pull to Deliver Innovation through Knowledge Management: A Case Study'. *Engineering Construction and Architectural Management*. **14**(1): 94–109.

Peansupap, V. (2004). An Exploratory Approach to the Diffusion of ICT Innovation a Project Environment. PhD, *School of Property, Construction and Project Management*. Melbourne: RMIT University.

Peansupap, V. and Walker, D. H. T. (2005a). 'Exploratory Factors Influencing ICT Diffusion and Adoption Within Australian Construction Organisations: A Micro Analysis'. *Journal of Construction Innovation*. **5**(3): 135–157.

Peansupap, V. and Walker, D. H. T. (2005b). 'Factors Affecting ICT Diffusion: A Case Study of Three Large Australian Construction Contractors'. *Engineering Construction and Architectural Management*. **12**(1): 21–37.

Peansupap, V. and Walker, D. H. T. (2005c). 'Factors Enabling Information and Communication Technology Diffusion and Actual Implementation in Construction Organisations'. *Electronic Journal of Information Technology in Construction*. **10**: 193–218.

Peansupap, V. and Walker, D. H. T. (2006a). 'Information Communication Technology (ICT) Implementation Constraints: A Construction Industry Perspective'. *Engineering Construction and Architectural Management*. **13**(4): 364–379.

Peansupap, V. and Walker, D. H. T. (2006b). 'Innovation Diffusion at the Implementation Stage of a Construction Project: A Case Study of Information Communication Technology'. *Construction Management and Economics*. **24**(3): 321–332.

Prather, C. W. (2000). 'Keeping Innovation Alive After the Consultants Leave.' *Research Technology Management*. **43**(5): 17.

Standing Committee on Industry, S. A. R. (1998). A sea of indifference – Australian industry participation in the North West Shelf project Report. Canberra, Australia: Parliament of Australia House of Representatives, ISBN 0 644 507888.

Szulanski, G. (1996). 'Exploring Internal Stickiness: Impediments to the Transfer of Best Practice within the Firm'. *Strategic Management Journal*. **17**(Winter Special Issue): 27–43.

Van de Ven, A. H. (1986). 'Central Problems in the Management of Innovation'. *Management Science*. **32**(5): 590–607.

Walker, D. H. T. (2005). *Having a Knowledge Competitive Advantage (K-Adv) A Social Capital Perspective*. Information and Knowledge Management in a Global Economy CIB W102, Lisbon, 19–20 May, Franciso L Ribeiro, Peter D. E Love, Colin H. Davidson, Charles O. Egbu and B. Dimitrijevic, DECivil, **1**: 13–31.

Walker, D. H. T. and Hampson, K. D. (2003a). Enterprise Networks, Partnering and Alliancing. In *Procurement Strategies: A Relationship Based Approach*. Walker D. H. T. and K. D Hampson, Eds. Oxford: Blackwell Publishing: Chapter 3, 30–73.

Walker, D. H. T. and Hampson, K. D. (2003b). Project Alliance Member Organisation Selection. *Procurement Strategies: A Relationship Based Approach*. Walker D. H. T. and K. D. Hampson, Eds. Oxford, Blackwell Publishing: Chapter 4, 74–102.

Walker, D. H. T., Maqsood, T. and Finegan, A. (2005). The Culture of The Knowledge Advantage (K-Adv) – An Holistic Strategic Approach to the Management of Knowledge. In *Knowledge Management in the Construction*

Industry: A Socio-Technical Perspective. Kazi A. S., Ed. Hershey, PA: Idea Group Publishing: 223–248.

Walker, D. H. T., Wilson, A. J. and Srikanathan, G. (2004). *The Knowledge Advantage (K-Adv) For Unleashing Creativity & Innovation in the Construction Industry.* Brisbane CRC in Construction Innovation. http://www.construction-innovation.info/images/pdfs/2001–004-A_Industry_Booklet.pdf.

Wood, D. J. and Gray, B. (1991). 'Toward a Comprehensive Theory of Collaboration'. *Journal of Applied Behavioral Science.* 27(2): 139–162.

Chapter 14

Business transformation through an innovative alliance

Alejandro C. Arroyo and Derek H. T. Walker

Chapter introduction

In Chapter 3, stakeholders and the place of supply chain partners in project procurement delivery was discussed. Chapter 5 linked strategy with creating value in projects through procurement choices, the translation of project vision into managing portfolios of projects and strategically choosing projects as well as aligning available resource combinations through strategic procurement decision making. Chapter 9 stressed the value of procurement choices relating to forms of collaboration as triggering innovation. These chapters underpin this case study chapter and provide a basis for extending concepts raised.

A key aim of this book is to introduce the reader to a wider range of concepts of project value delivery through procurement decision making that has been generally available in a number of the business disciples. The construction industry has had a growing focus on construction project procurement that other PM specialisations may learn valuable lessons from.

The aim of this chapter is to provide an innovative example – where the project involved transformation of the supply chain for a group of organisations in a realignment that allowed these companies to form a sustainable alliance centred on the construction, delivery and operation of a fleet of river barges. The way that these barges were procured and delivered represented a radical departure from 'business as usual' to enable participating firms to deliver a sustainable competitive advantage in their normal operational business activities. While the 'project' may seem superficially straightforward, its strategic nature, allowing the parties involved to transform the way they did business and find a means of synergistically sharing knowledge, resulted in this being a highly transformational project. It is as much an example of a change management project and business transformation as one involving building a fleet of barges. There is a dearth of readily available case studies that help us understand emerging PM practices from a Latin American context, and so this chapter also provides a useful source to help rectify that deficit.

This chapter is structured as follows. The next section will outline the general context and background of the project. This is followed by a discussion of the underpinning theory that the case study illustrates and presents – the 'what' issues. The following section then provides case study details explaining the 'how' and 'why' issues. The last section provides an evaluation and discussion of implications for PM procurement practice.

General context

An interest in PM techniques being used to enact a business transformation informed this chapter. One of the authors was intimately involved in an innovative logistics project in South America involving a complex business transformation. This occurred by moving from a standard business supply chain management (SCM) product delivery approach towards a PM approach that integrated organisations previously characterised by a hands-off relationship. The experience introduced interesting and valuable insights. All principal business participants described in this chapter are linked through experience of working together in another innovative major transformation – the 'Atlantic Corridor Project'.

The Atlantic Corridor Project stems from an innovative way in which to address problems associated with the logistics of managing projects and cross-border business in a recently emerged trading block in South America. South America is embarking on an integrated economic and political enterprise – the Southern Market (Mercosur).[1] Also, a way to overcome existing logistical and trading problems within the region was to link business interests across the Mercosur in the development of a community of practice (COP)[2] project in which participants were able to virtually meet and take part in 'integrating forums', initially established through an online community in various key points in Brazil and Argentina, linking industrial hubs with consumer centres along the continent. The aim of this project (the Atlantic Corridor Project) was to achieve, through an improved transportation logistics infrastructure, higher productivity and thus make participants accomplish greater economies of scale (Arroyo and Walker, 2004). Several participants of the project (including the two team leaders under study in this chapter) were members of that COP and have developed a measure of trust and sound business relationships as a consequence. This relationship and COP notwithstanding, the barriers to commerce are significant and real. Further, local and national media frequently concentrate on national sentiments and so awarding contracts and conducting business is subject to media pressure and interest that can negatively impact upon preferred business partnership formation in projects such as that described here.

Globalisation pressures and increasing demand for resources by the strongly emerging Chinese economy has been as fierce in Latin America as elsewhere in the world. The result of this emerging and sustained demand for raw materials, resources and guaranteed stream of supply by Chinese companies has not only driven up resource prices but also locked out many local recipients of these resources. This prompts a re-evaluation of the way that companies depending on local resources (in a region such as South America) approach their business dealings in existing supply chain networks. The rules of the game in South America for steel manufacturers have changed with the emergence of China as a major purchaser – for example, huge demands for iron ore and prevailing high prices/low availability for fuel used in transporting bulk commodities by barge. The geographical context of this case study adds to its complexity, involving more than 2,400 km of river navigation along one of the world's largest fluvial systems: the Paraná–Paraguay waterways linking Argentina, Brazil, Uruguay, Paraguay, and Bolivia. The project case study focuses on the construction and operation of fully dedicated barge convoys to carry iron ore supplies ex Brazil to Argentina. More details of this fascinating project follow in a later section.

Underpinning theory

This chapter draws upon SCM, leadership, business strategy and theory of the nature of organisations to reflect and make sense (Schön, 1983; Weick, 2001) of the experience described in the case study. Data were gathered from documented experiences gained from personal interaction during the first project phases, referring to project documentation generated during the project and reflections on critical project-external incidents.

SCM business relationships

SCM evolved from logistics management frameworks in the early 1980s (Lambert and Cooper, 2000). Its acceptance also benefited from the concept of value chains and the management of value-adding activities as being the way forward to achieve sustainable competitive advantage (Porter, 1985; Normann and Ramirez, 1993; Womack and Jones, 2000; Champion, 2001). Many standard manufacturing organisations have moved from *ad hoc* supply arrangements towards a more strategic lean SCM approach (Cox, 1999). Advantages are well documented from a number of industry sectors including oil and gas, computer component manufacturers, chemical industry, automotive and aerospace (Bessant *et al.*, 2003). SCM advantages include better supply reliability; improved quality culture alignment between supply and producers; and improved valuable knowledge transfer leading to waste reductions, improved competitive advantage to the participating chain of organisations and production simplification (Cox, 1999; Michaels, 1999; Croom *et al.*, 2000; Dubois and Gadde, 2000; Bessant *et al.*, 2003; Childerhouse *et al.*, 2003). Also a trend has emerged towards project managers taking a supply chain management approach, particularly in the construction industry (Akintoye *et al.*, 2000).

This has been evident from project managers concentrating on working more closely and openly with fewer competing suppliers so that mutual understanding can reduce transaction cost inefficiencies caused by excessive numbers of members of a supply chain wasting time developing quotes and tender proposals under extreme time pressures (Akintoye *et al.*, 2000). Reducing the number of competitors invited to quote for a contract (to a few well-qualified and credible suppliers) improves mutual understanding between parties and not only reduces risk and uncertainly relating to poor information and knowledge flows about the products to be delivered, but also results in improved relationship management leading to innovative mutual win–win problem solutions (Bessant *et al.*, 2003).

Thus, SCM has either influenced how projects are being managed or there has been a confluence of strategic thinking that has led project managers to embrace SCM principles not only on projects but across projects to programmes and portfolios of projects. Tan (2001: 41) links SCM through corporate vision directly to strategic planning, and from there to the development of processes designed to support tactical plans to execute an SCM strategy that delivers customer satisfaction, which in turn generates sustainable business results. This emerging theme has been explored and investigated in terms of how sophisticated organisations link their strategic management of projects, programmes of linked and related projects, and whole portfolios of programmes (Morris and Jamieson, 2004). Thus the link between formerly isolated manufacturing activities (that relate to delivering an product or service) and a project (that delivers a specific outcome) has become increasingly evident as a value-creating proposition delivered in a project management context using a variety of tools and approaches (such as SCM, strategic positioning and general project management methodologies).

What has driven this evolution? How can we better understand how this can be applied in practice? In answering the first question two general forces can be discerned as drivers. First, there are internal forces such as adoption of the evolving management theory and its practical application by senior managers. PM advances have provided fertile ground for project managers to experiment using theories (whether explicitly learned or discovered as part of osmosis from colleagues, educational institutions or general media sources) that they can test and refine – this is a form of knowledge or innovation 'push'. Second, there are strong knowledge innovation 'pull' forces at work. The need for sustainable business development provides an acute force for organisational and individual members' survival. This force is manifested by the current acute competition on a wider regional and global scale. Customer (and supply chain stakeholder) expectations have also been sharpened with increasing demands for greater value to be delivered. Naturally, force field analysis could unearth a lengthy factor breakdown structure; however, two principal forces (pull and push) can be generally used as a framework for understanding the evolution of this trend. Thus, using this simplified framework on an interesting case study, it is possible to better understand how this business transformation is taking place.

Strategic management influences and organisational responses

Disruptive regionalisation and globalisation forces have challenged the way that many organisations can viably maintain a sustainable business. These forces are organisation-external and must be responded to. Chapter 5 discussed the prescriptive strategic schools' focus on strategy through design, planning and positioning that partially explain how strategy evolves in a conscious way where it is sculpted from the situation. Descriptive strategy schools centre on motivation and what is going on in the mind of strategists be that entrepreneurial desire; a cognitive psychology approach to explain decision-making processes; a focus on learning with strategy as being an emerging experimental process; a power process of negotiation and political decision making; a cultural process by recognising collective decision making based on a variety of perceptions and assumptions of what needs responding to; or an environmental process where strategy responds to – rather than shapes – the environment in which strategic decision making needs to be made. The configuration school sees strategy as a process of transformation, dynamically responding to periods ranging from relative stability to high levels of turbulence.

In this case study the parties moved from a competitive stance where each party sought their part in any given project's supply chain through intense competition to one where they built a cluster/coalition/alliance. They did this by forming a joint venture of selected supply chain members who they felt they could cooperate with. This required significant change in business processes and leadership style. Through this action, the group decided to respond to the turbulent external environment by leveraging social capital generated through engagement on the Atlantic Corridor Project. Their aim was to transform the way they did business through their member's emergent learning and mutual adjustment. They also leveraged their power wherever possible to their advantage as a cluster to lock in materials supply for the steel manufacturing project partner and to provide synergistic business opportunities to other partners in the newly formed supply chain for the project. It was in some respects planned for but, as we explain, the strategy also emerged and was highly wrapped up in the region's culture.

The nature of the firm can be viewed from a number of perspectives. The resource-based view (RBV) sees organisations as a set of competencies and resources providing competitive advantage that can be deployed to successfully undertake business activities (Conner and Prahalad, 1996). Competencies have been argued as being a vital resource (Prahalad and Hamel, 1990) including knowledge (Conner and Prahalad, 1996). Further, organisations now need to respond to turbulence and market uncertainty by developing a set of dynamic capabilities that allow them to be flexible, lean, responsive to disruptive change and become learning organisations so that they can interact more effectively with both upstream and downstream supply chain members (Spender, 1996; Teece *et al.*, 1997; Eisenhardt and Brown, 1999; Eisenhardt and Martin, 2000).

Courtney *et al.* (1997) argue that in turbulent and rapidly changing circumstances, traditional strategic planning horizons are dangerously constricting and confining. They categorise four levels of uncertainty with corresponding strategic responses. Level 1, with a clear enough future that can be planned for using the traditional strategic toolkit as described by the positioning school of thought (Mintzberg *et al.*, 1998: chapter 4). Level 2, with alternative futures lends itself to concentrating on a few discrete outcomes that define the future using analytical tools game theory and so on. Level 3 comprises a range of possible outcomes but no natural scenarios where scenario planning can be used. Level 4 is in the realm of true ambiguity with no basis to forecast the future so the main tools used for planning would be emergent by drawing analogies, recognising patterns and using non-linear dynamic models. Transformational organisational change will respond to the perceived uncertainties and dynamic capabilities being developed by firms.

Fenton and Pettigrew (2000) provide a comprehensive summary of work based on the large-scale international INNFORM research programme (Pettigrew and Fenton, 2000: 37; Pettigrew *et al.*, 2003). They cite findings from their study of European organisations that during 1992–1996 there was a 65% increase in alliance formation and 65% increase in outsourcing. These turbulent changes have been driven by globalised competition and regionalised trading blocks. Thus a distinct pattern of business supply chain partner re-alignment and cooperative arrangements is becoming evident that responds to the emerging reality of global business. In this context, national boundaries seem to be increasingly irrelevant. Globalisation, particularly 'free' trade, has forced organisations to be competitive not just within their local context but to be able to withstand explicit and potential pressure from competitors completely outside national boundaries. Regionalisation mitigates, to some extent, against globalised competition through countries forming trading blocks where preferential treatment may be afforded to organisations in member states – for examples in the European Union and the Mercosur in Latin America. However, even within these trading blocks, stiff competition arises out of the inevitable rationalisation and reconfiguration of infrastructure that can then force a disruptive influence on firms previously comfortably protected by national monopolies or local advantages/ preferences.

One recommended response to deal with these kind of external threats and challenges is to form alliances or more closely working with clusters of organisations to be more competitive by working together rather than separately competing – as has been traditional in Japan (Anchordoguy, 1990). Clusters have naturally occurred within convenient geographical proximity exemplified by the electronics industry in Silicon Valley, the Californian wine industry and the fashion industry in Italy (Porter, 1998). Clustering has been effective because it allows firms to act is if they had greater scale, resources and capabilities whether that is through formal or informal relationships with others in their cluster (Porter, 1998: 80). However, this does require firms' leaders and managers to acquire and develop excellent relationship management skills because their new skills relate to effective persuasion and lobbying for collective action rather than use of command and control power. The formation of clusters, alliances or forms of joint ventures is congruent with recent moves towards better integrated SCM (Croom *et al.*, 2000; Lambert and Cooper, 2000) as highlighted earlier. Chapter 3 discussed how stakeholder management requires an acute appreciation of the nature of trust and distrust and how those concepts and the concept of commitment link to managing relationships.

The emergence of SCM and clustering of organisations to respond to global and regional trading challenges has forced a change in emphasis in what can be considered as the main competencies of project managers and organisational leaders. It is because these new structures require additional skills and competencies to previously reported leadership and PM characteristics (Kotter, 2001) that include impact and influence-based competences such as understanding power and influence processes (Bourne and Walker, 2004) as well self-control and flexibility characteristics (Dainty *et al.*, 2003; 2004; Dainty *et al.*, 2005). Principally amongst these is a transformational leadership style that matches the transformational business task that these firms face.

Transactional leaders motivate subordinates to perform as expected. They do this by helping them recognise task responsibilities, identify goals, acquire confidence about desired performance levels, and understand that their needs and the desired rewards are linked to goal achievement (Bass and Avolio, 1994: 53). There is an explicit or implicit understanding that the leader independently or with the follower, specifies expectations with a performance measurement system in place to reward desired behaviours and outputs. The goals are defined and specified, as are the means. Avolio (1996: 5) argues that transformational leadership requires 'the four I's – Idealised influence, Inspirational motivation, Intellectual stimulation and Individual consideration'. Transformational leaders motivate individuals to perform beyond expectations by inspiring them to focus on broader missions transcending their own immediate self-interest; concentrate on intrinsic higher-level goals (achievement and self-actualisation) and not lower-level goals (safety and security); and be confident in their abilities to achieve the missions given by the leader (Bass and Avolio, 1994). We can project this behaviour beyond individuals to SCM behaviour that respects each member's value proposition. A mix of transformational leadership by the lead agent in the supply chain can provide inspiration and vision and transactional behaviour and the detail can be observed by governance arrangement behaviour (discussed in Chapter 4) and trust–distrust issues (discussed in Chapter 3). SCM and stakeholder management can be explained in terms of leadership behaviours of the 'lead' partners in the supply chain and the maturity of followership of others in the supply chain.

This section reinforced theoretical discussion elsewhere in this book and provides us with a useful view to analyse the case study reported upon in this chapter. It helps explain how an extremely turbulent business environment triggered change and how the experiences and relationships of principal parties to the alliance shaped their strategic SCM response.

Case study details

The project comprised a highly customised design that was developed to suit a series of complex logistic constraints of a steel-manufacturing company, belonging to a large transnational industrial conglomerate. Project participants were located in Argentina and Paraguay. Geography influenced the project logistical design, as it is an extremely complex process to transport steelmaking raw materials by river barge over 2,400 km across national borders along one of the world's largest river systems. River barge transport remains the most financially viable mode of transporting such materials. An important facet is that case study participants shared a past history of business interaction and collaborated through the Atlantic Corridor Project e-collaboration[3] network (Arroyo and Walker, 2004). This collaboration cemented trust and the necessary relationship commitment and goodwill that provided the underpinning for attempting to transform the way the collaborating firms participated in a value/supply chain. This innovative approach represents an interesting example of a project (that could be seen as part of a programme or portfolio of projects) that helps explain how a business transformation was delivered through a series of projects in a portfolio of change initiatives.

Physical outcome project general description

The specific project focused upon in this study comprised the construction and operation of fully dedicated barge convoys to carry iron ore supplies from Brazil to Argentina commencing in 2003 and phased over six years. The first phase comprised development and acceptance of an innovative logistics design solution and its realisation through the construction of new barges and leasing of existing barges. The project had in-built replication possibilities to roll out a series of similar projects. The following proposal as illustrated in Table 14.1 was accepted to describe the broad scope of the six-year project (with the prospect of a larger follow-on project).

The first barge unit required 90 days to be built with a maximum of 30 days for every subsequent built barge. A group of 12 leased barges was sourced, surveyed, approved and chartered from one of the smaller barge operators. Figure 14.1 illustrates a typical barge convoy. A reliable river operator was introduced into the project team and persuaded to operate the convoys under strict efficiency standards throughout the six-year contract duration. Three potential shipyards were approached in Argentina, Brazil, Paraguay and invited to bid for the barge-building contract using steel sheets for the barges to be supplied by the steel-manufacturing company from one of its plants in Brazil.

Table 14.1 Project proposal scope

1 One fully dedicated barge convoy 12 months a year	7 12 days sailing up-river; 11 days sailing down-river; 2 days loading/ unloading; 1–2 days open
2 Building 12 barges of 2,500 tons each = 30,000 tons capacity	8 Gross operational time 25–27 days; 1 trip every month
3 One suitable tugboat of 'pusher' type basis 4,500 horse power	9 Total annual volume 250,000 tons + or – 10% basis 6 years contract
4 Port Loading at Ladario, Corumba, Sobramil/Brazil	10 24 hours loading/24 hours unloading
5 Speed 5 to 6 knots up-river and 8 to 9 knots down-river	11 Barge dimensions basis 65 m length by 15 m width
6 Port Discharge at San Nicolás/ Argentina	12 Building of new barges basis US$490,000 each

Figure 14.1 Typical barge convoy on the River Plate.

This project presented logistical challenges within an increasingly complex business environment that demanded high levels of leadership and strategy by participant executives (project team leaders). A high level of environmental contextual uncertainity was presented by the global nature of the steel company's supply sources; the impact upon global buyers and sellers of China's raw materials demand and the business practices of shipowners and port terminals. During the 1990s, the river transportation industry was sharply re-engineered with introduced capital concentration to substantially replace *ad hoc* casual service provision and lack of effective competition. Also, the river operators' behaviour made leasing agreements with a smaller number of dominant river operators highly problematic. Tactics used by one operator over its rivals included instigating prolonged legal challenges over dubious unpaid debts and seizing chartered barges. This illustrates the tough and antagonistic business negotiating style that exemplified the river operators' business culture.

Change management general description

Table 14.2 illustrates the project participant context. The steel-producing firm in the consortium had to quickly adapt to a changing business context to meet urgent steel production supply needs whilst meeting steel export demands. Achieving this project was of utmost importance to meet growing exports by supplying the steel company's recently opened production line as a way to capitalise on an increasing international price for steel products. The steel firm's vision was to transform itself through an innovative process, based on improving business performance and continued efficiency, as a way to compete in the global market while completely changing the way decision making was made.

Manu S was under intense pressure exerted by accelerating global market forces that were driving up costs of raw resources from Brazil used for steel manufacturing products in its Argentinean plant and thus radically changing demand patterns. China's rise as a major economy over recent years demanding greater iron ore and related resource supplies placed tremendous pressures on Manu S.

Iron ore and magnesium reserves (that are key raw materials utilised in the steel industry) in the Matto Grosso do Sul area of Brazil is estimated at 150m tonnes. Argentina lacks these strategic minerals and is therefore increasingly dependent on Brazil as a supply source for globally exported steel manufacturing products. Steel that is produced not only supplies Argentina's domestic market but also generates foreign currency. Traditionally, a single Argentine multinational group dominated this commercial activity with supply sources globally spread to avoid any supply concentration that might negatively impact on the production

Table 14.2 Project participant context

Critical players:
Steel manufacturing company (Manu S) is based in Buenos Aires with 4,000 employees
and showing an annual turnover exceeding US$1 bn, belonging to an industrial con-
glomerate with annual sales of US$3.2 bn and 16,000 employees worldwide. The
Argentine steel manufacturing company at present operates with global efficiency,
depends upon global cooperation, works on global basis is customer-focused and has
integrated its worldwide products and associated services supply chain
Logistic consulting company (Logi L) is a Buenos Aires-based firm specialising in the
formulation and implementation of complex logistic designs with offices all over
Latin America to service its client's logistic projects in areas of difficult physical access.
They are highly specialised in providing logistics advice in the oil, gas, power, infrastructure
and mining industry sectors. Customers are large transnational corporations undertaking
projects throughout the Latin American region. This company possesses a great volume
of tacit knowledge that is very difficult to imitate in an ever-increasing market position
Barge operator company (Barge B) is based in Asunción in Paraguay, this is one of the
various river transportation firms that went through a re-engineering process during
the 1990s by disposing of most of their floating assets and shore facilities. It maintained
a very efficient small operation based on its deep knowledge, solid planning and sound
strategy and was probably the only remaining operator skilful enough to undertake
this project and to deal with large, powerful and demanding transnational firms

Instrumental players:
Barge B's shipping agent, based in Buenos Aires and a barge owner himself during the
1990s and ex-partner of the Paraguayan barging company, acted as a cultural bridge
and managerial interpreter between the Argentine steel transnational and the
Paraguayan barge operator
Shipyards: competing shipyards in Argentina, Paraguay and Brazil actively influenced
through lobbying both critical and instrumental players. Their technical skills, social
contacts and community impact were some of the variables cleverly utilised by the
shipbuilders and the client
Competition: if competitors perceived that capital concentration provides too much
dominance over the waterways market in this part of South America, then there
was a potential threat that competitors would tend to retaliate to impede the
accomplishment of the project; however, this threat was not significant due to their
short-sighted strategy of relying excessively on their market dominance
Banks: these were purely instrumental in approving or rejecting the project from a
financial standpoint
Media: were instrumental in lobbying for the location where barge building was going
to take place

cycle or on their supply chain. Supply sources were utilised as needed by balancing the opportunity cost of stockpiling materials versus shipping it on a just-in-time basis from a limited set of sources. About one-quarter of the supply needs was thus allocated in a highly transitionally minded manner to historical supply origins of raw materials providers as indicated in Table 14.3.

Table 14.3 Project steel products supply source characteristics

- Matto Grosso do Sul – Brazil: rated as of very good product quality though of poor logistic performance with barge convoys sailing 2,400 km along the Paraná–Paraguay waterways. Reaching the final destination on time and on budget involved a number of difficult operational and commercial constraints. Critical stocks had to be increased while this supply source's relative share tended to diminish to accommodate other customers
- Newcastle and Fremantle – Australia: relatively lower quality iron ore was regularly supplied on ocean-going vessels in shipments of up to 40,000 tons ships of 60,000 tons deadweight. These vessels had high capacity but were underutilised due to River Plate draft limitations (meaning higher ocean freight rates/ton) though this was useful in offsetting supply shortfalls arising out of waterway logistic bottlenecks. A full shipment from Australia delivered 40,000–45,000 tons with 30 days transit time, whereas a full convoy on the waterways often represented a tonnage ranging from 18,000–30,000 tons and a transit time of 15–60 days. It became clear that the variability of volumes to be carried, and sailing times on the waterways, was disadvantageous even without considering port congestions at the Brazilian terminals. River draft variability was subject to a dynamic seasonality and further regulatory framework constraints
- Mobile, USG – Alabama United States: although the product was of high quality and the transit time and/or certainty of physical delivery was much better than those previously mentioned, draft limitations on the River Plate limited the size of vessels and so logistics economies of scale were not achieved. This supply source resulted in being similar to that of Australia. However, Mobile was an efficient supply source able to quickly offset any shortfall of other sources owing to its relative geographical proximity and ample ship tonnage supply. These variables were reinforced by existing easy access to the commercial or negotiating channels to reach both shipping companies and qualified shipbrokers. However, this supply source also started to diminish with an increasing degree of preference to serve customer's orders coming from China. This choice resulted in relatively larger volumes being available and higher prices being demanded for the product than was the case for being delivered via the River Plate
- Sepetiba, Tubarao, Carajas – North Eastern Brazil: quality was not the best though product delivery response by the producer and transit times resulted in being close to optimal. However, continuous labour strikes, unannounced port lock-outs, and the River Plate draft limit – affecting all ocean sources alike – effectively restricted this relative share of the supply source to no more than one-fourth, and often less than that

The situation represented by the case study is one of a highly volatile and turbulent environment. The impact of China's entry into the global procurement scene of acquiring steel production resources could have provided a tipping point (Gladwell, 2000; Kim and Mauborgne, 2003), tipping the balance between a past regime of one environment to another. Established relationship power balances and emerging ways of working globally have been changing to the extent that past ways of operating are challenged in today's global competition context. Corporate players see this situation as a business reality and need to adapt and align their actions to this reality for survival.

We will first argue that Logi L had a transformational impact upon Manu S but we also argue that the project-external environment plays a significant part in determining Manu S's behaviour and how this behaviour represented a significant change in the way its leadership group behaved and that it realigned its strategic direction.

The outcome was that Manu S accepted Logi L's proposal that required a radical change in Manu S's behaviours towards procuring its resources for its business operation. It had to move from a highly transactional approach that played raw material suppliers, barge operators and port operators in the logistical supply system against each other to get the cheapest price for service towards a more sustainable negotiated settlement of long-term contracts backed up by real commitments for performance. Given the river system's logistical supplier's historical short-termism, poor safety record and general unreliable performance, the transformation evidenced by Manu S can be regarded as remarkable and worthy of scrutiny through this case study analysis. The project was successfully delivered and provided a model for an extension to this approach. This has changed business practice and procurement decisions for Manu S for this aspect of its business activity.

Discussion of the case study context

One of the first issues to be explored in making sense of this case study is trying to understand Manu S's strategic quandary. During and before the early 1990s, Manu S enjoyed a commanding position in sourcing raw materials with a number of service and materials suppliers that could be played against each other.

Further, many river operators involved in transporting materials had been fragmented with their behaviours being highly opportunistic, with formal supply agreement contracts being broken when it suited them, risking legal redress or financial penalties that were deemed less damaging, for example, taking opportunities to undertake short-term transport work at higher rates than committed to by Manu S. This behaviour, locally translated in English as being 'informal', ranged from breaking contractual commitments, asserting losses from sinking or grounded barges and thus making insurance claims, and other practices that would be considered unethical, probably illegal and certainly not representative of behaviour that can forge alliances or sound relationships. This represented a pressure for Manu S to pursue a highly transactional approach to business relationships with its supply chain for the procurement of raw materials along that supply chain.

In addition, 'best value' represented cheapest price to Manu S. Maintaining 'safe' stock levels could be optimised by taking advantage of a balanced supply that offered options where one potential supply source failure could be compensated by another readily available source. While the costs of a short-term poor choice could be inconvenient it could nevertheless be mitigated against the organisation's purchasing power. Yukl (1998) describes various bases of power including legitimate power, in which the legitimacy of being the major client can force subservient behaviours. Table 14.3 indicates that Manu S had a number of options. Until the emergence of China as an economic power, affecting demand for raw materials to make steel, there was an equilibrium position where despite earlier highlighted difficulties, Manu S had the upper hand in any negotiations over any supply and logistics procurement system providers. This situation changed as the demand–supply equation tipped in the late 1990s to early 2000 period. Thus a shock was administered to a hitherto relatively stable situation, even though experience indicated that the supply of river transportation services could be characterised as an unprofessional and chaotic scramble, determined by the actions of many small-time wildcat operators seeking short-term gains. This *ad hoc* situation was gradually choked off by the consolidation of many of the wildcat barge operators and river port operators into a more dominant smaller number of operators who shaped their business potential.

Two driving forces can be identified. First, the force of China's global emergence tipped the balance of existing demand patterns to transform the nature and scale of resource supply and delivery demands. The cost of both steel raw materials and oil used as fuel for transportation and energy rose substantially. Second, there was a consolidation and rationalisation in river logistics distribution system suppliers. This left Manu S vulnerable in both areas of raw procurement and a reliable distribution system. Both factors had a detrimental impact upon the coercive and legitimate power position that Manu S could maintain.

The emergence of the Mercosur and the experience that Manu S shared with Logi L in being engaged in other cooperative ventures opened up possibilities for Manu S moving from a highly transactional business strategy to a relationship-based one. The important competitive advantage shifted from a cost-based advantage built upon price-dominant power to a shared network of contributors model where more sophisticated trade-offs occur that at one level are transactional (the 'what's in it for me?' syndrome). This led to a more relational and long-term strategy where the dominant directive is potential and realisable synergies in the near or long term with a trust that the potential will be not only realised but that it will be realised to a more fruitful extent than is likely in the purely transactional approach. Manu S has a track record of being very sophisticated in recognising opportunities and so it is not surprising that it would be open to transformational influences provided that benefits are apparent and perceived as real. Manu S showed a propensity to manage both transactional and transformational behaviour. The experience that Manu S developed by working on the project through the interaction with Logi L and Barge B opened up possibilities for an innovative approach to solving a supply chain problem in which Manu S's board approved changes that triggered transformational change to its project leadership team role.

Making sense of the case study

Stakeholder management behaviour of the supply chain was indicated by the mix of transactional and transformational leadership style. Transactional leaders motivate subordinates to perform as expected by helping them recognise task responsibilities, identify goals, acquire confidence about desired performance levels, and understand that their needs and the rewards desired are linked to goal achievement (Bass and Avolio, 1994: 53). There is an explicit or implicit understanding that the leader specifies independently or with the follower, a set of expectations and there is a performance measurement system in place to reward desired behaviours and outputs. The goals are defined and specified, as are the means.

In this case study, Logi L had transformational leadership behaviour in its role of proposing and facilitating innovative approaches to logistics and engaging Players S and B to support and experiment with this idea.

In many respects, management can be seen as transactional, that is getting things done through shifting balances of power towards an acceptable solution as a set of compromises amongst conflicting values. Leaders tend to be transformational in that they develop fresh approaches to long-standing problems. They project their ideas onto images that excite people to give these ideas substance (Zaleznik, 1977). Manu S clearly took an organisationally transformational approach to its PM approach on this project with its board supporting its continuance and growth towards a second and potential third contract in this form.

The fluctuations in oil prices, rises in raw material prices due to China's demand, and changes in the river transport operators' structure provided uncertainty, but these could be classified as presenting a Level 2 or Level 3 uncertainty (Courtney *et al.*, 1997). The required strategic responses to this lie between Level 2 a 'clear enough future', where a single forecast precise enough for determining strategy could be employed and Level 3 'alternate futures' where a few discrete outcomes that define the future could be employed. Uncertainty can prove a trigger that forces organisations to change their decision making and strategic approaches, forcing adaptation to changed circumstances in order to survive (Meyer *et al.*, 1990).

Transformation such as that experienced by case study participants over the past decade can be seen as typical of evolving environments where periods of relative equilibrium are punctuated by radical discontinuous changes. Gersick (1991) draws upon change theory occurring in individuals, groups, organisations, scientific fields, biological species and a grand theory approach to argue that change occurs in this way when the deeper structures that maintain equilibrium are challenged and forced to adapt to new conditions, and that these can be stabilised until that equilibrium is sufficiently challenged to require a revolutionary change. This links to the classical model put forward by Lewin (1947) in which study of the forces acting upon a system can lead to an action plan in which a three-step model of unfreezing, moving forward towards a new stable state and then re-freezing takes place. The change management approach advocated by Lewin can be seen as a transformational process. If we look at this case study from an industry level we can use the model developed by Meyer *et al.* (1990) to see adaptation taking place as an evolution process where industry players track small changes and adapt accordingly until the adjustments made cause internal tensions that create abrupt or revolutionary change because the previous system can no longer support the large numbers of small changes. At the firm level, managers and leaders instigate incremental changes often using a transactional leadership style to cope with evolutionary change until the situation reaches a revolutionary change point where, as proposed by Greiner (1998), metamorphosis is required. This requires organisational transformation as well as transformational leadership. The strategy of being 'king' and directly procuring and organising raw material purchasing, as well as completely outsourcing transportation logistics, was becoming increasingly difficult and uncertain in its outcomes.

Further, the situation was moving from a more loosely coupled system towards a more tightly coupled one because the slack and contingency raw materials storage capacity strategies that protected Manu S against poor logistics performers was no longer adequate and also the opportunity cost was commercially unviable. Manu S was being squeezed through globalisation to move towards being more competitive and was also pursuing a dual strategy of cost reduction and innovation in its SCM. This indicates a focus on strategic competitive advantage (Porter, 1990). The case study project with its relational based characteristics (of an integrated material delivery system of a dedicated barge convoy strictly compliant with safety and better-business practices as opposed to the previous informal *ad hoc* approach) indicated a transformational shift by Manu S influenced by Logi L's offered proposal. Manu S can be viewed as behaving in a transformational mode in response to being partially influenced by the transformational style of management by Logi L. However, Manu S still exhibited strong transactional behaviours with this being influenced by its organisational culture. Manu S senior managers tend to mainly come from an engineering background where analytical skills, attention to detail and highly cost conscious values prevail and where the observed culture was also very homogeneous in terms of this engineering background as the career path.

Much of the external pressures placed upon Manu S by the external environment led Manu S towards a managerial and transactional stance. Two examples of this are offered the impact exerted by a rising Chinese dominance on the demand and supply balance for business interactions that Manu S has with its supply chain; and the opportunities and challenges triggered by a changing landscape of consolidation of previous wild-cat barge and port operators along the river system that were used to transport vital resources. The problems being addressed and solved related to optimisation of returns for Manu S. Also, how to cope with a chaotic and turbulent business environment, where the relative stability of 'rules' or *modus operandi* was subject to power shift changes away from Manu S as 'king' to one where it was pitted against Chinese trading interests that began to dominate the change in preferred business partner.

Prior to the changed environment becoming a clear and present danger to Manu S, a relatively stable environment dominated this situation where Manu S had sufficient market power and range of suppliers to be able to optimise its raw material stockpiles between what was needed for immediate production and what was needed as a buffer stock. Finished steel prices for manufacturing, prior to the emergence of China as a mass market, increased from 2001 to 2004 by some 70% (see oil and gas journal www.ogjonline.com). Manu S responded to this changed environment by investing in a new plant to cope with growing demand. However, this triggered greater pressure to ensure raw material supplies and be able to transport that supply as well as transport finished goods to customers within the Mercosur. Thus the balance of power shifted for Manu S from a situation of dominance to being a contributing part of a broad supply chain for its products that included the important transportation link in that chain. This established a need for Manu S to improve its relationship and value chain management with participants in its supply chain.

The response that Manu S adopted was to join in this project as an active participant and to follow through with plans to repeat the experience for further similar projects – adopting organisational transformation. The business transformation triggered by the participation of Manu S was managed as a project. Risk management, planning, coordination, cost management and all the usual aspects of project management methodologies and body of knowledge (PMI, 2004) were applied.

Manu S saw no need to be intimately involved with the business of river logistics prior to the transformation, considering that river logistics was outside its core competencies. However, Manu S senior executives shifted their stance in agreeing with the view that outsourcing 'often means becoming hostage to providers who may feel tempted to gradually raise prices and reduce services' (Auguste et al., 2002: 54). They recognised that by participating in this project they could gain better value from the relationship in terms of reducing the uncertainties that they faced or at least to better cope with the market and physical uncertainties. They were able to better shape, in conjunction with Logi L and Barge B, a more attractive future including the business generated from

Table 14.4 Summary of aspects relating to this book

Chapter – perspective	Comments and discussion
Chapter 1 – Value proposition and make-or-buy decision	The value proposition of the supply chain members was extremely important to supply chain members. By choosing an alliance/joint venture procurement form, Manu S continued its outsourcing strategy but brought the players closer to its organisational core
Chapter 2 – Project type and procurement choice	The barge fleet aspect of the project was akin to an engineering project but it also involved a major change management project, with highly complex and evolving high relational and integrated design and delivery aspects and also making large-scale cultural adaptations
Chapter 3 – Stakeholder management	A high level of supply chain team stakeholder management effort was evident through undertaking the transformation from lowest cost to a relational basis. The project focused upon SCM
Chapter 4 – Governance	The trust and distrust issues needed to be considered and designed into the arrangement with an appropriate governance framework
Chapter 5 – Business strategy	The case study was centred on a strategic decision to change the way that this kind of project would be procured for Manu S. It went from a hand-off supply of steel raw materials supply to an integrated SCM using an emergent approach triggered by business turbulence
Chapter 6 – Performance measurement	There were performance measures built into the procurement system but space limitations inhibited us from discussing these in depth
Chapter 7 – E-business	There were no e-business issues raised here though prime players first collaborated through an e-informing interactive portal developed as a means of triggering collaboration (the Atlantic Corridor Project)
Chapter 8 – Innovation and organisational learning	The chosen procurement approach promoted knowledge sharing and organisational learning across the supply chain for the approach to work
Chapter 9 – Cultural dimensions	The change management aspects had many links to this chapter both from a cross-national and organisational culture perspective
Chapter 10 – Selecting and growing talented people	This case study did not address any of the issues raised in this chapter

supplying the steel to construct the barges to be used while helping to stabilise the uncertainty they faced. Thus, they were able to transform a severe business threat into an opportunity. They also positioned themselves through this project to be in a better position to learn from the experience and to fine-tune the next iteration.

Conclusions

The case study demonstrates the context and indicates the drivers that triggered a transformation in the organisation and management style of Manu S. This player moved from a totally transactional approach to one of forming a partnership, and while it is true that many transactional elements remain (such as requirements that indicate a concurrent healthy trust and distrust mitigated by governance by contracts and more tightly binding agreements with the project partners) the change in business practice could be classified as transformational. The leadership style indicates mixed transactional and transformational elements. Strategic decision making in the case study is consistent with a configuration strategic approach (Mintzberg *et al.*, 1998). This led to behaviours that supported improved SCM and stakeholder management that were designed into the procurement choice.

The trigger for the transformation appears to be mainly contextual. The changed environment demanded more than adaptation for survival with a required metamorphosis in business practice. Logi L's offering to design a logistics solution with the key Barge B's involvement facilitated this. Logi L could be seen as having a particular transformational leadership influence upon Manu S. Both Logi L and Barge B had important deep tacit knowledge to contribute to help Manu S make the necessary logistics business model transformation.

The purpose of this chapter is to provide a practical case study example of a project procurement decision and how it links to theory discussed in other chapters in this book. This chapter's focus centres on strategic decision making, SCM and to some extent stakeholder management.

Analysis of this case study helps us to make sense out of a messy and difficult situation that faced Manu S using a range of management theory and literature. Table 14.4 provides a summary of lessons from the case study linked to the chapters presented in the book.

Notes

1 The *Mercosur* was established, like the European Union, to better integrate the commercial, political and cultural life of its block of members. These member countries include Argentina, Bolivia, Brazil, Chile, Paraguay and Uruguay, with Peru, Venezuela, Mexico and South Africa as potential future members. Emerging trading blocks have to overcome a bewildering array of local laws and regulations (national as opposed to cross-country legislation as well as harmonising cross-country legal requirements) that impede their integration goal. While many of these laws and codes are explicit, their interpretation and application is subject to a great deal of tacit knowledge. A further Mercosur aim is to rationalise and improve integration of its infrastructure and logistical transport systems

2 See Chapter 8 for further details of COPs and their role in knowledge sharing and problem solving.

3 Participants from remote and central regions across the Mercosur agreed to meet online twice monthly in a virtual meeting place with the aim of finding ways to solve logistical problems within a 24-hour period.

References

Akintoye, A., McIntosh, G. and Fitzgerald, E. (2000). 'A Survey of Supply Chain Ccollaboration and Management in the UK Construction Industry'. *European Journal of Purchasing & Supply Management.* 6(3–4): 159–168.

Anchordoguy, M. (1990). 'A Brief History of Japan's Keiretsu'. *Harvard Business Review.* 68(4): 58–60.

Arroyo, A. C. and Walker, D. H. T. (2004). *A Latin American Strategic Community of Practice Experience.* 18th ANZAM Conference, Dunedin, New Zealand, 8–11 December, ANZAM: 11 pages on CD-Disk proceedings.

Auguste, B. G., Hao, Y., Singer, M. and Wiegand, M. (2002). 'The Other Side of Outsourcing'. *The McKinsey Quarterly.* (1): 52–64.

Avolio, B. (1996). What's All the Karping About Down Under? In *Leadership research and Practice.* Parry K. W., Ed. South Melbourne: Pitman Publishing: 3–15.

Bass, B. M. and Avolio, B. J. (1994). *Improving Organisational Effectiveness Through Transformational Leadership.* London: Sage.

Bessant, J., Kaplinski, R. and Lamming, R. (2003). 'Putting Supply Chain Learning Into Practice'. *International Journal of Operations & Production Management.* 23(2): 167–184.

Bourne, L. and Walker, D. H. T. (2004). 'Advancing Project Management in Learning Organizations'. *The Learning Organization, MCB University Press.* 11(3): 226–243.

Champion, D. (2001). 'Mastering the Value Chain'. *Harvard Business Review.* 79(6): 108–115.

Childerhouse, P., Hermiz, R., Mason-Jones, R., Popp, A. and Towill, D. R. (2003). 'Information Flow in Automotive Supply Chains – Identifying and Learning to Overcome Barriers to Change'. *Industrial Management & Data Systems.* 103(7): 491–502.

Conner, K. R. and Prahalad, C. K. (1996). 'A Resource-based Theory of the Firm: Knowledge Versus Opportunism'. *Organization Science: A Journal of the Institute of Management Sciences.* 7(5): 477–501.

Courtney, H., Kirkland, J. and Viguerie, P. (1997). 'Strategy Under Uncertainty'. *Harvard Business Review.* 75(6): 67–81.

Cox, A. (1999). 'Power, Value and Supply Chain Management'. *Supply Chain Management.* 4(4): 167–175.

Croom, S., Romano, P. and Giannakis, M. (2000). 'Supply Chain Management: An Analytical Framework for Critical Literature Review'. *European Journal of Purchasing & Supply Management.* 6(1): 67–83.

Dainty, A. R. J., Cheng, M.-I. and Moore, D. R. (2003). 'Redefining Performance Measures for Construction Project Managers: An Empirical Evaluation'. *Construction Management & Economics.* 21(2): 209–218.

Dainty, A. R. J., Cheng, M.-I. and Moore, D. R. (2004). 'A Competency-Based Performance Model for Construction Project Managers'. *Construction Management & Economics.* 22(8): 877–888.

Dainty, A. R. J., Cheng, M.-I. and Moore, D. R. (2005). 'A Comparison of the Behavioral Competencies of Client-Focused and Production-Focused Project Managers in the Construction Sector'. *Project Management Journal*. **36**(2): 39–49.

Dubois, A. and Gadde, L.-E. (2000). 'Supply Strategy and Network Effects – Purchasing Behaviour in the Construction Industry'. *European Journal of Purchasing & Supply Management*. **6**(3–4): 207–215.

Eisenhardt, K. M. and Brown, S. L. (1999). 'Patching Restitching Business Portfolios in Dynamic Markets'. *Harvard Business Review*. **77**(3): 73–82.

Eisenhardt, K. M. and Martin, J. A. (2000). 'Dynamic Capabilities: What Are They?' *Strategic Management Journal*. **21**(10/11): 1105–1121.

Fenton, E. M. and Pettigrew, A. (2000). Theoretical Perspectives on New Forms of Organizing. In *The Innovating Organization*. Pettigrew A. and A. M. Fenton, Eds. Thousand Oaks, CA: Sage: 1–46.

Gersick, C. J. G. (1991). 'Revolutionary Change Theories: A Multilevel Exploration of the Punctuated Equilibrium Paradigm'. *Academy of Management Review*. **16**(1): 10–36.

Gladwell, M. (2000). *The Tipping Point*. London: Abacus.

Greiner, L. E. (1998). 'Evolution and Revolution as Organizations Grow'. *Harvard Business Review*. **76**(3): 55–68.

Kim, W. C. and Mauborgne, R. (2003). 'Tipping Point Leadership'. *Harvard Business Review*. **81**(4): 60–70.

Kotter, J. P. (2001). 'What Leaders Really Do'. *Harvard Business Review*. **79**(11): 85–96.

Lambert, D. M. and Cooper, M. C. (2000). 'Issues in Supply Chain Management'. *Industrial Marketing Management*. **29**(1): 65–83.

Lewin, K. (1947). 'Frontiers in Group Dynamics'. *Human Relations*. **1**(1): 5–41.

Meyer, A. D., Brooks, G. R. and Goes, J. B. (1990). 'Environmental Jolts and Industry Revolutions: Organizational Responses to Discontinuous Change'. *Strategic Management Journal*. **11**(5): 93–110.

Michaels, L. M. J. (1999). 'The Making of a Lean Aerospace Supply Chain'. *Supply Chain Management*. **4**(3): 135–145.

Mintzberg, H., Ahlstrand, B. W. and Lampel, J. (1998). *Strategy Safari: The Complete Guide Through the Wilds of Strategic Management*. London: Financial Times/Prentice Hall.

Morris, P. W. G. and Jamieson, A. (2004). *Translating Corporate Strategy into Project Strategy*. Newtown Square, PA: PMI.

Normann, R. and Ramirez, R. (1993). 'From Value Chain to Value Constellation: Designing Interactive Strategy'. *Harvard Business Review*. **71**(4): 65–77.

Pettigrew, A. and Fenton, E. M., Eds. (2000). *The Innovating Organization*. Series The Innovating Organization. Thousand Oaks, CA: Sage.

Pettigrew, A., Whittington, R., Melin, L., Sánchez-Runde, C., van den Bosch, F. A. J., Ruigrok, W. and Numagami, T. (2003). *Innovative Forms of Organizing*. Thousand Oaks, CA: Sage.

PMI (2004). *A Guide to the Project Management Body of Knowledge*. Sylva, NC: Project Management Institute.

Porter, M. E. (1985). Competitive Advantage: Creating and Sustaining Superior Performance, New York: The Free Press.

Porter, M. E. (1990). *The Competitive Advantage of Nations*. New York: Free Press.

Porter, M. E. (1998). 'Clusters and the New Economics of Competition'. *Harvard Business Review*. **76**(6): 77–90.

Prahalad, C. K. and Hamel, G. (1990). 'The Core Competence of the Corporation'. *Harvard Business Review*. **68** (3): 79–91.

Schön, D. A. (1983) *The Reflective Practitioner – How Professionals Think in Action*. Aldershot: BasiAshgate ARENA.

Spender, J.-C. (1996). 'Making Knowledge the Basis of a Dynamic Theory of the Firm'. *Strategic Management Journal*. **17**(Winter Issue): 45–62.

Tan, K. C. (2001). 'A Framework of Supply Chain Management Literature'. *European Journal of Purchasing & Supply Management*. **7**(1): 39–48.

Teece, D., Pisano, G. and Shuen, A. (1997). 'Dynamic Capabilities and Strategic Management'. *Strategic Management Journal*. **18**(7): 509–533.

Weick, K. E. (2001) *Making Sense of the Organization*. Oxford: Blackwell Publishers.

Womack, J. P. and Jones, D. T. (2000). From Lean Production to Lean Enterprise. In *Harvard Business Review on Managing the Value Chain*. Review H. B, Ed. Boston, MA: Harvard Business School Press: 221–250.

Yukl, G. (1998). *Leadership in Organisations*. Sydney: Prentice-Hall.

Zaleznik, A. (1977). 'Managers and Leaders Are They Different'. *Harvard Business Review*. **55**(3): 126–135.

Index